European Workshop on Nonlinear Maps and Applications (NOMA 2017)

Journal of Physics: Conference Series Volume 990

Nizhni Novgorod, Russia
28 – 30 June 2017

ISBN: 978-1-5108-6070-4
ISSN: 1742-6588

Printed from e-media with permission by:

Curran Associates, Inc.
57 Morehouse Lane
Red Hook, NY 12571

Printed by Curran Associates, Inc. (2018)

For permission requests, please contact the Institute of Physics
at the address below.

Institute of Physics
Dirac House, Temple Back
Bristol BS1 6BE UK

Phone: 44 1 17 929 7481
Fax: 44 1 17 920 0979

techtracking@iop.org

Additional copies of this publication are available from:

Curran Associates, Inc.
57 Morehouse Lane
Red Hook, NY 12571 USA
Phone: 845-758-0400
Fax: 845-758-2633
Email: curran@proceedings.com
Web: www.proceedings.com

Table of contents

Volume 990

European Conference - Workshop "Nonlinear Maps and Applications"

28–30 June 2017, Nizhni Novgorod, Russian Federation

Accepted papers received: 15 March 2018
Published online: 29 March 2018

Preface

Papers

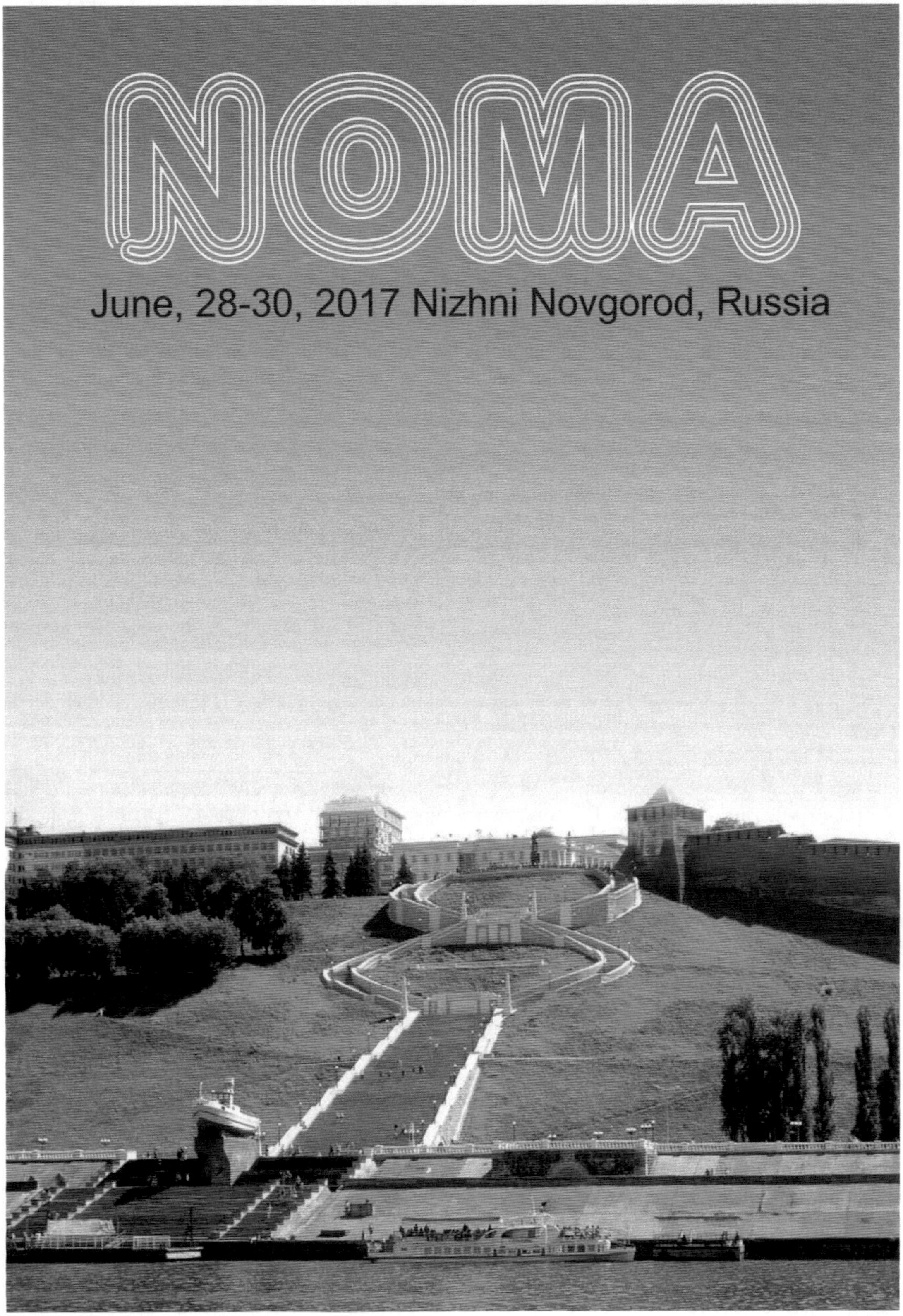
NOMA
June, 28-30, 2017 Nizhni Novgorod, Russia

IOP Conf. Series: Journal of Physics: Conf. Series **990** (2018) 011001 doi:10.1088/1742-6596/990/1/011001

European Workshop "Nonlinear Maps and Applications" – 2017, Nizhni Novgorod, Russia

Efremova L.S.[1], **Gracio M.C.**[2], **Lopez-Ruiz R.**[3], **Sakbaev V.Zh.**[4], **Stepin A.M.**[5]

[1] National Research University of Nizhni Novgorod, Russia
[2] University of Evora, CIMA - UE, Portugal
[3] University of Zaragoza, Spain
[4] Moscow Institute of Physics and Technology, Dolgoprudny, Russia
[5] Moscow State University, Russia

E-mail: [1]`lefunn@gmail.com`, [2]`mgracio@uevora.pt`, [3]`rilopez@unizar.es`, [4]`fumi2003@mail.ru`, [5]`anmist@inbox.ru`

Regular European Workshops "Nonlinear Maps and Applications" (NOMA) are held biannually in those European Universities where successful researchers in the area of nonlinear maps and their applications work.

In far 1973 year French scientist Christian Mira organized Colloquium "Point Mappings and Applications" in the University of Toulouse, where he worked. According to Christian Mira, his mathematical preferences were formed under the influence of works of the founder of the Nizhni Novgorod nonlinear oscillations school A.A. Andronov.

The idea of the regular European Workshops "Nonlinear Maps and Applications" belongs to French scientist D.Fournier-Prunaret and Japanese scientists T.Ueta, Y.Nishio. The first Conference NOMA was held in Toulouse in 2007. The following Conferences were held in Urbino (2009), Evora (2011), Zaragoza (2013) and Dublin (2015).

In 2017 the Steklov Mathematical Institute of the Russian Academy of Sciences and the Institute of Information Technologies, Mathematics and Mechanics of the National Research Nizhni Novgorod State University took up the torch to hold the 6-th European Workshop NOMA-2017.

50 participants of the Conference NOMA-2017 from France, Japan, Mexico, Poland, Portugal, Russia and Spain presented their results at the Sections "Topological and Differential Dynamics", "Ergodic Theory and Dynamical Systems", "Applications of Nonlinear Maps". All participants could listen 7 invited lectures and 4 plenary lectures on the contemporary problems of the theory of nonlinear maps and their applications.

In this Issue of JPCS we publish papers by the participants on the topics in the framework of 3 Sections of the Conference.

The Editors would like to thank all Conference Participants, Committees for their valuable contribution, as well as Reviewers for their time and expertise to review the works presented for this volume.

We thank also Dr. Elena V. Malkina for creation of the logotype of the Conference NOMA - 2017 and Dr. Andrey S. Filchenkov for technical assistance.

IOP Conf. Series: Journal of Physics: Conf. Series **990** (2018) 011002 doi:10.1088/1742-6596/990/1/011002

Scientific, Programme Committee:

Dmitrii V. Treschev (Russia) (Chairman)
Fransisco Balibrea (Spain)
Lyudmila S. Efremova (Russia) (Vice-Chairman)
Daniele Fournier-Prunaret (France)
Laura Gardini (Italy)
Maria Clara Gracio (Portugal)
Ricardo Lopez-Ruiz (Spain)
Mikhail I. Malkin (Russia)
Vladimir I. Nekorkin (Russia) (Vice-Chairman)
Yoshifumi Nishio (Japan)
Anatoly S. Stepin (Russia)

Local Organizing Committee:

Viktor P. Gergel (Chairman)
Dmitrii V. Balandin (Vice-Chairman)
Svetlana S. Belmesova
Lyudmila S. Efremova (Vice-Chairman)
Andrey S. Fil'chenkov
Elena N. Makhrova
Mikhail I. Malkin
Elena V. Malkina
Vsevolod Zh. Sakbaev
Ksenia S. Vaniukova

The Conference NOMA-2017 was partially supported by Russian Foundation for Basic Research, grant No 17-01-20264.

We would like to thank the National Research Nizhni Novgorod State University (Nizhni Novgorod, Russia) for hospitality and all kind of support during the Conference.

Peer review statement

All papers published in this volume of *Journal of Physics: Conference Series* have been peer reviewed through processes administered by the proceedings Editors. Reviews were conducted by expert referees to the professional and scientific standards expected of a proceedings journal published by IOP Publishing.

IOP Conf. Series: Journal of Physics: Conf. Series **990** (2018) 012001 doi:10.1088/1742-6596/990/1/012001

On the quadratic Morse-Smale endomorphisms

Svetlana S. Belmesova

National Research Nizhni Novgorod State University
Nizhny Novgorod, Russia

E-mail: belmesovass@mail.ru

Abstract. For the quadratic map $F_\mu = (xy, (x-\mu)^2)$ we indicate an interval of parameter values, such that map F_μ with a parameter value in the indicated interval is a (nonsingular) Morse-Smale endomophism.

1. Introduction

The bibliography of investigation of polynomial maps, and in particular, of quadratic maps is very extensive (see, for example, [1] – [10]). In this paper we consider the quadratic maps on the plane from the one-parameter family of the type (1)

$$F_\mu(x, y) = (xy, (x-\mu)^2), \tag{1}$$

where (x, y) is an arbitrary point of the plane xOy and $\mu \in (0, 1]$. Different aspects of dynamics of the maps from one-parameter family (1) are investigated in [7] – [10]. This paper is a continuation of [7] – [10].

We prove here, that the map F_μ is a (nonsingular) Morse-Smale endomorphism for any $\mu \in (0, 1)$ and the map F_1 is a singular Morse-Smale endomorphism.

We will extend the definition of a (nonsingular) Morse-Smale endomorphism acting on a compact manifold [11] to the case of maps of a plane.

Nonsingular Morse-Smale flows were considered in [12]. Let us define a nonsingular Morse-Smale endomorphism by analogy with [12].

In the case of endomorphisms, the global stable manifold of a point is not necessarily connected. Therefore, in Definition 1 we use the local stable manifold $W^s_{loc}(p)$ of a p; we understand it as a connected component of the global stable manifold W^s_p containing the point p [13].

Definition 1. An endomorphism $F : R^2 \to R^2$ is called a *(nonsingular) Morse-Smale endomorphism*, if

(i) the nonwandering set $\Omega(F)$ is finite and consists of hyperbolic periodic points;

(ii) the local stable manifold $W^s_{loc}(p)$ and the global unstable manifold $W^u(q)$ of different periodic points p, q intersect transversally, i. e., if $(x, y) \in W^s_{loc}(p) \cap W^u(q)$, then $T_{(x,y)} = T_{(x,y)}W^s_{loc}(p) \oplus T_{(x,y)}W^u(q)$ holds.

Definition 2. An endomorphism $F : R^2 \to R^2$ with a finite nonwandering set $\Omega(F)$ is called a *singular Morse-Smale endomorphism*, if at least one of the following conditions is fulfilled:

IOP Conf. Series: Journal of Physics: Conf. Series **990** (2018) 012001 doi:10.1088/1742-6596/990/1/012001

(i) the nonwandering set $\Omega(F)$ contains a nonhyperbolic periodic point;

(ii) there exist periodic points p', $q' \in \Omega(F)$ such that $W^s_{loc}(p')$ and $W^u(q')$ intersect nontransversally.

The main result of the paper work is the following.

Theorem 1. *Let F_μ be a map of type (1). Then for each $\mu \in (0,1)$ the map F_μ is a (nonsingular) Morse-Smale endomorphism, such that its nonwandering set $\Omega(F_\mu)$ consists of three fixed points (the sink $A_1(0;\mu^2)$, the source $A_2(\mu+1;1)$ and the saddle point $A_3(\mu-1;1)$) and periodic orbit B of period two, which is formed by two sources $B_1\left(\frac{\mu^2+1-\sqrt{\mu^2+1}}{\mu};\ \frac{\mu^2}{(1-\sqrt{\mu^2+1})^2}\right)$ and $B_2\left(\frac{\mu^2+1+\sqrt{\mu^2+1}}{\mu};\ \frac{\mu^2}{(1+\sqrt{\mu^2+1})^2}\right)$.*

The map F_1 is a singular Morse-Smale endomophism, such that its nonwandering set $\Omega(F_1)$ consists of two fixed points (the nonhyperbolic point $A_1(0;1)$, the source $A_2(2;1)$) and the periodic orbit B of periodic two, which is formed by two sources $B_1(2-\sqrt{2};\ \frac{1}{3-2\sqrt{2}})$, $B_2(2+\sqrt{2};\ \frac{1}{3+2\sqrt{2}})$.

This paper is organized as follows. Section 1 contains a main techical results. In Section 2 we construct the special partition of the first quadrant, which we use for the proof of Theorem 1. In Section 3 we give the description of nonwandering set $\Omega(F_\mu)$ of the map F_μ and complete the proof of Theorem 1.

2. The main techical results

In this part of the paper we give the preliminary information that will be used.

We begin this section with a formulation of the main properties of map F_μ.

Let us F^n_μ is the n-th iteration of map (1) and $f_{\mu,n}$ and $g_{\mu,n}$ are the first and the second coordinate functions of F^n_μ respectively, i.e. $F^n_\mu(x,y) = (f_{\mu,n}(x,y), g_{\mu,n}(x,y))$ for any $n \geq 1$.

Denote by $J(F_\mu(x,y)) = \frac{\partial F^n_\mu(x,y)}{\partial(x,y)}$ and $J(F^n_\mu(x,y))$ Jacobian matrix and Jacobian of F^n_μ respectively.

Proposition 1. *Map (1) for every $\mu \in (0,1]$ possesses the properties:*

(i) *The equality $J(F^n_\mu(x,y)) = 0$ $(n \geq 1)$ holds iff coordinates of a point $(x;y)$ satisfy one of the equations $f_{\mu,i}(x,y) = 0$ or $f_{\mu,i}(x,y) = \mu$, for some $0 \leq i \leq n-1$; in particular, the critical set of the map F_μ consists of two straight lines C_0: $x = 0$ and C_μ: $x = \mu$.*

(ii) *The following inclusions hold for every open quadrant K_i $(i = 1, 2, 3, 4)$ of the plane xOy: $F_\mu(\overline{K}_3) \subset \overline{K}_1,$, $F_\mu(\overline{K}_4) \subset \overline{K}_2$, $F_\mu(\overline{K}_2) \subset \overline{K}_2$, $F_\mu(\overline{K}_1) \subset \overline{K}_1$, where $\overline{(\cdot)}$ means the closure of a set.*

(iii) *Triangle $\Delta_\mu = \{(x,y) : x,\, y \geq 0, \frac{x}{2\mu} + \frac{y}{\mu^2} \leq 1\}$ is F_μ – invariant set [1] such that its legs $k_{\mu,x} = \{(x,0) : 0 \leq x \leq 2\mu\}$ and $k_{\mu,y} = \{(0;y) : 0 \leq y \leq \mu^2\}$ satisfy the equalities $F_\mu(k_{\mu,x}) = k_{\mu,y}$, $F_\mu(k_{\mu,y}) = (0;\mu^2)$; in addition the rectriction $F_\mu|_{h_\mu}$ on its hypotenuse h_μ is defined by the equality $F_\mu|_{h_\mu}(x,\,y) = (x\mu(\mu - \frac{x}{2}); (\mu - \frac{2y}{\mu})^2)$ such that the image $F_\mu(h_\mu)$ of hypotenuse h_μ is the closed interval of the line $2x + \mu y = \mu^3$, which lies in the triangle Δ_μ.*

(iv) *There are three fixed points $A_1(0;\mu^2)$, $A_2(\mu+1;1)$ and $A_3(\mu-1;1)$, such that $A_1(0;\mu^2)$ is a sink for any $\mu \in (0,1)$ and is a non-hyperbolic point when $\mu = 1$, $A_2(\mu+1;1)$ is a source for any $\mu \in (0,1]$ and point $A_3(\mu-1;1)$ is a saddle point for any $\mu \in (0,1]$.*

[1] We say that set A, $A \subset R^2$ is F_μ – invariant set if the inclusion $F_\mu(A) \subset A$ holds.

(v) *There exists a unique periodic orbit B with the least period two, which formed by two sources $B_1(\frac{\mu^2+1-\sqrt{\mu^2+1}}{\mu}; \frac{\mu^2}{(1-\sqrt{\mu^2+1})^2})$ and $B_2(\frac{\mu^2+1+\sqrt{\mu^2+1}}{\mu}, \frac{\mu^2}{(1+\sqrt{\mu^2+1})^2})$.*

(vi) *The unbounded set $D_{\mu,+\infty} = \{(x;\, y) \in K_1 : x \geq \mu+1,\, y \geq 1\}$ and the closed set $D_{\mu,A_1} = \{(x;y) \in K_1 : 0 < x \leq \mu+1, 0 < y \leq 1\}$ are F_μ–invariant sets; for every point $(x;\, y) \in D_{\mu,\infty} \setminus \{A_{2,\mu}(\mu+1;1)\}$ the following equalities hold:*

$$\lim_{n\to+\infty} f_{\mu,n}(x,\, y) = \lim_{n\to+\infty} g_{\mu,n}(x,\, y) = +\infty,$$

and for any point $(x;y) \in D_{\mu,A_1} \setminus \{A_2(\mu+1;1)\}$ the following equalities hold:

$$\lim_{n\to+\infty} f_{\mu,n}(x,y) = 0, \quad \lim_{n\to+\infty} g_{\mu,n}(x,y) = \mu^2;$$

in addition, the restriction $F_\mu|_{D_{\mu,\infty}}$ is diffeomorfism of set $D_{\mu,\infty}$ on its image $F_\mu(D_{\mu,\infty})$ such that $F_\mu(D_{\mu,\infty}) \subset D_{\mu,\infty}$.

To prove Theorem 1 we need also the following property of the map F_μ.

Proposition 2 [8]. *For any $\mu \in (0,1]$ there exists a C^1-smooth strictly decreasing function $y = \Gamma_\mu(x)$ defined on the interval $(\mu, +\infty)$, such that $\Gamma_\mu((\mu,\, +\infty)) = (0, +\infty)$; moreover, the graph Γ_μ of this function is a F_μ-invariant curve belonging to the unbounded noninvariant set $\{A_{2,\mu}(\mu+1;\, 1)\} \bigcup (K_1 \setminus (D_{\mu,+\infty} \cup D_{\mu,A_1}))$.*

The invariant curve Γ_μ makes it possible to construct the partition of the first quadrant and to prove that F_μ is a (nonsingular) Morse-Smale endomorphism for every $\mu \in (0,1)$ and the map F_1 is a singular Morse-Smale endomorphism.

Note, that the source $A_2(\mu+1;1)$ is the limit point of the preimages of unbounded rank $\{F_\mu^{-1}(C_\mu \cap K_1)\}_{i\geq 1}$ of the part of critical line C_μ in the first open quadrant K_1 and $\{F_\mu^{-1}(C_0 \cap K_1)\}_{i\geq 1}$ are preimages of unbounded rank of the part of critical line C_0 in K_1 ($\{F_\mu^{-1}(C_\mu \cap K_1)\}_{i\geq 1} \subseteq \{F_\mu^{-1}(C_0 \cap K_1)\}_{i\geq 1}$) since $F_\mu^2(C_\mu) = C_0$ for $\mu \in (0,1]$. It is means, that there not exist universal neighborhood $U(A_2)$ of the source A_2, such that the restriction $F_\mu^n|_{U(A_2)}$ is diffeomorphism for all $n \geq 1$.
Calculating eigenvectors of the differential $DF_\mu(A_2)$ and using claim 6 of Proposition 1, we obtain, that one of unit eigenvectors $p_{A_2} = \left(\frac{1+\sqrt{9+8\mu}}{\sqrt{26+8\mu+2\sqrt{9+8\mu}}}; \frac{4}{\sqrt{26+8\mu+2\sqrt{9+8\mu}}}\right)$ of $DF_\mu(A_2)$ of the map F_μ lies in the unbounded invariant set $D_{\mu,+\infty}$ and the other unit eigenvector $q_{A_2} = \left(\frac{1-\sqrt{9+8\mu}}{\sqrt{26+8\mu-2\sqrt{9+8\mu}}}; \frac{4}{\sqrt{26+8\mu-2\sqrt{9+8\mu}}}\right)$ lies in the unbounded noninvariant set $K_1 \setminus D_{\mu,+\infty}$.
Therefore, the standard technique of cones and normal forms can not be applied for the proof of existence of the invariant curve Γ_μ.

For the proof of the Proposition 2 we use the nonlocal theorem of existence of unlocal strictly decreasing implicit functions $y = \eta_{\mu,n}(x)$, $n \geq 1$ [8]. The function $y = \eta_{\mu,n}(x)$ is the solution of equation $f_{\mu,n}(x,y) = \mu+1$ on the interval $(\mu, +\infty)$ for any $n \geq 2$ and the function $y = \eta_{\mu,1}(x)$ is the solution of equation $f_{\mu,1}(x,y) = \mu+1$ on the interval $(0, +\infty)$. The graph $\eta_{\mu,n}$ of this function for all $n \geq 1$ contains the unique fixed point – source $A_2(\mu+1;1)$ and doest not contain the common points with the periodic orbit B of the period two. The properties of the function $y = \eta_{\mu,n}(x)$, $n \geq 1$ are formulated below, in Proposition 3.

Proposition 3. *Let F_μ be the map (1), $\mu \in (0;1]$. Then:*

(i) *for any $n \geq 1$ the following relations hold:*

(a) $\eta_{\mu,2n-1} \prec \eta_{\mu,2n+1}$, $\eta_{\mu,2n+2} \prec \eta_{\mu,2n}$ *for any $x \in (\mu, \mu+1]$*[2].

[2] We say that $\eta_{\mu,k}$ is preceded $\eta_{\mu,l}$ for $k \neq l$, $k,l \geq 1$ and denote by $\eta_{\mu,k} \prec \eta_{\mu,l}$ if the inequality $\eta_{\mu,k}(x) < \eta_{\mu,l}(x)$ holds for any $x \in (\mu, \mu+1]$ (for any $x \in (\mu+1, +\infty)$).

(b) $\eta_{\mu,2n} \prec \eta_{\mu,2n+2}$, $\eta_{\mu,2n+1} \prec \eta_{\mu,2n-1}$ *for any* $x \in (\mu+1, +\infty)$.

(ii) *the sequence of the restrictions* $\{\eta_{\mu,2n}|_{[a,\mu+1]}(x)\}_{n\geq 1}$ $(\{\eta_{\mu,2n-1}|_{[\mu+1,b]}(x)\}_{n\geq 1})$ *of* C^1 *– smooth strictly decreasing functions* $y = \eta_{\mu,2n}(x)$ $(y = \eta_{\mu,2n-1}(x))$, $n \geq 1$ *on the interval* $[a, \mu+1] \subset (\mu, \mu+1]$ $([\mu+1, b] \subset [\mu+1, +\infty))$ *converges in* C^1 *norm on the interval* $[a, \mu+1]$ *(resp.* $[\mu+1, b]$*) to the strictly decreasing function* $y = \Gamma_\mu(x)$ *restricted on the interval* $[a, \mu+1]$ *(resp.*$[\mu+1, b]$*).*

Note also, that the preimages of the different orders of the straight line $x = \mu + 1$ and the critical line C_μ intersect each other under curve Γ_μ. Therefore, for the proof Theorem 1 we need the sets:

$$D^a_{\Gamma_\mu} = \{(x; y) : x \in (\mu, +\infty), y > \Gamma_\mu\};$$

$$D^u_{\Gamma_\mu} = \{(x; y) : x \in (0, +\infty), 0 < y < \Gamma_\mu\} \text{ (see Fig. 1).}$$

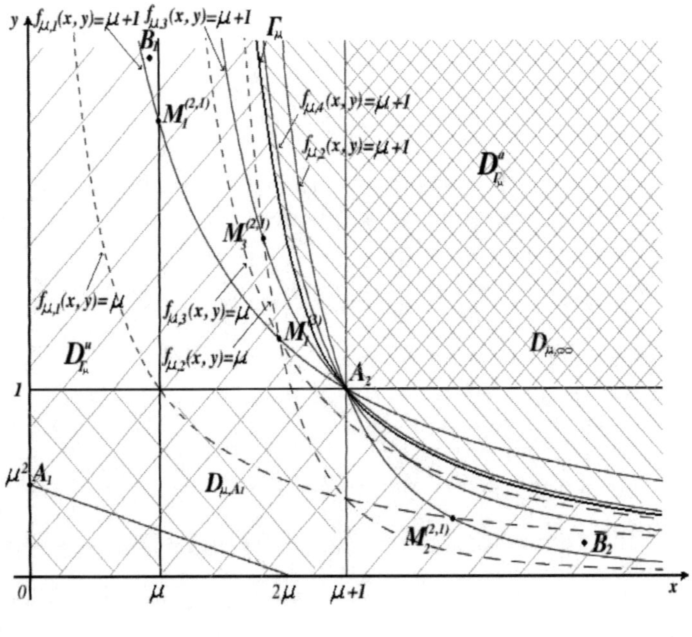

Figure 1. Intersections of preimages of the straight line $x = \mu + 1$ with the preimages of the critical line C_μ in double and triple points, which lie in the K_1.

The set of intersection points of the above preimages is described in the following lemma.

Lemma 1. *Let* F_μ *be the map (1),* $\mu \in (0, 1]$. *Then*

(i) *for any odd number* $n \geq 1$ *and for any* $x \in (0, \mu+1)$ *the curve* $f_{\mu,n}(x, y) = \mu + 1$ *has in* $D^u_{\Gamma_\mu}$ *at least one triple point* $M^{(3)}_n$ *(i. e. a point of the intersection of three curves* $f_{\mu,n}(x, y) = \mu + 1$, $f_{\mu,n+1}(x, y) = \mu$ *and* $f_{\mu,n+2}(x, y) = \mu$*(* $n \geq 1$*)), at least one double point* $M^{(2,1)}_n$ *of the first type (i.e. a point of the intersection of two curves* $f_{\mu,n}(x, y) = \mu + 1$ *and* $f_{\mu,n-1}(x, y) = \mu$ $(n \geq 1)$*) and a countable set of double points* $M^{(2,2)}_{n,j}$ *of the second type (i.e. the points of intersection of the curve* $f_{\mu,n}(x, y) = \mu + 1$ *with every curve* $f_{\mu,j}(x, y) = \mu$ *for any* $j \geq n + 3$, $n \geq 1$*);*

(ii) *for any even number $n \geq 1$ and for any $x \in (\mu+1,+\infty)$ the curve $f_{\mu,n}(x,y) = \mu+1$ has in $D^u_{\Gamma_\mu}$ at least one triple point $M^{(3)}_n$, at least one double point $M^{(2,1)}_n$ of the first type and countable set of double points $M^{(2,2)}_{n,j}$ of the second type;*

(iii) *the set $D^u_{\Gamma_\mu}$ doesn't contain points of intersection of the curves $f_{\mu,n}(x,y) = \mu+1$ and $f_{\mu,m}(x,y) = \mu$ $(m,n \geq 1)$ other than above.*

3. The special partition of the first quadrant K_1 on the plane

In this section we construct the partition in K_1 using the properties of the invariant curve Γ_μ, the properties of strictly decreasing functions $y = \eta_{\mu,n}(x)$ and the preimages of the different orders of the critical line C_μ. This partition help us to investigate the asymptotic behaviour of F_μ – trajectories in the sets $D^a_{\Gamma_\mu}$ and $D^u_{\Gamma_\mu}$.

Proposition 4. *Let F_μ be a map of type (1), $\mu \in (0,1]$. Then*

(i) *the set $D^a_{\Gamma_\mu}$ is invariant set and for every point $(x;y) \in D^a_{\Gamma_\mu}$ the equalities hold*

$$\lim_{n\to+\infty} f_{\mu,n}(x,\,y) = +\infty, \qquad \lim_{n\to+\infty} g_{\mu,n}(x,\,y) = +\infty.$$

(ii) *the set $D^u_{\Gamma_\mu}$ is invariant set and for every point $(x;y) \in D^u_{\Gamma_\mu}$ with exeption of the periodic orbit B of period two and all of its preimages the equalities hold*

$$\lim_{n\to+\infty} f_{\mu,n}(x,\,y) = 0, \qquad \lim_{n\to+\infty} g_{\mu,n}(x,\,y) = \mu^2.$$

We devide the proof of Proposition 4 in some steps.

I. Let us consider the set $D^a_{\Gamma_\mu}$. For the proof of the claim 1 of Proposition 4 we define the sets:

$$T_{D^a_{\Gamma_\mu},2n} = \{(x;y) \in D^a_{\Gamma_\mu} : \eta_{\mu,2n+2}|_{[a,\mu+1]}(x) \leq y \leq \eta_{\mu,2n}|_{[a,\mu+1]}(x)\} \text{ where } a \in (\mu,\mu+1);$$

$$T_{D^a_{\Gamma_\mu},2n-1} = \{(x;y) \in D^a_{\Gamma_\mu} : \eta_{\mu,2n+1}|_{[\mu+1,b)}(x) \leq y \leq \eta_{\mu,2n-1}|_{[\mu+1,b)}(x)\} \text{ where } b \in (\mu+1,+\infty);$$

$$T^l_{D_{\mu,+\infty}} = \{(x;y) \in D^a_{\Gamma_\mu} : y \geq \eta_{\mu,2}\}; \; T^r_{D_{\mu,+\infty}} = \{(x;y) \in D^a_{\Gamma_\mu} : y \geq \eta_{\mu,1}\}.$$

Then from Proposition 3 we obtain the following equality

$$D^a_\Gamma = D_{\mu,+\infty} \cup T^l_{D_{\mu,+\infty}} \cup T^r_{D_{\mu,+\infty}} \cup \Big(\bigcup_{n\geq1} T_{D^a_{\Gamma_\mu},2n}\Big) \cup \Big(\bigcup_{n\geq1} T_{D^a_{\Gamma_\mu},2n-1}\Big), \tag{2}$$

such that the inclusions are valid

$$F_\mu(T_{D^a_{\Gamma_\mu},2}) \subset T^l_{D_{\mu,+\infty}}, \; F_\mu(T^l_{D_{\mu,+\infty}}) \subset T^r_{D_{\mu,+\infty}} \text{ and } F_\mu(T^r_{D_{\mu,+\infty}}) \subset D_{\mu,+\infty}. \tag{3}$$

Lemma 2. *For all $\mu \in (0,1]$ and any $n \geq 1$ the map F_μ is the C^1 – diffeomorphism of the set $T_{D^a_{\Gamma_\mu},2n}$ on the set $T_{D^a_{\Gamma_\mu},2n-1}$.*

Proof. Indeed, by Proposition 3 and Lemma 1 the restriction $F_\mu|_{T_{D^a_{\Gamma_\mu},2n}}$ is surjective local diffeomorphism of the set $T_{D^a_{\Gamma_\mu},2n}$ on the set $T_{D^a_{\Gamma_\mu},2n-1}$. We show, that $F_\mu : T_{D^a_{\Gamma_\mu},2n} \to T_{D^a_{\Gamma_\mu},2n-1}$ is injective map. Let us $(x_1;y_1)$ and $(x_2;y_2)$ are an arbitrary points in $T_{D^a_{\Gamma_\mu},2n}$, such that $(x_1;y_1) \neq (x_2;y_2)$. The coordinate function $f_{\mu,1}(x,y) = xy$ is the strictly monotone function for each variable and the coordinate function $g_{\mu,1}(x,y) = (x-\mu)^2$ is the strictly monotone function for variable x $(x > \mu)$. We have

$$(f_{\mu,1}(x_1,y_1), g_{\mu,1}(x_1,y_1)) \neq (f_{\mu,1}(x_2,y_2), g_{\mu,1}(x_2,y_2)).$$

IOP Conf. Series: Journal of Physics: Conf. Series **990** (2018) 012001 doi:10.1088/1742-6596/990/1/012001

Then $F_\mu : T_{D^a_{\Gamma_\mu},2n} \to T_{D^a_{\Gamma_\mu},2n-1}$ is bijection and there exists the inverse map $(F_\mu|_{T_{D^a_{\Gamma_\mu},2n-1}})^{-1}$ to the map $F_\mu|_{T_{D^a_{\Gamma_\mu},2n}}$ with the Jacobian matrix

$$J((F_\mu|_{T_{D^a_{\Gamma_\mu},2n-1}})^{-1}(x,y)) = \begin{pmatrix} 0 & \frac{1}{2\sqrt{y}} \\ \frac{1}{\mu+\sqrt{y}} & \frac{x}{2\sqrt{y}} \end{pmatrix} |_{T_{D^a_{\Gamma_\mu},2n-1}}.$$

The partial derivatives of coordinate functions of map $(F_\mu|_{T_{D^a_{\Gamma_\mu},2n-1}})^{-1}$ are continuous on the set $T_{D^a_{\Gamma_\mu},2n-1}$. Therefore, $F_\mu : T_{D^a_{\Gamma_\mu},2n} \to T_{D^a_{\Gamma_\mu},2n-1}$ is C^1 – diffeomophism. Lemma 2 is proved.

Using Lemma 2, formula (2), inclusions (6) and claim 6 of Proposition 1 we obtain the following result.

Lemma 3. *For any $\mu \in (0,1]$ set $D^a_{\Gamma_\mu}$ is F_μ – invariant set and for every point $(x;y) \in D^a_{\Gamma_\mu}$ the following equalities hold:*

$$\lim_{n\to+\infty} f_{\mu,n}(x,\, y) = +\infty, \qquad \lim_{n\to+\infty} g_{\mu,n}(x,\, y) = +\infty.$$

Thus, the claim 1 of Proposition 4 is proved.

II. Our next step is to prove, that the set $D^u_{\Gamma_\mu}$ is invariant set, such that the trajectories of the points under Γ_μ satisfy equalities

$$\lim_{n\to+\infty} f_{\mu,n}(x,\, y) = 0, \qquad \lim_{n\to+\infty} g_{\mu,n}(x,\, y) = \mu^2.$$

II (a). We begin from the construction the partition of the set $D^u_{\Gamma_\mu}$. For this goal we need the sets:

$$T_{D^u_{\Gamma_\mu},2n-1} = \{(x;y) \in D^u_{\Gamma_\mu} : \eta_{\mu,2n-1}|_{[a,b]}(x) \le y \le \eta_{\mu,2n+1}|_{[a,b]}(x)\}, \text{ where } [a,b] \subset (\mu, \mu{+}1), n \ge 1;$$

$$T_{D^u_{\Gamma_\mu},2n} = \{(x;y) \in D^u_{\Gamma_\mu} : \eta_{\mu,2n}|_{[a,b]}(x) \le y \le \eta_{\mu,2n+2}|_{[a,b]}(x)\}, \text{ where } [a,b] \subset (\mu{+}1, +\infty), n \ge 1;$$

$$T^l_{D_{\mu,A_1}} = \{(x;y) \in K_1 : 0 < x < \mu+1, 1 < y \le \frac{\mu+1}{x}\};$$

$$T^r_{D_{\mu,A_1}} = \{(x;y) \in K_1 : x > \mu+1, 0 < y \le \frac{\mu+1}{x(x-\mu)^2}\}.$$

Lemma 4. *For any $\mu \in (0,1]$ the inclusions*

$$F_\mu(T^l_{D^u_{\Gamma_\mu}}) \subset D_{\mu,A_1}, F_\mu(T^r_{D^u_{\Gamma_\mu}}) \subset T^l_{D_{\mu,A_1}}$$

are valid.

Proof. We prove the first of the inclusions. Note that, the critical line C_μ intersect the set $T^l_{D_{\mu,A_1}}$. Therefore, we represent set $T^l_{D_{\mu,A_1}}$ as the union of two sets:

$$T^{l,1}_{D_{\mu,A_1}} = \{(x;y) \in T^l_{D_{\mu,A_1}} : x \in (0,\mu)\};$$

$$T^{l,2}_{D_{\mu,A_1}} = \{(x;y) \in T^l_{D_{\mu,A_1}} : x \in (\mu, \mu+1)\};$$

i. e. $T^l_{D_{\mu,A_1}} = T^{l,1}_{D_{\mu,A_1}} \cup T^{l,2}_{D_{\mu,A_1}}$.

We show, that the following inclusions $F_\mu(T^{l,1}_{D_{\mu,A_1}}) \subset D_{\mu,A_1}$, $F_\mu(T^{l,2}_{D_{\mu,A_1}}) \subset D_{\mu,A_1}$ are valid. Let $(x;y)$ is an arbitrary point of the set $T^{l,1}_{D_{\mu,A_1}}$. By definition of the set $T^{l,1}_{D_{\mu,A_1}}$ the following inequlities hold: $0 < xy < \mu+1$, $(x-\mu)^2 < 1$. Hence, $F_\mu(T^{l,1}_{D_{\mu,A_1}}) \subset D_{\mu,A_1}$.
If $(x;y) \in T^{l,2}_{D_{\mu,A_1}}$ then $\mu < xy < \mu+1$, $(x-\mu)^2 < 1$ and the inclusion $F_\mu(T^{l,2}_{D_{\mu,A_1}}) \subset D_{\mu,A_1}$ is valid. Therefore, $F_\mu(T^{l}_{D_{\mu,A_1}}) \subset D_{\mu,A_1}$. Analogously above we prove, that $F_\mu(T^{r}_{D_{\mu,A_1}}) \subset T^{l}_{D_{\mu,A_1}}$.
Lemma 4 is proved.
Using Lemma 4 and claim 6 of the Proposition 1 we obtain the following result.

Lemma 5. *The set* $D_{\mu,A_1} \cup T^{l}_{D_{\mu,A_1}} \cup T^{r}_{D_{\mu,A_1}}$ *is* F_μ *– invariant set and for every point* $(x;y) \in D_{\mu,A_1} \cup T^{l}_{D_{\mu,A_1}} \cup T^{r}_{D_{\mu,A_1}} \setminus \{A_2(\mu+1;1)\}$ *the following equalities hold:*

$$\lim_{n\to+\infty} f_{\mu,n}(x,\,y) = 0, \qquad \lim_{n\to+\infty} g_{\mu,n}(x,\,y) = \mu^2.$$

II (b). In this part of the work we investigate the properties of the set

$$G = \{(x;y) \in K_1 : x \in (0,\mu+1), \eta_{\mu,1}(x) < y < \Gamma_\mu\} \cup \{(x;y) \in K_1 : x > \mu+1, \eta_{\mu,2}(x) < y < \Gamma_\mu\}.$$

Using Lemma 5 we obtain

Corollary 1. *Let* $(x;y)$ *is an arbitrary point of the set* G*. Then the preimages of the point* $(x;y)$ *with respect to the restriction* $F_\mu|_{K_1}$ *does not belong to the set* $D_{\mu,A_1} \cup T^{l}_{D_{\mu,A_1}} \cup T^{r}_{D_{\mu,A_1}}$.
By definitions of the sets G and $T_{D^u_{\Gamma_\mu},2n-1}$, $T_{D^u_{\Gamma_\mu},2n}$ the following equality hold

$$\bigcup_{n=1}^{+\infty} (T_{D^u_{\Gamma_\mu},2n-1} \cup T_{D^u_{\Gamma_\mu},2n}) = G.$$

Our next step is the investigate of F_μ – trajectories of the points of the sets $T_{D^u_{\Gamma_\mu},2n-1}$ and $T_{D^u_{\Gamma_\mu},2n}$. For this goal, we consider the existence problem for implicit functions defined by the equations

$$f_{\mu,n}(x,y) = \mu \tag{4}$$

with initial condition

$$f_{\mu,n}(x^*;y^*) = \mu \tag{5}$$

on the interval $(\mu,+\infty)$, where $(x^*;y^*)$ is the point of one of the types described in Lemma 1. Using the existence theorem for a nonlocal C^1-smooth implicit function (see [7]) we obtain the following claim.

Lemma 6. *Let* F_μ *be a map (1),* $\mu \in (0,1]$*. Then for every* $n \geq 1$ *there exists the* C^1 *– smooth function* $y = \psi_{\mu,n}(x)$ *which is the solution of the problem (4) with the initial condition (5) on the interval* $(\mu,+\infty)$*. Moreover, the graph* $\psi_{\mu,n}$ *of the function* $y = \psi_{\mu,n}(x)$ *passes through the triple point* M^3_{n-1} *and the double point of the first type* $M^{(2,1)}_{n+1}$ *for* $n \geq 2$*; for* $n \geq 5$ *and* $k \geq 3$ *the graph* $\psi_{\mu,n}$ *of the function* $y = \psi_{\mu,n}(x)$ *passes through the double points of the second type* $M^{(2,2)}_{n-k}$.
Let us note, that the ray $x = 2\mu$ for $\mu \in (0,1]$ and its preimages of the different orders plays an important role in investigation of F_μ – trajectories from the set G.

Thus, we consider also the existence problem for implicit functions defined by the equations

$$f_{\mu,n}(x,y) = 2\mu \tag{6}$$

with initial conditions

$$f_{\mu,n}(x_{Q_n}, y_{Q_n}) = 2\mu \tag{7}$$

on the interval $(\mu, +\infty)$.
The point $Q_n(x_{Q_n}; y_{Q_n})$ is the unique point of intersection of the connected component of the full preimage of the ray $x = 2\mu$ of the $(n-1)$ order with the unique smooth strictly monotone branch of the nth preimage of the ray $x = \mu+1$, which passes through the fixed point $A_2(\mu+1; 1)$ $(n \geq 1)$.
Let $Q_1 = (2\mu; \frac{\mu+1}{2\mu})$ is the unique point of intersection of the ray $x = 2\mu$ with unique smooth strictly decreasing branch of the first preimage of the ray $x = \mu + 1$, which passes through the fixed point $A_2(\mu + 1; 1)$. Note that, for $x > \mu$ there exist the unique point $Q_2(\mu + \frac{\mu+1}{2\mu}; \frac{2\mu}{\mu+\frac{\mu+1}{2\mu}})$ of intersection of the first preimage of the ray $x = 2\mu$ with the unique smooth strictly monotone branch of the second preimage of the ray $x = \mu + 1$, which passes through the point A_2 and the following equality hold $F_\mu(Q_2) = Q_1$. Denote by Q_n $(n \geq 1)$ the unique point of intersection of connected component of the full preimage of the ray $x = 2\mu$ of the $(n-1)$ order with the unique continuos strictly monotone branch of the nth preimage of the ray $x = \mu + 1$, which passes through the point A_2 such that the following equality hold $F_\mu(Q_{n+1}) = Q_n$ for any $n \geq 1$. The point Q_n for any $n \geq 1$ is the correctly defined.

Analogously above we apply the existence theorem of a nonlocal C^1 – smooth implicit function [7] and obtain the following result.

Lemma 7. *Let F_μ be a map (1), $\mu \in (0, 1]$. Then there exists C^1 – smooth monotone function $y = \phi_{\mu,n}(x)$, which defined on the interval $(\mu, +\infty)$ by equality (6). The graph $\phi_{\mu,n}$ of this function passes through the point Q_n.*
Lemmas 1, 6, 7 and definition of the set $\{Q_n, n \geq 1\}$ makes it possible to construct the partition of the set $T_{D^u_{\Gamma_\mu},2n-1} \subset G$, $n \geq 1$, such that boundaries of elements of this partition may contain the curves $\eta_{\mu,n}$, $\phi_{\mu,m}$, $\psi_{\mu,k}$ (where $m \geq 0$, $k \geq 0$ are defined by Lemma 1 and the set $\{Q_n, n \geq 1\}$), the critical line C_μ and the ray $x = 2\mu$.

For determination, we construct the partition of the set $T_{D^u_{\Gamma_\mu},2n-1}$. Let $[a, b] \subset (\mu, \mu + 1)$ is an arbitrary closed interval. For $n = 1$ we have

$$T_{D^u_{\Gamma_\mu},1} = \{(x; y) \in D^u_{\Gamma_\mu} : \eta_{\mu,1}|_{[a,b]}(x) \leq y \leq \eta_{\mu,3}|_{[a,b]}(x)\}.$$

The critical line C_μ and the line $x = 2\mu$ intersect set $T_{D^u_{\Gamma_\mu},1}$ in the double point of the first type $M_1^{(2,1)}(x_{M_1^{(2,1)}}, y_{M_1^{(2,1)}})$ and in the point Q_1 respectively, such that the equality hold

$$T_{D^u_{\Gamma_\mu},1} = \bigcup_{i=1}^{3} T^i_{D^u_{\Gamma_\mu},1},$$

where

$$T^1_{D^u_{\Gamma_\mu},1} = \{(x; y) \in T_{D^u_{\Gamma_\mu},1} : x \in (0, x_{M_1^{(2,1)}})\};$$

$$T^2_{D^u_{\Gamma_\mu},1} = \{(x; y) \in T_{D^u_{\Gamma_\mu},1} : x \in (x_{M_1^{(2,1)}}, x_{Q_1})\};$$

$$T^3_{D^u_{\Gamma_\mu},1} = \{(x; y) \in T_{D^u_{\Gamma_\mu},1} : x \in (x_{Q_1}, \mu + 1)\}.$$

From Lemmas 1, 6 follows, that the boundary of the set $T_{D^u_{\Gamma_\mu},1}$ contains the triple point $M_1^{(3)}(x_{M_1^{(3)}}; y_{M_1^{(3)}})$, which is the point of intersection of three branches: the monotone branch $\eta_{\mu,1}$ of the first preimage of the line $x = \mu + 1$ and the monotone branches $\psi_{\mu,2}$ and $\psi_{\mu,3}$ of the second and the third preimages of the critical line C_μ respectively; in addition, the abscissa $x_{M_1^{(3)}}$ of the point $M_1^{(3)}$ satisfies inequality

$$\mu < x_{M_1^{(3)}} < 2\mu \text{ if } \mu \in (0, \frac{\sqrt{5}-1}{2})$$

or satisfies inequality

$$2\mu < x_{M_1^{(3)}} < \mu + 1 \text{ if } \mu \in (\frac{\sqrt{5}-1}{2}, 1).$$

For the case $\mu \in (0, \frac{\sqrt{5}-1}{2})$ we define the sets

$$T^{2,1}_{D^u_{\Gamma_\mu}} = \{(x; y) \in T_{D^u_{\Gamma_\mu},1} : x \in (x_{M_1^{(2,1)}}, x_{M_1^{(3)}}), y < \psi_{\mu,3}(x)\};$$

$$T^{2,2}_{D^u_{\Gamma_\mu}} = \{(x; y) \in T_{D^u_{\Gamma_\mu},1} : x \in (x_{M_1^{(2,1)}}, x_{M_1^{(3)}}), \psi_{\mu,3}(x) < y < \psi_{\mu,2}(x)\};$$

$$T^{2,3}_{D^u_{\Gamma_\mu}} = \{(x; y) \in T_{D^u_{\Gamma_\mu},1} : x \in (x_{M_1^{(3)}}, x_{Q_1}), y > \psi_{\mu,2}(x)\},$$

such that $T^2_{D^u_{\Gamma_\mu},1} = \bigcup_{i=1}^3 T^{2,i}_{D^u_{\Gamma_\mu},1}$. From this follows the equality

$$T_{D^u_{\Gamma_\mu},1} = T^1_{D^u_{\Gamma_\mu},1} \cup T^3_{D^u_{\Gamma_\mu},1} \cup (\bigcup_{i=1}^3 T^{2,i}_{D^u_{\Gamma_\mu},1}) \text{ (see Figure 2)}$$

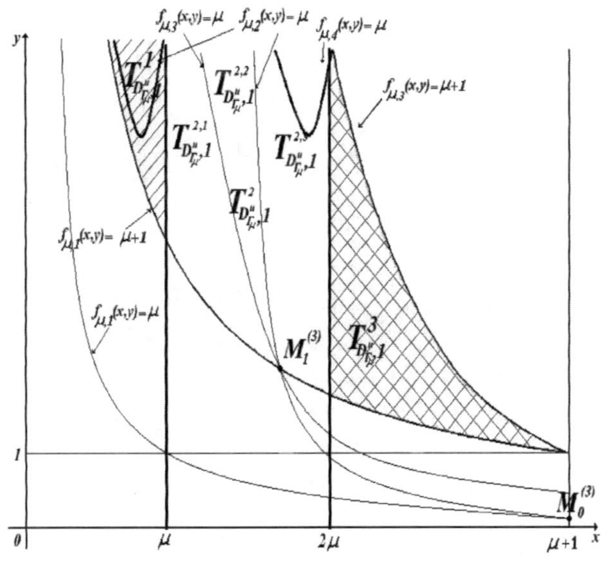

Figure 2. The partition of the set $T_{D^u_{\Gamma_\mu},1}$ with using critical line C_μ, ray $x = 2\mu$ and the monotone branches $\psi_{\mu,2}$, $\psi_{\mu,3}$ for $\mu \in (0, \frac{\sqrt{5}-1}{2})$.

If $\mu \in (\frac{\sqrt{5}-1}{2}, 1)$ we need the following sets

$$T^{2,1}_{D^u_{\Gamma_\mu},1} = \{(x; y) \in T_{D^u_{\Gamma_\mu},1} : x \in (x_{M_1^{(2,1)}}, x_{Q_1}), y < \psi_{\mu,3}(x)\};$$

$$T^{2,2}_{D^u_{\Gamma_\mu},1} = \{(x;y) \in T_{D^u_{\Gamma_\mu},1} : x \in (x_{M_1^{(2,1)}}, x_{Q_1}), \psi_{\mu,3}(x) < y < \psi_{\mu,2}(x)\};$$

$$T^{2,3}_{D^u_{\Gamma_\mu},1} = \{(x;y) \in T_{D^u_{\Gamma_\mu},1} : x \in (x_{M_1^{(2,1)}}, x_{Q_1}), y > \psi_{\mu,2}(x)\};$$

$$T^{3,1}_{D^u_{\Gamma_\mu},1} = \{(x;y) \in T_{D^u_{\Gamma_\mu},1} : x \in (x_{Q_1}, x_{M_1^{(3)}}), y < \psi_{\mu,3}(x)\};$$

$$T^{3,2}_{D^u_{\Gamma_\mu},1} = \{(x;y) \in T_{D^u_{\Gamma_\mu},1} : x \in (x_{Q_1}, x_{M_1^{(3)}}), \psi_{\mu,3}(x) < y < \psi_{\mu,2}(x)\};$$

$$T^{3,3}_{D^u_{\Gamma_\mu},1} = \{(x;y) \in T_{D^u_{\Gamma_\mu},1} : x \in (x_{Q_1}, \mu+1), y > \psi_{\mu,2}(x)\}.$$

Then we obtain

$$T_{D^u_{\Gamma_\mu},1} = T^1_{D^u_{\Gamma_\mu},1} \cup (\bigcup_{i=1}^{3} T^{2,i}_{D^u_{\Gamma_\mu},1}) \cup (\bigcup_{j=1}^{3} T^{3,j}_{D^u_{\Gamma_\mu},1}) \text{ (see Figure 3).}$$

Further, we consider the second case in more detail.

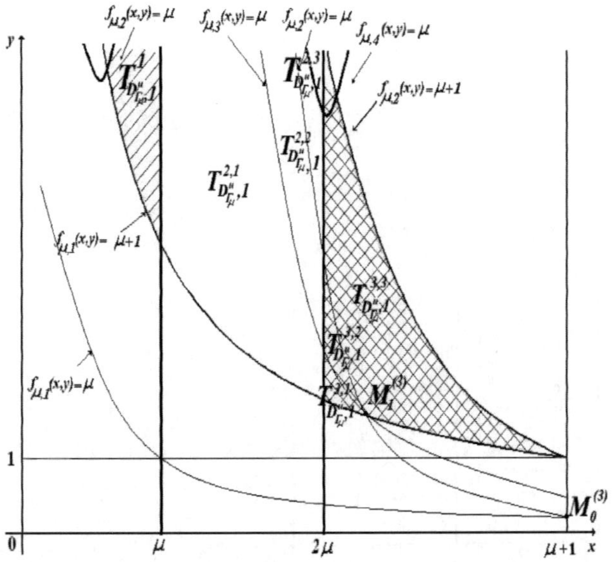

Figure 3. The partition of the set $T_{D^u_{\Gamma_\mu},1}$ with using critical line C_μ, ray $x = 2\mu$ and the monotone branches $\psi_{\mu,2}$, $\psi_{\mu,3}$ for $\mu \in (\frac{\sqrt{5}-1}{2}, 1)$.

Let $\mu \in (\frac{\sqrt{5}-1}{2}, 1)$. On the first step, describe above, we obtain the partition of the set $T_{D^u_{\Gamma_\mu},1}$. In this partition there exist elements, which contains the preimages of critical line C_μ of the order $k \geq 2$ (for example, the set $T^1_{D^u_{\Gamma_\mu}}$ and the set $T^{(3,3)}_{D^u_{\Gamma_\mu}}$ are contain the connected components of the preimages of critical line C_μ of the order 2 and 4 respectively, see Figure 3).

Suppose that after n steps, we constructed the partition of the set $T_{D^u_{\Gamma_\mu},2n-1}$ on the following subsets:

$$T^1_{D^u_{\Gamma_\mu},2n-1} = \{(x;y) \in T_{D^u_{\Gamma_\mu},2n-1} : x \in (\mu, x_{M^{(2,1)}_{2n-1}}), y < \psi_{\mu,2n-2}(x)\};$$

$$T^{2,1}_{D^u_{\Gamma_\mu},2n-1} = \{(x;y) \in T_{D^u_{\Gamma_\mu},2n-1} : x \in (x_{M^{(2,1)}_{2n-1}}, x_{Q_{2n-1}}), \psi_{\mu,2n-2}(x) < y < \psi_{\mu,2n+1}(x)\};$$

$$T^{2,2}_{D^u_{\Gamma_\mu},2n-1} = \{(x;y) \in T_{D^u_{\Gamma_\mu},2n-1} : x \in (x_{M^{(2,1)}_{2n-1}}, x_{Q_{2n-1}}), \psi_{\mu,2n+1}(x) < y < \psi_{\mu,2n}(x)\};$$
$$T^{2,3}_{D^u_{\Gamma_\mu},2n-1} = \{(x;y) \in T_{D^u_{\Gamma_\mu},2n-1} : x \in (x_{M^{(2,1)}_{2n-1}}, x_{Q_{2n-1}}), \psi_{\mu,2n}(x) < y < \phi_{\mu,2n-2}(x)\};$$
$$T^{3,1}_{D^u_{\Gamma_\mu},2n-1} = \{(x;y) \in T_{D^u_{\Gamma_\mu},2n-1} : x \in (x_{Q_{2n-1}}, x_{M^{(3)}_{2n-1}}), \phi_{\mu,2n-2}(x) < y < \psi_{\mu,2n+1}(x)\};$$
$$T^{3,2}_{D^u_{\Gamma_\mu},2n-1} = \{(x;y) \in T_{D^u_{\Gamma_\mu},2n-1} : x \in (x_{Q_{2n-1}}, x_{M^{(3)}_{2n-1}}), \psi_{\mu,2n+1}(x) < y < \psi_{\mu,2n}(x)\};$$
$$T^{3,3}_{D^u_{\Gamma_\mu},2n-1} = \{(x;y) \in T_{D^u_{\Gamma_\mu},2n-1} : x \in (x_{Q_{2n-1}}, \mu+1), y > \psi_{\mu,2n}(x)\}.$$

Let us note that every element of this partition can be contain the connected components of the preimages of critical line C_μ of the order $k \geq 2n$.

Let us describe nth step. Using definition of the points $Q_n, n \geq 1$, and definition of C^1 – smooth decreasing monotone functions $y = \eta_{\mu,n}(x)$, $n \geq 1$, we obtain that the monotone branches $\psi_{\mu,2n}$ and $\phi_{\mu,2n}$ intersects set $T_{D^u_{\Gamma_\mu},2n+1}$; in addition, the following equality

$$T_{D^u_{\Gamma_\mu},2n+1} = T^1_{D^u_{\Gamma_\mu},2n+1} \cup T^2_{D^u_{\Gamma_\mu},2n+1} \cup T^3_{D^u_{\Gamma_\mu},2n+1}$$

hold, where the sets $T^1_{D^u_{\Gamma_\mu},2n+1}$, $T^2_{D^u_{\Gamma_\mu},2n+1}$, $T^3_{D^u_{\Gamma_\mu},2n+1}$ are defined by equalities

$$T^1_{D^u_{\Gamma_\mu},2n+1} = \{(x;y) \in T_{D^u_{\Gamma_\mu},2n+1} : x \in (0, x_{M^{(2,1)}_{2n+1}})\};$$
$$T^2_{D^u_{\Gamma_\mu},2n+1} = \{(x;y) \in T_{D^u_{\Gamma_\mu},2n+1} : x \in (x_{M^{(2,1)}_{2n+1}}, x_{Q_{2n+1}})\};$$
$$T^3_{D^u_{\Gamma_\mu},2n+1} = \{(x;y) \in T_{D^u_{\Gamma_\mu},2n+1} : x \in (x_{Q_{2n+1}}, \mu+1)\}.$$

For all values of parameter μ from the interval $(\frac{\sqrt{5}-1}{2}, 1)$ the inequalities are fulfilled $x_{Q_{2n+1}} < x_{M^{(2,1)}_{2n+1}} < \mu + 1$. Hence, the equality

$$T_{D^u_{\Gamma_\mu},2n+1} = T^1_{D^u_{\Gamma_\mu},2n+1} \cup \bigcup_{i=1}^{3} T^{2,i}_{D^u_{\Gamma_\mu},2n+1} \cup \bigcup_{j=1}^{3} T^{3,j}_{D^u_{\Gamma_\mu},2n+1}$$

holds, where

$$T^{2,1}_{D^u_{\Gamma_\mu},2n+1} = \{(x;y) \in T_{D^u_{\Gamma_\mu},2n+1} : x \in (x_{M^{(2,1)}_{2n+1}}, x_{Q_{2n+1}}), \psi_{\mu,2n}(x) < y < \psi_{\mu,2n+3}(x)\};$$
$$T^{2,2}_{D^u_{\Gamma_\mu},2n+1} = \{(x;y) \in T_{D^u_{\Gamma_\mu},2n+1} : x \in (x_{M^{(2,1)}_{2n+1}}, x_{Q_{2n+1}}), \psi_{\mu,2n+3}(x) < y < \psi_{\mu,2n+2}(x)\};$$
$$T^{2,3}_{D^u_{\Gamma_\mu},2n+1} = \{(x;y) \in T_{D^u_{\Gamma_\mu},2n+1} : x \in (x_{M^{(2,1)}_{2n+1}}, x_{Q_{2n+1}}), \psi_{\mu,2n+2}(x) < y < \phi_{\mu,2n}(x)\};$$
$$T^{3,1}_{D^u_{\Gamma_\mu},2n+1} = \{(x;y) \in T_{D^u_{\Gamma_\mu},2n+1} : x \in (x_{Q_{2n+1}}, x_{M^{(3)}_{2n+1}}), \phi_{\mu,2n}(x) < y < \psi_{\mu,2n+3}(x)\};$$
$$T^{3,2}_{D^u_{\Gamma_\mu},2n+1} = \{(x;y) \in T_{D^u_{\Gamma_\mu},2n+1} : x \in (x_{Q_{2n+1}}, x_{M^{(3)}_{2n+1}}), \psi_{\mu,2n+3}(x) < y < \psi_{\mu,2n+2}(x)\};$$
$$T^{3,3}_{D^u_{\Gamma_\mu},2n+1} = \{(x;y) \in T_{D^u_{\Gamma_\mu},2n+1} : x \in (x_{Q_{2n+1}}, \mu+1), y > \psi_{\mu,2n+2}(x)\}.$$

Thus, the partition of the set $T_{D^u_{\Gamma_\mu},2n-1}$ (for every $n \geq 1$) is constructed.
Analogously above we construct the partition of the set $T_{D^u_{\Gamma_\mu},2n}$ for every $n \geq 1$.

The properties of elements of the above partitions of the sets $T_{D^u_{\Gamma_\mu},2n-1}$ and $T_{D^u_{\Gamma_\mu},2n}$ for the case $\mu \in (\frac{\sqrt{5}-1}{2}, 1]$ are formulated in following claim.

Lemma 8. *Let $\mu \in (\frac{\sqrt{5}-1}{2}; 1]$. Then map F_μ is the diffeomorphism:*

IOP Conf. Series: Journal of Physics: Conf. Series **990** (2018) 012001 doi:10.1088/1742-6596/990/1/012001

(i) *of the set* $T^{2,i}_{D^u_{\Gamma_\mu},2n}$ *on the set* $T^{2,i}_{D^u_{\Gamma_\mu},2n-1}$, $i = 1, 2, 3$;

(ii) *of the set* $T^{3,j}_{D^u_{\Gamma_\mu},2n}$ *on the set* $T^{3,j}_{D^u_{\Gamma_\mu},2n-1}$, $j = 1, 2, 3$;

(iii) *of the set* $T^{1}_{D^u_{\Gamma_\mu},2n}$ *on the set* $T^{1}_{D^u_{\Gamma_\mu},2n-1}$.

Proof. For determination, we give the proof of the claim 1. As it follows from the definition of sets $T^{2,i}_{D^u_{\Gamma_\mu},2n}$ and $T^{2,i}_{D^u_{\Gamma_\mu},2n-1}$ $(i = 1, 2, 3)$ the restriction $F_\mu|_{T_{D^u_{\Gamma_\mu},2n}}$ – is local diffeomorphism.

Let us prove, that $F_\mu : T^{2,i}_{D^u_{\Gamma_\mu},2n} \to T^{2,i}_{D^u_{\Gamma_\mu},2n-1}$ is surjection for every $i = 1, 2, 3$. In fact, by the formula (1) the connected component of the full preimage of the closed interval $\{(x; y) : y = c, x \in [x_1, x_2]\}$ contains the closed interval $\{(x; y) : x = c^*, y \in [y_1, y_2]\}$. Arbitrarily take and fix a leave $y = c$, where $c \in T^{2,i}_{D^u_{\Gamma_\mu},2n-1}$, $c > 0$. From definition of the set $T^{2,i}_{D^u_{\Gamma_\mu},2n-1}$ follows, that the abscisses of the points $(x_1; c)$ and $(x_2; c) \in T^{2,i}_{D^u_{\Gamma_\mu},2n-1}$ satisfy the inequalities: $x_{M^{(2,1)}_{2n-1}} < x_1 < x_2 < x_{M^{(3)}_{2n-1}}$. Using formula (1), we obtain, that there exist the points $(c^*; y_1)$ and $(c^*; y_2) \in T^{2,i}_{D^u_{\Gamma_\mu},2n}$, such that the equalities $F_\mu(c^*, y_1) = (x_1; c)$, $F_\mu(c^*; y_2) = (x_2, c)$ are fulfilled. It means, that F_μ is surjection of the set $T^{2,i}_{D^u_{\Gamma_\mu},2n}$ on the set $T^{2,i}_{D^u_{\Gamma_\mu},2n-1}$ for $i = 1, 2, 3$.

We show, that $F_\mu : T^{2,i}_{D^u_{\Gamma_\mu},2n} \to T^{2,i}_{D^u_{\Gamma_\mu},2n-1}$ is injection for every $i = 1, 2, 3$. Using the monotonicity of the function $f_{\mu,1}(x, y)$ by the variable x (resp. y), we obtain, that for the arbitrary two points $(x_1; y_1), (x_2; y_2) \in T^{2,i}_{D^u_{\Gamma_\mu},2n}$, such that $x_1 < x_2$ (resp. $y_1 < y_2$) the inequality is fulfilled $F_\mu((x_1, y_1)) \neq F_\mu(x_2, y_2)$, where $F_\mu(x_1, y_1)$, $F_\mu(x_2, y_2) \in T^{2,i}_{D^u_{\Gamma_\mu},2n-1}$ $(i = 1, 2, 3,.)$ Hence, $F_\mu : T^{2,i}_{D^u_{\Gamma_\mu},2n} \to T^{2,i}_{D^u_{\Gamma_\mu},2n-1}$ is injection. Therefore, we obtain, that $F_\mu : T^{2,i}_{D^u_{\Gamma_\mu},2n} \to T^{2,i}_{D^u_{\Gamma_\mu},2n-1}$ is bijection $(i = 1, 2, 3)$.

Then, the inverse map $F_\mu^{-1} : T^{2,i}_{D^u_{\Gamma_\mu},2n-1} \to T^{2,i}_{D^u_{\Gamma_\mu},2n}$ is defined, and the partial derivaties of its coordinate functions are continuous in $T^{2,i}_{D^u_{\Gamma_\mu},2n-1}$. The claim 1 of Lemma 8 is proved.
Lemma 8 is proved.
By Lemmas 8 the following statement hold.

Corollary 2. *The set* G *is* F_μ – *invariant set for every* $\mu \in (0, 1]$.

II (c) In this part of the paper we describe asymptotic behaviour of F_μ – trajectories of the points of the set G.

Note that set G contains periodic orbit B of period 2. Prove the following statement.

Lemma 9. *The map* F_μ *have the unique periodic orbit* B *with the least period two, which formed by two sources* $B_1\left(\frac{\mu^2+1-\sqrt{\mu^2+1}}{\mu}; \frac{\mu^2}{(1-\sqrt{\mu^2+1})^2}\right)$ *and* $B_2\left(\frac{\mu^2+1+\sqrt{\mu^2+1}}{\mu}; \frac{\mu^2}{(1+\sqrt{\mu^2+1})^2}\right)$ *such that* $B_1 \in T^1_{D^u_{\Gamma_\mu},1}$, $B_2 \in T^1_{D^u_{\Gamma_\mu},2}$.

Proof.

(i) Existence of unique periodic orbit B is proved in [8].

(ii) Let us prove, that $B_1 \in T^1_{D^u_{\Gamma_\mu},1}$. Note, that for any $\mu \in (0, 1]$ for the coordinate of point B the equalities holds:

$$\frac{\mu^2 + 1 - \sqrt{\mu^2 + 1}}{\mu} = \frac{(\sqrt{\mu^2 + 1} - 1)\sqrt{\mu^2 + 1}}{\mu} = \frac{\mu\sqrt{\mu^2 + 1}}{\sqrt{\mu^2 + 1} + 1}$$

and

$$\frac{\mu^2}{(1-\sqrt{\mu^2+1})^2} = \frac{(1+\sqrt{1+\mu^2})^2}{\mu^2} = \left(\frac{1}{\mu}+\sqrt{1+\frac{1}{\mu^2}}\right)^2.$$

Then the following inequalities are fulfilled

$$0 < \frac{\mu\sqrt{\mu^2+1}}{\sqrt{\mu^2+1}+1} < \mu, \quad \left(\frac{1}{\mu}+\sqrt{1+\frac{1}{\mu^2}}\right)^2 > 1. \tag{8}$$

Moreover, for any $\mu \in (0,1]$ we have:
$\frac{\mu^2+1-\sqrt{\mu^2+1}}{\mu} \cdot \frac{\mu^2}{(1-\sqrt{\mu^2+1})^2} - \mu - 1 = \frac{\mu\sqrt{\mu^2+1}}{\sqrt{\mu^2+1}-1} - \mu - 1 = -1 + \sqrt{1+\frac{1}{\mu^2}} + \frac{1}{\mu} > 0.$

Hence, we obtain that $f_{\mu,1}(x,y)|_{B_1} > \mu + 1$.
Using the last formula and the inequalities (8), we obtain, that point B_1 belongs to the unbounded curvilinear sector, which boundary formed by ray $\{(x;y) : x = \mu, y > \frac{\mu+1}{\mu}\}$ and by unbounded continuous arc $\{(x;y) : 0 < x < \mu, y = \frac{\mu+1}{x}\}$. Consequently, $B_1 \in T^1_{D^u_{\Gamma_\mu},1}$ (see definition of set $T^1_{D^u_{\Gamma_\mu},1}$). Using Lemma 6 for any $\mu \in (0,1]$ we obtain, that $B_2 \in T^1_{D^u_{\Gamma_\mu},2}$.
Lemma 9 is proved.
By Lemma 9 and Corollary 2 the preimages of periodic orbit B of the arbitrary order $n \geq 1$ (with resrect to restriction $F_\mu|_{K_1}$) belong to the set $\bigcup_{n=1}^{+\infty}(T^1_{D^u_{\Gamma_\mu},2n-1} \cup T^1_{D^u_{\Gamma_\mu},2n})$.

Denote by $\{(F_\mu|_{K_1})^{-n}(x_{B_1};y_{B_1})\}$ the subset of the full nth preimage of the point B_1 which does not contain the periodic orbit B. From Lemmas 8, 9 follows, that $\{(F_\mu|_{K_1})^{-n}(x_{B_1};y_{B_1})\}$ is single point set. The unique point of this set we denote by $(x^{-n}_{B_1}; y^{-n}_{B_1})$.

By the claim 3 of Proposition 3, Lemmas 8, 9 and Corollary 2 the following statement holds.

Lemma 10. *Let F_μ be a map of type (1), $\mu \in (0,1]$. Then for any $\mu \in (0,1]$ the limit equalities $\lim_{n\to+\infty} x^{-n}_{B_1} = \mu+1$, $\lim_{n\to+\infty} y^{-n}_{B_1} = 1$. are valid.*
Note, that the trajectory $\{(F_\mu|_{K_1})^{-2n}(x_{B_1};y_{B_1})\}_{n\in Z}$ $(\{(F_\mu|_{K_1})^{-2n}(x_{B_2};y_{B_2})\}_{n\in Z})$ is heteroclinic, such that it is attracted to the point B_1(B_2) and it is repeled from the point A_2.

By definition of sets $T^1_{D^u_{\Gamma_\mu},1}$ and $T^r_{D^u_{\Gamma_\mu}}$ the following inclusion

$$F_\mu(T^1_{D^u_{\Gamma_\mu},1}) \subset T^r_{D^u_{\Gamma_\mu}} \tag{9}$$

holds.

From Lemmas 5 and 8 we obtain the following statement.

Corollary 3. *The set $D^u_{\Gamma_\mu} = G \cup D_{\mu,A_1} \cup T^r_{D^u_{\Gamma_\mu}} \cup T^l_{D^u_{\Gamma_\mu}}$ is invariant set.*
Using inclusion (9) and Lemmas 4, 5, 8–10 we obtain the following result.

Lemma 11. *Let F_μ be a map of type (1), $\mu \in (0;1]$. Then for every point $(x;y) \in G$ with exeption of periodic orbit B of period two and all of its preimages the equalities hold*

$$\lim_{n\to+\infty} f_{\mu,n}(x,y) = 0, \quad \lim_{n\to+\infty} g_{\mu,n}(x,y) = \mu^2.$$

Claim 2 of Proposition 4 follows immediately from Lemmas 5, 11 and Corollary 3.
Proposition 4 is proved.

4. Completion of the proof of Theorem 1

In this part we describe nonwandering set $\Omega(F_\mu)$ of map F_μ and complete the proof of Theorem 1.

From Proposition 4 follows, that for any $\mu \in (0,1]$ nonwandering set $\Omega(F_\mu|_{K_1})$ of the map $F_\mu|_{K_1}$ is finite and consists of the fixed point $A_2(\mu+1;1)$ and the periodic orbit B of period two.

Let us describe behaviour of F_μ – trajectories in K_2. As it follows from the claims 1 and 2 of Proposition 1, $F_\mu|_{K_2} : K_2 \to F_\mu(K_2)$ is diffeomorphism.
By the claim 4 of Proposition 4 we obtain, that for any $\mu \in (0,1)$ the set $\overline{K}_2$ contains two fixed points: the saddle point $A_3(\mu-1;1)$ and the sink $A_1(0;\mu^2)$. Therefore, there exist continuous functions $y = \gamma_i^u$, $y = \gamma_i^s$, $i = 1,2$, such that the graphs γ_1^u, γ_2^u are the nonstable separatrix of saddle point A_3 and the graphs γ_1^s, γ_2^s are the stable separatrix of saddle point A_3 [14]. Moreover, the equalities holds $\gamma_1^u \cup \gamma_2^u = W^u(A_3)$, $\gamma_1^s \cup \gamma_2^s = W^s(A_3)$. Let us define the sets

$$K_{21} = \{(x:y) \in K_2 : y > \gamma_1^u(x), y > \gamma_2^s(x)\};$$

$$K_{22} = \{(x;y) \in K_2 : \gamma_2^u(x) < y < \gamma_1^s(x)\};$$

$$K_{23} = \{(x;y) \in K_2 : y < \gamma_2^u(x), y < \gamma_2^s(x)\};$$

$$K_{24} = \{(x;y) \in K_2 : \gamma_2^s(x) < y < \gamma_1^u(x)\},$$

such that $\bigcup_{i=1}^4 K_{2i} = K_2$.

Lemma 12. *Let F_μ be a map of type (1), $\mu \in (0,1]$. Then*

(i) *for any point $(x;y) \in K_{22} \cup K_{23}$ the following equlities holds*

$$\lim_{n\to+\infty} f_{\mu,n}(x,y) = 0, \quad \lim_{n\to+\infty} g_{\mu,n}(x,y) = \mu^2;$$

(ii) *for any point $(x;y) \in K_{21} \cup K_{24}$ the following equalities holds*

$$\lim_{n\to+\infty} f_{\mu,n}(x,y) = -\infty, \quad \lim_{n\to+\infty} g_{\mu,n}(x,y) = +\infty.$$

Let us describe the nonwandering set $\Omega(F_\mu|_{\overline{K}_2})$ of the map $F_\mu|_{\overline{K}_2}$, $\mu \in (0,1]$.

From Lemma 12 follows, that for any $\mu \in (0,1)$ set $\Omega(F_\mu|_{\overline{K}_2})$ is finite and consists of fixed points $A_3(\mu-1;1)$, $A_1(0;\mu^2)$; for $\mu = 1$ the equality is fulfilled $A_3(\mu-1;1) = A_1(0;\mu^2)$ and the set $\Omega(F_1|_{\overline{K}_2})$ consists of the single point $A_1(0;1)$.

By formula (1) we obtain, that for any $\mu \in (0,1]$ the following equalities holds

$$F_\mu(Ox) = \{0\} \times [0,+\infty), \quad F_\mu(Oy) = (0;\mu^2).$$

Therefore, for any $\mu \in (0,1]$ the equality is fulfilled

$$(Ox \cup Oy) \cap \Omega(F_\mu) = A_1(0;\mu^2).$$

In addition, for any $\mu \in (0,1)$ the global stable manifold $W^s(A_1)$ of sink A_1 and the global unstable manifold $W^u(A_3)$ of saddle point A_3 intersect transversally (see Figure 4). Therefore, we obtain, that for any $\mu \in (0,1)$ the map F_μ is a (nonsingular) Morse-Smale endomorphism. For $\mu = 1$ there exist the nonhyperbolic point $A_1(0;1)$ and map F_1 is the singular Morse-Smale endomorphism.

Thus, Theorem 1 is proved.

Acknowledgments
The author thanks Professor Efremova L. S. for helpful discussions.

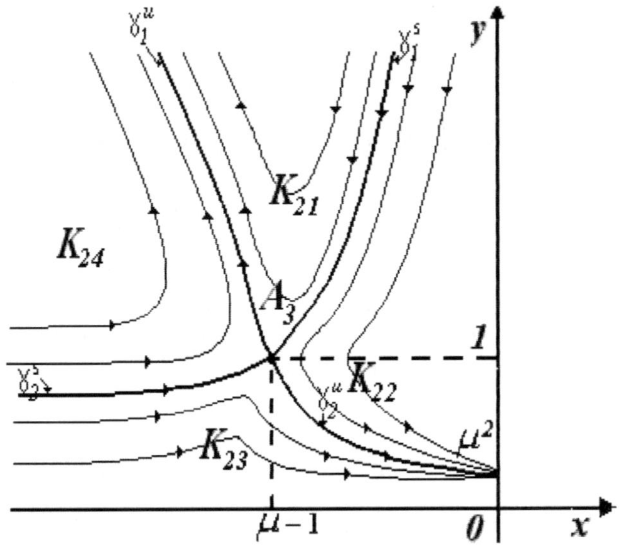

Figure 4. Asymptotic behaviour of F_μ – trajectories in K_2.

References

[1] C Mira 1987 *Chaotic dynamics from the One-Dimensional Endomophism to the Two-Dimensional Diffeomorphism* (Singapore: World Scientific)
[2] C E Frouzakis, L Gardini, I G Kevrekidis, G Millereoux, C Mira 1997 On some properties of invariant sets of two-dimensional noninvertible maps *Intern. Journ. Bifurcations and Chaos* **7** (6) 1167–1194
[3] M-C Li and M Malkin 2004 Bounded nonwandering sets for polynomial mappings *Dynam. Control Syst.* **10** (3) 377-389
[4] M Lyubich 1997 Dynamics of quadratic polynomials I-II *Acta Mathematica* **178** 185-297
[5] M Henon 1969 Numerical study of quadratic area preserving mapping *Appl. Math.* **27** 291-312
[6] S Friedland, J Milnor 1989 Dynamical properties of plane polynomial automorphisms, *Ergod. Theory Dyn. Syst.* **9** 67–99
[7] S S Bel'mesova, L S Efremova 2015 On the Concept of Integrability for Discrete Dynamical Systems *Nonlinear maps and their applications. Springer Proceedings in Mathematics and Statistics* **112** 127–158
[8] S S Belmesova, L S Efremova 2013 A one-parameter family of quadratic maps of a plane including Morse Smale endomorphisms *Russian Math.* **8** 70–74
[9] S Belmesova, L Efremova 2010 On Quadratic Maps from Some One-Parameter Family Closed to Unperturbed Map *Proceed. of MIPT* **2** (2) 46–57 (Russian)
[10] S S Belmesova, L S Efremova, D Fournier-Prunaret 2014 Invariant curves of quadratic maps on the plane from the one-parameter family containing the trace map *ESAIM: Proceedings and Surveys* **46** 98-110
[11] M Brin, Ya Pesin 1996 On Morse-Smale endomorphisms *American Math. Soc. Transl.* **171** (2) 35–45
[12] D Azimov 1975 Round handles and non-singular Morse-Smale flows, *Ann. Math.* **102** 41-54
[13] A N Sharkovskii, Yu L Maistrenko and E Yu Romanenko 1986 *Difference Equations and Their Applications* (Nauk. Dumka, Kiev) (Ukrainian)
[14] A A Andronov, E A Leontovich, I I Gordon, A G Maier 1966 *Qualitative Theory of Second-Order Dynamic Systems* (M: Science) (Russian)
[15] N P Erugin 1956 *Implicit Functions* (Leningrad, Leningrad University Press) (Russian)
[16] A Katok, B Hasselblat 1999 *Introduction to the Modern Theory of Dynamical Systems* (Factorial, Moscow) (Russian)

IOP Conf. Series: Journal of Physics: Conf. Series **990** (2018) 012002 doi:10.1088/1742-6596/990/1/012002

Superposition principle for the continuity equation in a bounded domain

P Bonicatto[1] and N A Gusev[2]

[1] SISSA, Via Bonomea, 265, Trieste TS, 34136, Italy

[2] Steklov Mathematical Institute of Russian Academy of Sciences 8 Gubkina St, Moscow, 119991, Russia *and* RUDN University 6 Miklukho-Maklay St, Moscow, 117198

E-mail: [1] paolo.bonicatto@sissa.it, [2] n.a.gusev@gmail.com

Abstract. We consider an initial–boundary value problem for the continuity equation in a class of non-negative measure-valued solutions. We prove that any solution in the considered class can be represented as a superposition of elementary solutions, associated with the solutions of the corresponding ordinary differential equation.

1. Introduction

Let $b\colon [0,T] \times \mathbb{R}^d \to \mathbb{R}^d$ be a Borel measurable uniformly bounded time-dependent vector field, where $T > 0$ and $d \in \mathbb{N}$.

Given $x \in \mathbb{R}^d$ let δ_x denote the Dirac measure concentrated at x. With the term *measure* in this work we will, by default, refer to a Borel measure on some underlying space, which will be always clear from the context.

It is well-known that if an absolutely continuous function $\gamma\colon [0,T] \to \mathbb{R}^d$ solves

$$\partial_t \gamma(t) = b(t, \gamma(t)) \quad \text{a.e. } t \in (0,T) \tag{ODE}$$

then the family $\{\mu_t\}_{t\in[0,T]}$ of measures on $\mathbb{R}^d$

$$\mu_t := \delta_{\gamma(t)}, \quad t \in [0,T], \tag{1}$$

solves the continuity equation

$$\partial_t \mu_t + \operatorname{div}_x(b\mu_t) = 0 \quad \text{in } \mathscr{D}'(Q) \tag{CE}$$

where $Q = (0,T) \times \mathbb{R}^d$. Solutions of (CE) having the form (1) will be called *elementary solutions* (associated with solutions γ of (ODE)).

Let

$$\Gamma_b := \{\gamma \in \Gamma \colon \gamma \text{ is absolutely continuous and solves (ODE)}\}. \tag{2}$$

Note that the set Γ_b is Borel since b is Borel, see [13], Proposition 2.

By linearity of (CE) any linear combination of elementary solutions will solve (CE) as well. More generally, suppose that η is a (possibly signed) Borel measure on

$$\Gamma := C([0,T]; \mathbb{R}^d) \tag{3}$$

such that η is concentrated on the set Γ_b (i.e. satisfies $|\eta|(\Gamma \setminus \Gamma_b) = 0$, where $|\cdot|$ denotes the total variation). Then family $\{\mu_t\}_{t\in[0,T]}$ of measures

$$\mu_t := \int_\Gamma \delta_{\gamma(t)}\, d\eta(\gamma) \tag{4}$$

solves (CE). (Eq. (4) can be understood in the following sense: for any Borel set $E \subset \mathbb{R}^d$ it holds that $\mu_t(E) = \int_\Gamma 1_E(\gamma(t))\, d\eta(\gamma)$, where 1_E is the indicator of E.)

In other words, any *superposition* of the elementary solutions of (CE) solves (CE) as well. It is therefore natural to ask whether *any* measure-valued solution of (CE) can be represented as a superposition of elementary solutions.

Recall that a family $\{\mu_t\}_{t\in[0,T]}$ of finite measures on $\mathbb{R}^d$ is called a *measure-valued solution* of (CE) if

(i) $\{\mu_t\}_{t\in[0,T]}$ is a Borel family, i.e. the map $t \mapsto \mu_t(E)$ is Borel measurable for any Borel set $E \subset \mathbb{R}^d$;

(ii) $\int_0^T |\mu_t|(\mathbb{R}^d)\, dt < \infty$;

(iii) Eq. (CE) holds.

Points 1 and 2 here imply that for any bounded Borel function $g\colon [0,T] \times \mathbb{R}^d$ the function $t \mapsto \int_{\mathbb{R}^d} g(t,x)\, d\mu_t(x)$ is Borel measurable (see e.g. [1]), therefore the distributional formulation of (CE) is well-defined.

For the class of *non-negative* measure-valued solutions of (CE) the question stated above was positively answered by Ambrosio, Gigli and Savare [11, 2]:

Theorem 1 (Superposition principle). *Let $\{\mu_t\}_{t\in[0,T]}$ be a non-negative measure-valued solution of* (CE). *Then there exists a non-negative measure η on Γ such that* (4) *holds for a.e. $t \in [0,T]$.*

Apart from providing a precise characterisation of non-negative measure-valued solutions of (CE), this result can be used together with additional assumptions. For instance, in [5] it was involved in the proof of uniqueness of bounded weak solutions of the continuity equation when $d = 2$ and b is nearly incompressible, autonomous and has bounded variation. This result was recently generalized in [12] for multidimensional non-autonomous nearly incompressible BV vector fields.

In the present work we provide a generalization of Theorem 1 for the case of a bounded domain (instead of the whole space). Solvability of the initial–boundary value problem for the continuity equation with a non-smooth vector field b was studied in [6] (under some weak differentiability assumptions on b).

Let $\Omega \subset \mathbb{R}^d$ be an open set and let $Q := (0,T) \times \Omega$. Suppose that $b\colon \Omega \times \mathbb{R}^d \to \mathbb{R}^d$ is a uniformly bounded Borel time-dependent vector field. We will assume that b is extended with 0 to $\mathbb{R}^{d+1} \setminus Q$. Similarly, for any family $\{\mu_t\}_{t\in[0,T]}$ of Borel measures on Ω we will extend the measure μ_t (for each $t \in [0,T]$) with 0 outside of Ω.

Let

$$\Gamma_\Omega := \{\gamma \in \Gamma \colon \gamma^{-1}(\Omega) \neq \emptyset\}$$

denote the set of curves intersecting Ω. Clearly Γ_Ω is an open subset of Γ.

In order to take into account the initial and boundary conditions for the continuity equation we will introduce some auxiliary maps. (In Lemma 4 below we prove that these maps are Borel measurable.)

Definition 1. *Given $\gamma \in \Gamma$ let*

$$S_\gamma := \inf(\gamma^{-1}(\Omega) \cup \{T\}), \qquad E_\gamma := \sup(\gamma^{-1}(\Omega) \cup \{0\})$$

If $\gamma \in \Gamma_\Omega$ then S_γ and E_γ are the *times when* γ *starts* and, respectively, *ends* in Ω.

Now we are ready to state our main result:

Theorem 2. *Let $\{\mu_t\}_{t\in[0,T]}$ be a Borel family of non-negative measures on $\mathbb{R}^d$ such that $\int_0^T \mu_t(\Omega)\,dt < \infty$ and $\int_0^T \mu_t(\mathbb{R}^d \setminus \Omega)\,dt = 0$. Let ν be a signed measure on $\mathbb{R}^{d+1}$ concentrated on ∂Q. Suppose that*

$$\partial_t \mu_t + \operatorname{div}(b\mu_t) = \nu \quad \text{in } \mathscr{D}'(\mathbb{R}^{d+1}) \tag{5}$$

Then there exists a non-negative measure η on Γ such that

(i) for η-a.e. γ we have $S_\gamma < E_\gamma$ and
- *$\gamma(t) = \gamma(0)$ for $t \in [0, S_\gamma]$,*
- *$\gamma(t) \in \Omega$ for a.e. $t \in [S_\gamma, E_\gamma]$,*
- *$\gamma(t) = \gamma(T)$ for $t \in [E_\gamma, T]$,*
- *γ solves the ODE $\dot\gamma(t) = b(t, \gamma(t))$ a.e. on $[S_\gamma, E_\gamma]$;*

(ii) for a.e. $t \in [0,T]$ and any $\varphi \in \mathscr{D}(\mathbb{R}^d)$

$$\int_{\mathbb{R}^d} \varphi(x)\,d\mu_t(x) = \int_\Gamma \varphi(\gamma(t)) 1_\Omega(\gamma(t))\,d\eta(\gamma);$$

(iii) for any $\Phi \in \mathscr{D}(\mathbb{R}^{d+1})$

$$\int_{\mathbb{R}^{d+1}} \Phi(t,x)\,d\nu(t,x) = \int_\Gamma (\Phi(S_\gamma, \gamma(S_\gamma)) - \Phi(E_\gamma, \gamma(E_\gamma)))\,d\eta(\gamma), \tag{6a}$$

$$\int_{\mathbb{R}^{d+1}} \Phi(t,x)\,d|\nu|(t,x) = \int_\Gamma (\Phi(S_\gamma, \gamma(S_\gamma)) + \Phi(E_\gamma, \gamma(E_\gamma)))\,d\eta(\gamma). \tag{6b}$$

Note that (5) is not a definition of the solution of the initial–boundary value problem for the continuity equation, but a property of such solutions. In particular, this property (i.e. existence of the measure ν) holds for the distributional solutions introduced in [6, Definition 3.4]. The motivation for such setting is the following. Let $U \subset \mathbb{R}^n$ be a bounded open set with a C^1 boundary. Suppose that $V \colon U \to \mathbb{R}^n$ is a bounded measurable vector field and $\operatorname{div} V = 0$ (in sense of distributions) inside U. Let us extend V with 0 to $\mathbb{R}^n \setminus U$. Then by [7] there exists a bounded Borel measurable function $g \colon \partial U \to \mathbb{R}$ such that

$$\operatorname{div}(1_U V) = g\mathscr{H}^{n-1}|_{\partial U} \quad \text{in } \mathscr{D}'(\mathbb{R}^n),$$

where $\mathscr{H}^k$ denotes the k-dimensional Hausdorff measure and $\mu|_E$ denotes the restriction of a measure μ to the set E. From this viewpoint the measure ν in (5) represents both initial and boundary conditions.

2. Currents and their properties

The proof of Theorem 2 is based on the structure of *normal 1-currents*, therefore first we would like to recall some relevant notation.

2.1. Polyvectors and covectors

Let $\Lambda_m(\mathbb{R}^n) \equiv \Lambda_m\mathbb{R}^n$ denote the space of m-vectors (polyvectors) on $\mathbb{R}^n$. Let $\Lambda^m(\mathbb{R}^n) \equiv \Lambda^m\mathbb{R}^n$ denote the space of m-forms (covectors) on $\mathbb{R}^n$.

The *inner product* on $\Lambda_m\mathbb{R}^n$ is defined as follows. For simple polyvectors we set

$$(v_1 \wedge \cdots \wedge v_m, w_1 \wedge \cdots \wedge w_m) := \det \|(v_i, w_j)\| \tag{7}$$

Then we extend bilinearly the inner product $(\cdot,\cdot)$ for non-simple polyvectors.

The inner product defined above induces the *euclidean norm* on $\Lambda_m\mathbb{R}^n$:

$$|v| := \sqrt{(v,v)}. \tag{8}$$

2.2. Differential forms and currents

Let $U \subset \mathbb{R}^n$ be an open set. Let $\mathscr{D}^m(U) := \mathscr{D}(U; \Lambda^m \mathbb{R}^n)$ denote the space of smooth m-forms. By definition, an *m-dimensional current* T is a linear continuous functional on $\mathscr{D}^m(U)$. Let

$$\mathscr{D}_m(U) := \mathscr{D}(U; \Lambda^m \mathbb{R}^n)' \tag{9}$$

denote the space of m-dimensional currents. It is clear that $\Lambda_m(\mathbb{R}^n) \subset \mathscr{D}_m(\mathbb{R}^n)$. More generally, any polyvector field is a current.

The *mass* of $T \in \mathscr{D}_m(\mathbb{R}^n)$ is defined as

$$M(T) := \sup\langle T, \omega \rangle, \tag{10}$$

where the supremum is taken over all differential forms $\omega \in \mathscr{D}^m(\mathbb{R}^n)$ such that

$$|\omega|_\infty := \sup_{x \in \mathbb{R}^d} |\omega(x)| \leq 1. \tag{11}$$

(We use here the euclidean norm of $\Lambda^m(\mathbb{R}^n)$.)

Definition 2. *A current S is called a* subcurrent *of a current T with $M(T) < \infty$ if*

$$M(T - S) + M(S) = M(T). \tag{12}$$

If (12) *holds then we write $S \subset T$.*

2.3. Operations with currents

The *boundary* $\partial T \in \mathscr{D}_{m-1}(\mathbb{R}^n)$ of $T \in \mathscr{D}_m(\mathbb{R}^n)$ is defined by

$$\langle \partial T, \omega \rangle := \langle T, d\omega \rangle, \qquad \omega \in \mathscr{D}^{m-1}(\mathbb{R}^n) \tag{13}$$

if $m > 0$. If $m = 0$ then we set $\partial T = 0$.

Let $U \subset \mathbb{R}^n$, $V \subset \mathbb{R}^k$ be open and let $f \colon U \to V \subset \mathbb{R}^k$ be a smooth proper map (i.e. preimage of any compact is compact). The *pushforward* $f_\# T \in \mathscr{D}_m(V)$ of a current $T \in \mathscr{D}_m(U)$, is defined by

$$\langle f_\# T, \omega \rangle := \left\langle T, f^\# \omega \right\rangle \qquad \omega \in \mathscr{D}^m(\mathbb{R}^n), \tag{14}$$

where $f^\# \omega$ is the pullback of ω with respect to f.

Recall the following definition [8, Definition 3.7]:

Definition 3. *Let $T \in \mathscr{D}_m(\mathbb{R}^n)$. A current $C \in \mathscr{D}_m(\mathbb{R}^n)$ is called a* cycle *of T if $C \subset T$ and $\partial C = 0$. The current T is called* acyclic *if $C = 0$ is the only cycle of T.*

2.4. Examples of currents

The simplest example of a current is an oriented segment. Let $[a, b] \subset \mathbb{R}$. We define

$$\langle [\![a, b]\!], \omega \rangle := \int_a^b \omega \tag{15}$$

for any $\omega \in \mathscr{D}^1(\mathbb{R})$ (smooth compactly supported 1-form). Clearly $[\![a, b]\!] \in \mathscr{D}_1(\mathbb{R})$.

More generally, suppose that $v \in L^1(\mathbb{R}^n; \mathbb{R}^n)$. Given $\omega \in \mathscr{D}^1(\mathbb{R}^n)$ define

$$\langle P, \omega \rangle := \int_{\mathbb{R}^n} \langle \omega(x), v(x) \rangle \, dx.$$

Then P is a 1-current on $\mathbb{R}^n$.

If $\theta\colon [a,b] \mapsto \mathbb{R}^n$ is Lipschitz and $\theta' \neq 0$ a.e., then we define

$$[\![\theta]\!] := \theta_\# [\![a,b]\!] \tag{16}$$

Clearly $[\![\theta]\!] \in \mathscr{D}_1(\mathbb{R}^n)$. Let us show that if, in addition, θ is a simple curve with fixed orientation (or even if it intersects itself in finitely many points) then $[\![\theta]\!]$ does not depend on the choice of the parametrization θ.

Let $\mathscr{L}^n$ denote the Lebesgue measure and let $\mathscr{H}^k$ denote the k-dimensional Hausdorff measure on $\mathbb{R}^n$.

From the definition of the pullback of a differential form (see e.g. [3], def. 6.2.7) it follows that

$$(\theta^\# \omega)(t) = \big\langle \omega(\theta(t)), \theta'(t) \big\rangle \, dt \tag{17}$$

Then using Area formula we get

$$\begin{aligned} \langle [\![\theta]\!], \omega \rangle &= \left\langle [\![a,b]\!], \theta^\# \omega \right\rangle \\ &= \int_a^b \big\langle \omega(\theta(t)), \theta'(t) \big\rangle \, dt \\ &= \int_a^b \langle \omega(x), \tau(x) \rangle \, d\mathscr{H}^1(x), \end{aligned} \tag{18}$$

where $\tau(\theta(t)) = \theta'(t)/|\theta'(t)|$ for a.e. $t \in [a,b]$. In other words, we have proved that

$$[\![\theta]\!] = \tau \mathscr{H}^1 \llcorner \theta([a,b]), \tag{19}$$

where $\mu \llcorner A$ denotes the restriction of the measure μ on the set A. Hence $[\![\theta]\!]$ does not depend on the choice of the parametrization θ (which preserves the orientation).

2.5. Polar decomposition of a current with finite mass

Given a topological space $\mathfrak{X}$ let $\mathcal{M}(\mathfrak{X})$ denote the set of finite Borel measures on $\mathfrak{X}$. Let $\mathcal{M}_+(\mathfrak{X}) \subset \mathcal{M}(\mathfrak{X})$ denote the set of non-negative measures.

Let $\mu \in \mathcal{M}_+(\mathbb{R}^n)$. Let $f\colon \mathbb{R}^n \to \Lambda_m(\mathbb{R}^n)$ be a Borel polyvector field on $\mathbb{R}^n$ such that $f \in L^1(\mu)$. Then one can define an m-current $T := f\mu$ as follows:

$$\langle T, \omega \rangle = \int_{\mathbb{R}^n} \langle \omega, f \rangle \, d\mu, \qquad \forall \omega \in \mathscr{D}^m(\mathbb{R}^n). \tag{20}$$

Lemma 1. *If $f \in L^1(\mathbb{R}^n, \mu; \Lambda_m \mathbb{R}^n)$ then*

$$M(f\mu) = \int_{\mathbb{R}^n} |f| \, d\mu. \tag{21}$$

Proof. First, notice that $T\colon \omega \mapsto \langle f\mu, \omega \rangle$ can be extended by continuity for any $\omega \in C_b(\mathbb{R}^n; \Lambda^m \mathbb{R}^n)$. Using Lusin's theorem we extend T further to any bounded Borel $\omega\colon \mathbb{R}^n \to \Lambda^m \mathbb{R}^n$. Then in the definition of mass (10) we can take sup over all bounded Borel differential forms.

By (10) $M(f\mu) \le \int_{\mathbb{R}^n} |f| \, d\mu$.

Let us define a Borel differential form $\omega\colon \mathbb{R}^n \to \Lambda^m \mathbb{R}^n$ by

$$\langle \omega, v \rangle := \frac{(f, v)}{|f|}. \tag{22}$$

IOP Conf. Series: Journal of Physics: Conf. Series **990** (2018) 012002 doi:10.1088/1742-6596/990/1/012002

(Without loss of generality we may assume that $f \neq 0$.) Clearly $|\omega| = 1$, and

$$\langle f\mu, \omega\rangle = \int_{\mathbb{R}^n} \langle \omega, f\rangle \, d\mu = \int_{\mathbb{R}^n} |f| \, d\mu. \tag{23}$$

□

All currents with finite mass have the form (20) (see e.g. [3, §7.2]):

Proposition 1 (Polar decomposition). *Let $T \in \mathscr{D}_m(\mathbb{R}^n)$ and suppose that $M(T) < \infty$. Then there exists a Borel polyvector field $f\colon \mathbb{R}^n \to \Lambda_m(\mathbb{R}^n)$ such that $\|f(x)\| = 1$ for all $x \in \mathbb{R}^n$ and a finite non-negative Borel measure μ on $\mathbb{R}^n$ such that (20) holds.*

(Proposition 1 essentially follows from the Riesz representation theorem.)

It is easy to see that this decomposition is unique in the following sense: if $T = g\nu$ for some other non-negative Borel measure ν on $\mathbb{R}^n$ and Borel polyvector field $g\colon \mathbb{R}^n \to \Lambda_m(\mathbb{R}^n)$ such that $\|g(x)\| = 1 \ \forall x \in \mathbb{R}^n$ then $\nu = \mu$ (as measures) and the equality $f = g$ holds μ-a.e. We will therefore denote

$$|T| := \mu \tag{24}$$

where μ is the measure provided by Proposition 1.

Clearly for the current $[\![\theta]\!]$ defined in (16) one has

$$|[\![\theta]\!]| = \mathscr{H}^1 \llcorner \theta([a,b]) \quad \text{and} \quad |\partial[\![\theta]\!]| = \begin{cases} \delta_{\theta(0)} + \delta_{\theta(1)}, & \theta(0) \neq \theta(1); \\ 0, & \theta(0) = \theta(1). \end{cases} \tag{25}$$

Lemma 2. *Suppose that $\mu = \mu_1 + \mu_2$ and $\mu_1 \perp \mu_2$ are non-negative finite Borel measures. Let $f \in L^1(\mathbb{R}^n, \mu; \Lambda^m\mathbb{R}^n)$. Then*

$$M(\mu f) = M(\mu_1 f) + M(\mu_2 f). \tag{26}$$

Proof. The proof follows immediately from (21).

□

The generic form of subcurrent of a current with finite mass can be described as follows:

Proposition 2. *Let $\mu \in \mathcal{M}_+(\mathbb{R}^n)$ and $f \in L^1(\mathbb{R}^n, \mu; \Lambda^m\mathbb{R}^n)$. Then $S \in \mathscr{D}_m(\mathbb{R}^n)$ is a subcurrent of $T := f\mu$ if and only if*

$$S = g\mu \tag{27}$$

where $g \in L^1(\mathbb{R}^n, \mu; \Lambda^m\mathbb{R}^n)$ satisfies

$$g(x) \in \operatorname{conv}\{0, f(x)\} := \{tf(x) \mid t \in [0,1]\} \tag{28}$$

for μ-a.e. $x \in \mathbb{R}^n$.

Proof. By Proposition 1 there exists $\nu \in \mathcal{M}_+(\mathbb{R}^n)$ and $g \in L^1(\mathbb{R}^n, \mu; \Lambda^m\mathbb{R}^n)$ such that $S = g\nu$. By Radon-Nikodym theorem there exist mutually singular $\nu^a, \nu^s \in \mathcal{M}_+(\mathbb{R}^n)$ such that $\nu = \nu^a + \nu^s$, $\nu^a \ll \mu$ and $\nu^s \perp \mu$. Then

$$\begin{aligned} M(f\mu) &\overset{(12)}{=} M(f\mu - g\nu) + M(g\nu) \\ &\overset{(26)}{=} M(f\mu - g\nu^a - g\nu^s) + M(g\nu^a) + M(g\nu^s) \\ &\overset{(26)}{=} M(f\mu - g\nu^a) + M(g\nu^a) + 2M(g\nu^s). \end{aligned}$$

IOP Conf. Series: Journal of Physics: Conf. Series **990** (2018) 012002 doi:10.1088/1742-6596/990/1/012002

On the other hand by triangle inequality

$$M(f\mu) \leq M(f\mu - g\nu^a) + M(g\nu^a)$$

hence $2M(g\nu^s) = 0$, i.e. $\nu^s = 0$.

Therefore without loss of generality we can assume that $S = g\mu$.

Again using (12) we get

$$\int_{\mathbb{R}^n} (|f-g| + |g|)\, d\mu = \int_{\mathbb{R}^n} |f|\, d\mu. \tag{29}$$

Let $E \subset \mathbb{R}^n$ be a Borel set. By triangle inequality $|f(x)| \leq |f(x) - g(x)| + |g(x)|$ for any $x \in \mathbb{R}^n$ hence

$$\int_{\mathbb{R}^n \setminus E} (|f-g| + |g|)\, d\mu \geq \int_{\mathbb{R}^n \setminus E} |f|\, d\mu. \tag{30}$$

Subtracting (30) from (29) we get

$$\int_E (|f-g| + |g|)\, d\mu \leq \int_E |f|\, d\mu. \tag{31}$$

Again by triangle inequality the opposite inequality holds, hence in fact (31) is an equality. By arbitrariness of E we conclude that

$$|f(x) - g(x)| + |g(x)| = |f(x)| \tag{32}$$

for μ-a.e. $x \in \mathbb{R}^n$. Let $f_1 := \frac{(g,f)}{|f|^2} f$, $f_2 := \frac{(f-g,f)}{|f|^2} f = f - f_1$. Let $h := g - f_1$. Then $f - g - f_2 = -h$. Since the norm is euclidean,

$$|f-g| + |g| = |f_1| + |f_2| + 2|h|.$$

By triangle inequality $|f| \leq |f_1| + |f_2|$ hence using (32) and the equality above we get $h = 0$. Consequently $|f_1| \leq |f|$.

□

Corollary 1. *Suppose that $\mu \in \mathcal{M}_+(\mathbb{R}^n)$, $f \in L^1(\mathbb{R}^n, \mu; \Lambda_1\mathbb{R}^n)$ has the components $(f_1, \ldots, d_n)$. If $f_1 > 0$ μ-a.e. then the current $P := f\mu$ is acyclic.*

Proof. Let $C \subset P$ be a cycle of P. By Proposition 2 $C = g\mu$ with some g such that $g_1 \geq 0$. For every $k \in \mathbb{N}$ set $\omega^k(x) := x_1 \vee k \wedge (-k)$. Then ω^k is continuous and bounded, so we can test and deduce that

$$0 = \langle \partial C, \omega^k \rangle = \langle C, d\omega^k \rangle = \int_{\{x_1 \in [-k,+k]\}} g_1(x)\, d\mu(x) = \int_{\mathbb{R}^n} 1_{\{x_1 \in [-k,+k]\}}(x) g_1(x)\, d\mu(x) \to \int_{\mathbb{R}^n} g_1 d\mu(x)$$

where the passage to the limit is allowed by dominated convergence ($g_1 \in L^1(\mu)$).

□

2.6. Decomposition of normal 1-currents

Our main tool is the decomposition result for normal 1-currents obtained originally in [9] for finite-dimensional Euclidean spaces, and generalized in [8] for arbitrary complete separable metric spaces. Given metric spaces M, N let $\mathrm{Lip}(M, N)$ denote the metric space of bounded Lipschitz continuous maps from M to N, endowed with the standard sup norm. We state the decomposition result for $\mathbb{R}^n$ following [8]:

Theorem 3 (Smirnov; Paolini, Stepanov). *Let P be an acyclic normal 1-current on $\mathbb{R}^n$. Then there exists a finite measure $\xi \geq 0$ on $X := \mathrm{Lip}([0,1]; \mathbb{R}^n)$ such that*

(i) ξ-a.e. θ is simple;

(ii) for any $\omega \in \mathscr{D}^1(\mathbb{R}^n)$

$$\langle P, \omega\rangle = \int_X \langle [\![\theta]\!], \omega\rangle \, d\xi(\theta),$$

where $[\![\cdot]\!]$ was introduced in (16);

(iii) for any $\varphi \in \mathscr{D}^1(\mathbb{R}^n)$

$$\int_{\mathbb{R}^n} \varphi \, d|P| = \int_X \left(\int_{\mathbb{R}^n} \varphi(x) \, d|[\![\theta]\!]|(x) \right) d\xi(\theta),$$

where $|[\![\theta]\!]|$ is the total variation of the current $[\![\theta]\!]$, see (24) and (25);

(iv) for any $\varphi \in \mathscr{D}^1(\mathbb{R}^n)$

$$\langle |\partial P|, \omega\rangle = \int_X \langle |\partial[\![\theta]\!]|, \omega\rangle \, d\xi(\theta).$$

We will refer to any measure ξ given by Theorem 3 as to a *decomposition of P*.

Similarly to the proof of Lemma 1 one can observe that the equalities (1)–(3) of Theorem 3 hold not only for the smooth test functions and forms φ and ω, but also for merely bounded and Borel ones.

Let $\varphi \in \mathscr{D}(\mathbb{R}^n)$ and $\omega := d\varphi$. By Theorem 3(2)

$$\langle \partial P, \varphi\rangle = \int_X \langle \partial[\![\theta]\!], \varphi\rangle \, d\xi(\theta). \tag{33}$$

The following result was obtained in [10]:

Proposition 3. *Let $\mu \in \mathcal{M}_+(\mathbb{R}^n)$. Suppose that $B\colon \mathbb{R}^n \to \mathbb{R}^n$ is Borel measurable, $B \neq 0$ on $\mathbb{R}^n$ and $B \in L^1(\mathbb{R}^n, \mu)$. Suppose that 1-current $P := B\mu$ is normal. Then for any decomposition $\xi \in \mathcal{M}(X)$ of P (provided by Theorem 3) for ξ-a.e. $\theta \in X$ it holds that*

$$\dot\theta(s) = |\dot\theta(s)| \frac{B(\theta(s))}{|B(\theta(s))|} \quad \textit{for a.e. } s \in [0,1]. \tag{34}$$

For convenience of the reader let us present the proof of this result.

Proof. We apply Theorem 3 (2) with $\omega := |B|^{-1} B_i dx^i$:

$$\int_{\mathbb{R}^n} 1 \, d\mu = \int_X \int_0^1 \theta'(s) \cdot \frac{B}{|B|}(\theta(s)) \, ds \, d\xi(\theta).$$

Since $|P| = \mu$, from Theorem 3 (3) with $\varphi \equiv 1$ it follows that

$$\int_{\mathbb{R}^n} 1 \, d\mu = \int_X \int_0^1 |\theta'(s)| \, ds \, d\xi(\theta).$$

IOP Conf. Series: Journal of Physics: Conf. Series **990** (2018) 012002 doi:10.1088/1742-6596/990/1/012002

The equalities above imply that

$$\int_X \int_0^1 (|\theta'(s)| - \theta'(s) \cdot \frac{B}{|B|}(\theta(s)))\, ds\, d\xi(\theta) = 0.$$

By Cauchy–Bunyakovsky inequality the integrand is non-negative. Hence for ξ-a.e. $\theta \in X$

$$|\theta'(s)| - \theta'(s) \cdot \frac{B}{|B|}(\theta(s)) = 0$$

holds for a.e. $s \in [0,1]$. Clearly this equality is only possible when $\theta'(s) = |\theta'(s)|\frac{B}{|B|}(\theta(s))$.

□

Corollary 2. *In addition to the assumptions of Proposition 3 suppose that* $B_1 \equiv 1$ *on* $\mathbb{R}^n$ *and there exists* $C > 0$ *such that* $|B_j| \le C$ *for any* $j \in 2, 3, \ldots, n$. *Then for any* $\tau, \tau' \in [0,1]$ *and any* $j \in 2, 3, \ldots, n$

$$|\theta_1(\tau') - \theta_1(\tau)| \ge \frac{1}{C}|\theta_j(\tau') - \theta_j(\tau)| \tag{35}$$

Proof. Since $B_1 = 1$ and $|B_j| \le C$ the following inequalities hold:

$$\begin{cases} B_1(z) + \frac{1}{C}B_j(z) & \ge 0, \\ B_1(z) - \frac{1}{C}B_j(z) & \ge 0. \end{cases}$$

Substituting $z = \theta(s)$ and multiplying these inequalities by $|\dot\theta(s)|\frac{1}{|B(\theta(s))|}$ in view of (34) we get

$$\begin{cases} \dot\theta_1(s) + \frac{1}{C}\dot\theta_j(s) & \ge 0, \\ \dot\theta_1(s) - \frac{1}{C}\dot\theta_j(s) & \ge 0. \end{cases}$$

Integrating with respect to s from τ to τ' (observe that $B_1 \ge 0$) we get the desired result.

□

3. Continuity equation and 1-currents

In this section, similar to section 2.6, we denote $X := \mathrm{Lip}([0,1]; \mathbb{R}^n)$, where $n = d+1$ is the dimension of the spacetime.

Before proving Theorem 2 it remains to state several technical propositions.

Let

$$X_Q := \{\theta \in X : \theta^{-1}(Q) \neq 0\}.$$

Clearly X_Q is an open subset of X.

Given $c > 0$ let X_c denote the closed subspace of X consisting of $\theta\colon [0,1] \to \mathbb{R}^{d+1}$ with the components $(\theta_1, \ldots, \theta_{d+1})$ such that

- $s \mapsto \theta_1(s)$ is non-decreasing;
- for any $\tau, s \in [0,1]$ and any $j \in 2, \ldots, d+1$

$$|\theta_j(\tau) - \theta_j(s)| \le c|\theta_1(\tau) - \theta_1(s)|. \tag{36}$$

One can say that the elements of X_c are *timelike* curves in the spacetime, which are confined by a family of cones with a fixed shape.

Consider $\theta \in X_c$ with the components $(\theta_1, \ldots, \theta_{d+1})$. By definition of X_c for any $t \in \theta_1([0,1])$ the equation

$$\theta_1(s) = t$$

has at least one solution, i.e. $\theta_1^{-1}(t) \neq \emptyset$. Let

$$s_\theta(t) := \begin{cases} 0, & t < \theta_1(0); \\ \min \theta_1^{-1}(t), & t \in \theta_1([0,1]); \\ 1, & t > \theta_1(1). \end{cases}$$

Let us define the function $\gamma_\theta \colon [0,T] \to \mathbb{R}^d$ as the composition

$$\gamma_\theta := (\theta_2, \ldots, \theta_{d+1}) \circ s_\theta.$$

In general θ_1 may be not injective and therefore s_θ might be discontinuous. Nevertheless γ_θ is Lipschitz continuous by (36).

Lemma 3. *The map*

$$g \colon X_c \ni \theta \mapsto \gamma_\theta \in \Gamma$$

is continuous.

Proof. Consider $\theta, \vartheta \in X_c$ and suppose that $\varepsilon := \|\gamma_\vartheta - \gamma_\theta\| > 0$. Let $t \in [0,T]$ be such that $|\gamma_\vartheta(t) - \gamma_\theta(t)| = \varepsilon$. Let $z := s_\vartheta(t)$. By (36) the image $\theta([0,1])$ is contained in the cone

$$K_t := \{(\tau, \xi) \in \mathbb{R} \times \mathbb{R}^d : \|\xi - \gamma_\theta(t)\| \leq c|\tau - t|\}.$$

Hence

$$\|\vartheta - \theta\| \geq \operatorname{dist}(\vartheta(z), \theta(z)) \geq \operatorname{dist}(\vartheta(z), K_t) \geq \varepsilon \sin\varphi,$$

where $\varphi = \arctan \frac{1}{c}$. Therefore g is Lipschitz continuous with the Lipschitz constant $\frac{1}{\sin\varphi}$. □

Lemma 4. *The maps $\Gamma_\Omega \ni \gamma \mapsto S_\gamma \in \mathbb{R}$ and $\Gamma_\Omega \ni \gamma \mapsto E_\gamma \in \mathbb{R}$ are Borel measurable.*

Proof. Let $\bar\gamma \in \Gamma_\Omega$. For any $\varepsilon > 0$ there exists $t \in [S_{\bar\gamma}, S_{\bar\gamma} + \varepsilon)$ such that $\gamma(t) \in \Omega$. Since Ω is open we can always find $\delta > 0$ such that $B_\delta(\bar\gamma(t)) \subset \Omega$. Hence for any $\gamma \in B_\delta(\bar\gamma)$ it holds that

$$S_\gamma \leq t < S_{\bar\gamma} + \varepsilon$$

Therefore $\gamma \mapsto S_\gamma$ is upper semicontinuous. By an analogous argument the map $\gamma \mapsto E_\gamma$ is lower semicountinuous. Consequently, both maps are Borel measurable. □

Now we are ready to prove our main result:

of Theorem 2. Let $B \colon \mathbb{R}^{d+1} \to \mathbb{R}^{d+1}$ be the vector field with components

$$B(z) := (1, b_1(z), b_2(z), \ldots, b_d(z)), \qquad z \in \mathbb{R}^{d+1}.$$

Let us introduce the measure μ on $\mathbb{R}^{d+1}$ as follows:

$$\mu(E) = \int_0^T \int_{\mathbb{R}^d} 1_E(t,x)\, d\mu_t(x)\, dt \tag{37}$$

for any Borel set $E \subset \mathbb{R}^{d+1}$, where $1_E(z) = 1_E(z_1, z_2, \dots, z_{d+1})$ is the indicator function for the set E. Notice that μ is a Borel measure, because the family $\{\mu_t\}_{t\in[0,T]}$ is Borel (see e.g. [1]).

It is easy to check that by (5)

$$\operatorname{div}(B\mu) = \nu \quad \text{in } \mathscr{D}'(\mathbb{R}^{d+1}), \tag{38}$$

where the divergence is taken with respect to the space-time $(t, x) = (t, x_1, \dots, x_d)$. (Recall that both μ_t and b are extended with 0 outside of Q.)

Therefore by Corollary 1 the current $P := B\mu$ is acyclic. Hence by Theorem 3 there exists a decomposition ξ of the current P.

By Proposition 3 the measure ξ is concentrated on the solutions of (34). In view of Corollary 2 there exists a constant $C > 0$ such that ξ is concentrated on X_C (which was introduced above in Section 3). Therefore for ξ-a.e. θ we can define $\gamma_\theta := g(\theta)$, where g is the map introduced in Lemma 3.

The measure μ by definition is concentrated on Q. Hence by the equality (3) of Theorem 3 for ξ-a.e. θ it holds that $|[\![\theta]\!]|(\mathbb{R}^{d+1} \setminus Q) = 0$. For any such θ let $S := \{s \in [0,1] : \theta(s) \notin Q\}$ and $\gamma := g(\theta)$. Then $\mathscr{H}^1(\theta(S)) = 0$. It is obvious that $\{t \in \theta_1([0,1]) : \gamma(t) \notin \Omega\} \subset \pi(\theta(S))$, where $\pi\colon [0,T] \times \mathbb{R}^d \ni (t,x) \mapsto t$ is the standard projection. Consequently $\mathscr{L}^1(t \in \theta_1([0,1]) : \gamma(t) \notin \Omega) \le \mathscr{H}^1(\pi(\theta(S))) \le \mathscr{H}^1(\theta(S)) = 0$ (see e.g. (1.4.6) in [4]). Therefore for a.e. $t \in \theta_1([0,1])$ it holds that $\gamma(t) \in \Omega$. Hence $S_\gamma = \theta_1(0)$ and $E_\gamma = \theta_1(1)$. Since ξ-a.e. θ is injective and belongs to X_C, we also have $S_\gamma < E_\gamma$.

By Area formula and (34) for any $t \in [S_\gamma, E_\gamma]$ we have

$$\begin{aligned}
\int_{S_\gamma}^{t} b_j(\tau, \gamma(\tau))\, d\tau &= \int_{\theta_1(0)}^{\theta_1(t)} b_j(\theta_1(s), \gamma(\theta_1(s)))\, \theta_1'(s)\, ds \\
&= \int_{\theta_1(0)}^{\theta_1(t)} b_j(\theta_1(s), \gamma(\theta_1(s)))\, \frac{|\theta'(s)|}{|B(\theta(s))|}\, ds = \int_{\theta_1(0)}^{\theta_1(t)} b_j(\theta(s))\, \frac{|\theta'(s)|}{|B(\theta(s))|}\, ds \\
&= \int_{\theta_1(0)}^{\theta_1(t)} \theta_j'(s)\, ds = \theta_j(\theta_1(t)) - \theta_j(\theta_1(S_\gamma)) = \gamma_j(t) - \gamma_j(S_\gamma),
\end{aligned}$$

where $j = 1, 2, \dots, d$. Hence γ satisfies the ODE $\gamma'(t) = b(t, \gamma(t))$ for a.e. $t \in [S_\gamma, E_\gamma]$.

Let $\Phi = (1_Q \varphi, 0, \dots, 0)$, where $\varphi\colon \mathbb{R}^{d+1} \to \mathbb{R}$ is a bounded Borel function. Then

$$\begin{aligned}
\int_0^T \int_{\mathbb{R}^d} \varphi\, d\mu_t\, dt &= \int_Q B \cdot \Phi\, d\mu = \langle P, \Phi \rangle = \int_{X_C} \langle [\![\theta]\!], \Phi \rangle\, d\xi(\theta) \\
&= \int_{X_C} \int_0^1 \Phi(\theta(s)) \cdot \theta'(s)\, ds\, d\xi(\theta) \\
&= \int_{X_C} \int_0^1 (1_Q \cdot \varphi)(\theta(s)) \theta_1'(s)\, ds\, d\xi(\theta) \\
&= \int_{X_C} \int_0^1 (1_Q \cdot \varphi)(\theta_1(s), \gamma_\theta(\theta_1(s))) \theta_1'(s)\, ds\, d\xi(\theta) \\
&= \int_{X_C} \int_0^T (1_Q \cdot \varphi)(t, \gamma_\theta(t))\, dt\, d\xi(\theta) && \text{(by Area formula)} \\
&= \int_0^T \int_{X_C} (1_Q \cdot \varphi)(t, \gamma_\theta(t))\, d\xi(\theta)\, dt && \text{(by Fubini's theorem)} \\
&= \int_0^T \int_\Gamma (1_Q \cdot \varphi)(t, \gamma(t))\, d\eta(\gamma)\, dt,
\end{aligned}$$

where $\eta := g_{\#}\xi$ is the pushforward of the measure ξ under the map g. The measure η is well-defined by Lemma 3. Hence the claim (3) of Theorem 2 is proved.

It remains to prove claim (4) of Theorem 2. Arguing as above and using (33) we obtain (6a). By Theorem 3(4)

$$\begin{aligned}\int_{\mathbb{R}^{d+1}} \varphi\, d\nu &= \langle |\partial P|, \varphi\rangle \\ &= \int_{X_C} [\varphi(\theta(0)) + \varphi(\theta(1))]\, d\xi(\theta) \\ &= \int_{X_C} [\varphi(\theta_1(0), \gamma_\theta(\theta_1(0))) + \varphi(\theta_1(1), \gamma_\theta(\theta_1(1)))]\, d\xi(\theta) \\ &= \int_{X_C} [\varphi(S_{\gamma_\theta}, \gamma_\theta(S_{\gamma_\theta})) + \varphi(E_{\gamma_\theta}, \gamma_\theta(E_{\gamma_\theta}))]\, d\xi(\theta) \\ &= \int_\Gamma [\varphi(S_\gamma, \gamma(S_\gamma)) + \varphi(E_\gamma, \gamma(E_\gamma))]\, d\eta(\gamma)\end{aligned}$$

Hence (6b) is proved. □

Acknowledgments
The publication was prepared with the support of the "RUDN University Program 5-100".

References
[1] Bogachev V I 2007 *Measure theory* vol 2 (Springer)
[2] Ambrosio L, Gigli N and Savare G 2008 *Gradient Flows* (Springer Nature)
[3] Krantz S G and Parks H R 2008 *Geometric Integration Theory* (Birkhauser)
[4] Ziemer W P 1989 *Weakly differentiable functions* (Springer)
[5] Bianchini S, Bonicatto P and Gusev N A 2016 Renormalization for Autonomous Nearly Incompressible BV Vector Fields in Two Dimensions *SIAM Journal on Mathematical Analysis* **48:1** pp 1–33
[6] Crippa G, Donadello C and Spinolo L V 2014 Initial–boundary value problems for continuity equations with BV coefficients *Journal de mathématiques pures et appliquées* **102:1** pp 79–98
[7] Anzellotti G 1983 Pairings between measures and bounded functions and compensated compactness *Annali di Matematica Pura ed Applicata* **135:1** pp 293–318.
[8] Paolini E and Stepanov E O 2013 Structure of metric cycles and normal one-dimensional currents *J. Funct. Anal.* **264:6** pp 1269–1295.
[9] Smirnov S K 1994 Decomposition of solenoidal vector charges into elementary solenoids, and the structure of normal one-dimensional flows *St. Petersburg Math. J.* **5:4** pp 841–867
[10] Ambrosio L and Bernard P 2008 Uniqueness of signed measures solving the continuity equation for Osgood vector fields *Rendiconti Lincei - Matematica e Applicazioni* **19:3** pp 237–245
[11] Ambrosio L 2005 Transport equation and Cauchy problem for non-smooth vector fields, in *International Mathematical Summer Course "Calculus of variations and non linear partial differential equations"* (Cetraro)
[12] Bianchini S and Bonicatto P 2017 A uniqueness result for the decomposition of vector fields in $\mathbb{R}^d$ (*Preprint* SISSA 15/2017/MATE)
[13] Bernard P 2008 Some remarks on the continuity equation, *Séminaire: Équations aux Dérivées Partielles* (Ecole Polytechnique, Palaiseau)

IOP Conf. Series: Journal of Physics: Conf. Series **990** (2018) 012003 doi:10.1088/1742-6596/990/1/012003

The Trace Map and Integrability of the Multifunctions

Lyudmila S. Efremova

National Research University of Nizhni Novgorod, Gagarin ave. 23a, Nizhni Novgorod, Russia, 603950

E-mail: lefunn@gmail.com

Abstract. In this paper we construct the quotient map (with respect to the special equivalence relation) for the restriction of the trace map $F(x, y) = (xy, (x-2)^2)$ on the invariant subset of the plane that consists of unbounded curves. There is the unique curve in every point of which (with the exception of the unique point) the quotient map is two-valued and upper semicontinuous (but not lower semicontinuous). In other points this map is single-valued and continuous. We introduce also the concept of integrability for multifunctions of the above type, formulate and prove the (necessary) conditions for integrability. These conditions are based on the reduction of the above multifunctions to the multivalued skew products.

1. Introduction

In 90-s years of XX-th century the papers were published where investigation of discrete Schrödinger equation was reduced to investigation of special trace maps (see, e.g., [1, 2, 3, 4]). For example, as it is shown in [2], study of passing and reflection coefficients of the plane wave with a given impulse in the field of the crystal lattice with knots formed Thue-Morse chain, can be reduced to study of the trace map

$$F(x, y) = (xy, (x-2)^2). \tag{1}$$

Geometric properties and the unbounded subset of the set of wandering points of the quadratic trace map given by formula (1), were described in [5]. Different dynamical properties of maps of the one-parameter family $F_\mu(x, y) = (xy, (x-\mu)^2)$ ($\mu \in [0, 2]$) were investigated in [6, 7, 8]. Point out the papers [9, 10] devoted to symbolic aspects of dynamics of the trace map for the weakly coupled Fibonacci hamiltonian.

This paper is the direct continuation of [5]. We need the main result of the work [5] here. For this goal we consider the following subsets of the closure $\overline{K}_1$ of first quadrant K_1:

$$\Delta_2 = \{(x; y) : x; y \geq 0;\ x + y \leq 4\};$$

$$G_{\Delta_2} = \{(x; y) \in \overline{K}_1 : x + y \geq 4\};$$
$$G^{\circ}_{\Delta_2} = \{(x; y) \in \overline{K}_1 : x + y > 4\};$$
$$D_{2,\infty} = \{(x; y) : x \geq 3,\ y \geq 1\};$$

$$\widetilde{G} = G^{\circ}_{\Delta_2} \bigcap \left(\bigcup_{i=0}^{\infty} F^{-i}(D_{2,\infty}) \right),$$

where $F^{-i}(\cdot)$ is i-th complete preimage of a set;

$$G' = G_{\Delta_2} \setminus \widetilde{G}.$$

Theorem. *The set $\widetilde{G}$ is open and everywhere dense in G_{Δ_2} and consists of F-wandering points.*

The set G' is the union of continuous unbounded curves (homeomorphic to an unbounded interval); the cardinality of the set of these curves equals continuum. Moreover, these curves satisfy the following conditions:

- *every two curves either do not intersect each other in G_{Δ_2}, or intersect each other in a unique point on the hypothenuse of the triangle Δ_2; in addition, second property holds iff these two curves form a boundary of a connected component of $\widetilde{G}$;*
- *every tube neighborhood of a curve from G' contains a set of curves from G' of continuum cardinality;*
- *each curve from G' intersects a closed interval $x + y = C$, $0 \le x, y \le C$, where $C > 4$, in a unique point.*

The set G' is F-invariant and nowhere dense in G_{Δ_2} [5].

In this paper we construct the quotient map for the restriction of the trace map F on G' with respect to the special equivalence relation. There is the unique curve of the set G' that consists of points (with the exception of the unique point on the hypothenuse of the triangle Δ_2) where the quotient map is two-valued and upper semicontinuous (but not lower semicontinuous), in other points this map is single-valued and continuous. We introduce also the concept of integrability for multifunctions of the above type, formulate and prove the (necessary) conditions for integrability.

2. Construction of the Quotient Map

2.1. Preliminaries

To construct the quotient map for the trace map F (with respect to the special equivalence relation) we need detailed information about the connected components of the set $\widetilde{G}$ and their boundaries. Let $A_2 = (3, 1)$ be F-fixed point (A_2 is a source), and A^i_k be a homoclinic point of F on the hypothenuse of the triangle Δ_2 of an order $k \ge 1$ ($i = 1, 2, \dots, 2^k - 1$) to the fixed point A_2. (Remind that by *the order* of a homoclinic point (x, y) of the trace map F on the hypothenuse of the triangle Δ_2 to a fixed point (x^*, y^*) we mean the minimal natural number k such that $F^k(x, y) = (x^*, y^*)$.)

In [5] the following unbounded closed domains were correctly defined:

(i) $H^{max}_0(A_2)$ satisfying $D_{2,\infty} \subset H^{max}_0(A_2)$ and $\partial H^{max}_0(A_2) = \eta^{A_2}_l \bigcup \eta^{A_2}_r$, where $\partial(\cdot)$ is the boundary of a set; $\eta^{A_2}_l$ is the graph of the unbounded C^1-smooth strictly decreasing function $y = \eta^{A_2}_l(x)$, which is defined on the interval $(2, 3]$ and has the vertical asymptote $x = 2$; $\eta^{A_2}_r$ is the graph of C^1-smooth strictly decreasing function $y = \eta^{A_2}_r(x)$, which is defined on the unbounded interval $[3, +\infty)$ and has the horizontal asymptote $y = 0$;

(ii) $H^{max}_k(A^i_k)$, where $H^{max}_k(A^i_k)$ is a connected component of k-th complete preimage of H^{max}_0 with respect to $F_{|G_{\Delta_2}}$ containing a homoclinic point A^i_k. Moreover, $\partial H^{max}_k(A^i_k) = \eta_l^{A^i_k} \bigcup \eta_r^{A^i_k}$, where $\eta_l^{A^i_k}$, $\eta_r^{A^i_k}$ are C^1- smooth curves (with the unique common point A^i_k) which have asymptotes such that every of them belongs to the union of the set of connected components of m-th ($0 \le m \le k$) complete preimage (with respect to the map $F_{|G_{\Delta_2}}$) of the critical lines $x = 2$, $x = 0$; and also the axis $y = 0$. (Note that there are curves of the set

$\{\eta_l^{A_k^i},\ \eta_r^{A_k^i}\}$ for $k \geq 1, 1 \leq i \leq 2^k - 1$, which are not graphs of a function of the variable x or a function of the variable y.)

The unbounded closed domain $H_0^{max}(A_2)$ is called *the principal maximal generative domain*, and the unbounded closed domains $H_k^{max}(A_k^i)$ $(k \geq 1,\ 1 \leq i \leq 2^k - 1)$ are called *the maximal generative domains of k-th order.*

Consider, for example, the maximal generative domain $H_1^{max}(A_1^1)$ of first order. Then, by items (i) - (ii), the curve $\eta_l^{A_1^1}$ has the curvelinear asymptote $y = 2/x$ (and, consequently, the vertical asymptote $x = 0$), and the curve $\eta_r^{A_1^1}$ has the vertical asymptote $x = 2$. Note that $y = 2/x$ for $y \geq 2 + \sqrt{2}$ ($x = 2$ for $y \geq 2$) is one of the connected components of the complete preimage of the part of the critical line $x = 2$ for $y \geq 2$ (of the part of the axis $y = 0$ for $x \geq 4$) in G_{Δ_2}.

Remark 1. The trace map F is C^1-diffeomorphism on the set of interior points $(int(\cdot))$ of every maximal generative domain $H_0^{max}(A_2)$, $H_k^{max}(A_k^i)$ $(k \geq 1,\ 1 \leq i \leq 2^k - 1)$, and F is homeomorphism on every maximal generative domain. Hence, the following equalities are valid (see [5]):

$$F(int(H_0^{max}(A_2))) = int(H_0^{max}(A_2)), \qquad F(\eta_l^{A_2}) = \eta_r^{A_2}, \qquad F(\eta_r^{A_2}) = \eta_l^{A_2};$$

$$F(int(H_1^{max}(A_1^1))) = int(H_0^{max}(A_2)), \qquad F(\partial(H_1^{max}(A_1^1))) = \partial(H_0^{max}(A_2));$$

$$F(int(H_k^{max}(A_k^i))) = int(H_{k-1}^{max}(A_{k-1}^{i'})), \qquad F(\partial(H_k^{max}(A_k^i))) = \partial(H_{k-1}^{max}(A_{k-1}^{i'})),\ \ k \geq 2.$$

The map F preserves the order of points on the curves $\eta_l^{A_2}$, $\eta_r^{A_2}$; $\eta_l^{A_k^i}$, $\eta_r^{A_k^i}$ $(k \geq 1, 1 \leq i \leq 2^k - 1)$.

Remark 2. Since curves $\eta_l^{A_2}$, $\eta_r^{A_2}$; $\eta_l^{A_k^i}$, $\eta_r^{A_k^i}$ are C^1-smooth then, bounded arcs of these curves are rectifiable [5].

By previous Remark 2 all curves $\eta_l^{A_2}$, $\eta_r^{A_2}$; $\eta_l^{A_k^i}$, $\eta_r^{A_k^i}$ admit parametrization with use of the natural parameter $s \in [0, +\infty)$. Take a number $s_* \in (0, +\infty)$. Let $2^{\Gamma_{s*}}$ be the space (endowed with Hausdorff metric *dist*) of compact rectifiable arcs of the set G_{Δ_2} which start from all points of the hypothenuse $x + y = 4$, $x \in [0, 4]$, of the triangle Δ_2; in addition, these arcs are parametrized with the natural parameter $s \in [0, s_*]$.

Remind that Hausdorff metric *dist* (see [11, ch. 2, § 21, VII]) for arcs $\gamma'_{|[0,\, s^*]}$, $\gamma''_{|[0,\, s^*]} \in 2^{\Gamma_{s*}}$ is defined by the equality

$$dist(\gamma'_{|[0,\, s^*]},\ \gamma''_{|[0,\, s^*]}) = \max\{ \sup_{(x,\, y) \in \gamma'_{|[0,\, s^*]}} \rho((x,\, y),\ \gamma''_{|[0,\, s^*]}),\ \sup_{(x,\, y) \in \gamma''_{|[0,\, s^*]}} \rho(\gamma'_{|[0,\, s^*]},\ (x,\, y))\}.$$

Here ρ is a metric in the plane.

Proposition 1. *The metric space* $(2^{\Gamma_{s*}}, dist)$ *is complete for any* $s_* \in (0, +\infty)$ *[5].*

Proposition 2. *The set of compact arcs*

$$\eta_l^{A_2,\, s_*} = \eta_l^{A_2}{}_{|[0,\, s_*]},\quad \eta_r^{A_2\, s_*} = \eta_r^{A_2}{}_{|[0,\, s_*]},\quad \eta_l^{A_k^i,\, s_*} = \eta_l^{A_k^i}{}_{|[0,\, s_*]},\quad \eta_r^{A_k^i,\, s_*} = \eta_r^{A_k^i}{}_{|[0,\, s_*]},$$

where $k \geq 1,\ 1 \leq i \leq 2^k - 1$, *is dense in itself with respect to Hausdorff metric dist for any* $s_* \in (0, +\infty)$; *moreover, every set* $\{\eta_l^{A_2,\, s_*},\ \eta_l^{A_k^i,\, s_*}\}$ *and* $\{\eta_r^{A_2,\, s_*},\ \eta_r^{A_k^i,\, s_*}\}$ *is dense in itself too [5].*

Proposition 3. *Let* $\{(x_n,\, y_n)\}_{n \geq 1}$ *be a convergent sequence of points in the plane which belong to compact arcs of the set* $\{\eta_l^{A_2,\, s_*},\ \eta_l^{A_k^i,\, s_*},\ \eta_r^{A_2,\, s_*},\ \eta_r^{A_k^i,\, s_*}\}$ $(k \geq 1,\ 1 \leq i \leq 2^k - 1)$. *Then corresponding sequence of arcs (containing the above points) converges in Hausdorff metric dist [5].*

Proposition 4. *Between any two different curves of the set G' which do not belong to the boundary of the same maximal generative domain there exist a maximal generative domain and a curve of G' which does not belong to the boundary of a maximal generative domain [5].*

Proposition 5. *Functions $y = \eta_l^{A_2}(x)$ and $y = \eta_r^{A_2}(x)$ are convex down. The straight line $x+y = 4$ is the tangent to the graphs $\eta_l^{A_2}$ and $\eta_r^{A_2}$ of these functions respectively at the point A_2; moreover, $x+y = 4$ is the tangent to curves $\eta_l^{A_k^i}$, $\eta_r^{A_k^i}$ at the point A_k^i $(k \geq 1,\ 1 \leq i \leq 2^k - 1)$ [5].*

Proposition 6. *The following equality is valid for the open set $\widetilde{G}$:*

$$\widetilde{G} = int(H_0^{max}(A_2)) \bigcup \Big(\bigcup_{k=1}^{+\infty} \bigcup_{i=1}^{2^k-1} int(H_k^{max}(A_k^i)) \Big) \ [5].$$

2.2. The Equivalence Relation and the Quotient Map

We use the mathematical induction principle to define the equivalence relation on the maximal generative domains.

1. Begin to define the equivalence relation from the principal maximal generative domain $H_0^{max}(A_2)$. Let $M_l(A_2)$ and $M_r(A_2)$ be arbitrary points of the curves $\eta_l^{A_2}$ and $\eta_r^{A_2}$ respectively, where $M_l(A_2) \neq A_2$, $M_r(A_2) \neq A_2$.

We identify points $M_l(A_2)$ and $M_r(A_2)$ iff

$$s([A_2,\ M_l(A_2)]) = s([A_2,\ M_r(A_2)]), \tag{2}$$

where $s([\cdot])$ is the length of an arc.

Let points $M_l(A_2)$ and $M_r(A_2)$ be identified by the equality (2). As it follows from Proposition 5, the closed interval $[M_l(A_2),\ M_r(A_2)]$ belongs to the domain H_0^{max}. Let a point $M_l'(A_2) \in \eta_l^{A_2}$ be so that $s([A_2,\ M_l'(A_2)]) > s([A_2,\ M_l(A_2)])$, and a point $M_r'(A_2) \in \eta_r^{A_2}$ be identified with $M_l'(A_2)$. Then, the inequality $s([A_2,\ M_r'(A_2)]) > s([A_2,\ M_r(A_2)])$ holds. It means that two closed intervals $[M_l(A_2),\ M_r(A_2)]$ and $[M_l'(A_2),\ M_r'(A_2)]$, where $M_l(A_2) \in \eta_l^{A_2}$ and $M_r(A_2) \in \eta_r^{A_2}$ ($M_l'(A_2) \in \eta_l^{A_2}$ and $M_r'(A_2) \in \eta_r^{A_2}$) are identified, either do not intersect each other (in H_0^{max}) or coincide. Therefore, it is correct to identify all points of the closed interval $[M_l(A_2),\ M_r(A_2)]$ with the point $M_l(A_2)$.

2. Define the equivalence relation on the maximal generative domain $H_1^{max}(A_1^1)$ of first order. By item (ii) of the Subsection 2.1 the domain $H_1^{max}(A_1^1)$ is the connected component of first complete preimage of the principal maximal generative domain $H_0^{max}(A_2)$ with respect to $F_{|G_{\Delta_2}}$, moreover, $H_1^{max}(A_1^1)$ contains the homoclinic point A_1^1. By Remark 1 the map $F_{|H_1^{max}(A_1^1)}$ is homeomorphism of $H_1^{max}(A_1^1)$ on $H_0^{max}(A_2)$, moreover, $F_{|int(H_1^{max}(A_1^1))}$ is diffeomorphism of $int(H_1^{max}(A_1^1))$ on $int(H_0^{max}(A_2))$. Hence, there exists a regular arc $\gamma[M_l(A_1^1),\ M_r(A_1^1)] \subset H_1^{max}(A_1^1)$ satisfying

$$F(\gamma[M_l(A_1^1),\ M_r(A_1^1)]) = [M_l(A_2),\ M_r(A_2)]. \tag{3}$$

Identify all points of the arc $\gamma[M_l(A_1^1),\ M_r(A_1^1)]$ with the point $M_l(A_1^1) \in \eta_l^{A_1^1}$.

3. Suppose that we have defined the equivalence relation on maximal generative domains $H_m^{max}(A_m^i)$ of m-th order, where $1 \leq m \leq k$, $k \geq 1$, $1 \leq i \leq 2^m - 1$. It means that we have identified all points of a regular arc $\gamma[M_l(A_m^i),\ M_r(A_m^i)]$ with the point $M_l(A_m^i) \in \eta_l^{A_i^m}$. Here

$$F(\gamma[M_l(A_m^i),\ M_r(A_m^i)]) = \gamma[M_l(A_{m-1}^{i'}),\ M_r(A_{m-1}^{i'})]$$

for $k \geq 2$, $2 \leq m \leq k$, $1 \leq i' \leq 2^{m-1} - 1$.

Define the equivalence relation on a maximal generative domain $H^{max}_{k+1}(A^j_{k+1})$ of $(k+1)$-st order, $k \geq 1$, $1 \leq j \leq 2^{k+1}-1$. By Remark 1 the map $F_{|H^{max}_{k+1}(A^j_{k+1})}$ is homeomorphism of $H^{max}_{k+1}(A^j_{k+1})$ on $H^{max}_k(A^{j'}_k)$, and $F_{|int(H^{max}_{k+1}(A^j_{k+1}))}$ is diffeomorphism of $int(H^{max}_{k+1}(A^j_{k+1}))$ on $int(H^{max}_k(A^{j'}_k))$, where a maximal generative domain $H^{max}_{k+1}(A^j_{k+1})$ is the connected component of first complete preimage of $H^{max}_k(A^{j'}_k)$ $(1 \leq j' \leq 2^k - 1)$ with respect to $F_{|G_{\Delta_2}}$, moreover, the homoclinic point A^j_{k+1} belongs to $H^{max}_{k+1}(A^j_k)$. Therefore, there exists a regular arc $\gamma[M_l(A^j_{k+1}),\, M_r(A^j_{k+1})] \subset H^{max}_{k+1}(A^j_k)$ satisfying the equality

$$F(\gamma[M_l(A^j_{k+1}),\, M_r(A^j_{k+1})]) = \gamma[M_l(A^{j'}_k),\, M_r(A^{j'}_k)] \tag{4}$$

Identify all points of the arc $\gamma[M_l(A^j_{k+1}),\, M_r(A^j_{k+1})]$ with the point $M_l(A^j_{k+1}) \in \eta_l^{A^j_{k+1}}$.

Thus, using equalities (2) – (4) we constructed the equivalence relation on the set

$$H^{max}_0(A_2) \bigcup \Big(\bigcup_{k=1}^{+\infty} \bigcup_{i=1}^{2^k-1} H^{max}_k(A^i_k) \Big)$$

and, in particular, on the set of curves

$$L_H = \{\eta_l^{A_2},\, \eta_r^{A_2}\} \bigcup \Big(\bigcup_{k=1}^{+\infty} \bigcup_{i=1}^{2^k-1} \{\eta_l^{A^i_k},\, \eta_r^{A^i_k}\} \Big).$$

By the above, in the both cases (see Remark 1 and Proposition 6) the set of equivalence classes (with respect to the introduced equivalence relation) coincides with the set of curves

$$L_{H,l} = \eta_l^{A_2} \bigcup \Big(\bigcup_{k=1}^{+\infty} \bigcup_{i=1}^{2^k-1} \eta_l^{A^i_k} \Big).$$

Further, we consider the equivalence relation only on the set $L_H \subset G'$. Denote by π^{L_H} the constructed equivalence relation on the set of curves L_H.

4. Extend the equivalence relation π^{L_H} to the equivalence relation π defined on all curves of the set G'. Let $\gamma(A)$ be an unbounded curve from the set $G' \setminus L_H$ (here A is the original point of $\gamma(A)$ on the hypothenuse of the triangle Δ_2) and $M_{\gamma(A)}$ be a point of $\gamma(A)$. By the Theorem and Proposition 3 there is a sequence of points $M_n \in \eta_p^{A^i_{k_n}}$ $(p = l$ or r; $n \geq 1)$ on curves of the set L_H satisfying

$$M_{\gamma(A)} = \lim_{n\to+\infty} M_n. \tag{5}$$

Let $M_{l,n}(A^i_{k_n}) \in \eta_l^{A^i_{k_n}}$ be a point identified with M_n under equivalence relation π^{L_H}. Using Proposition 1 and the equality (5) we find a positive number s_* satisfying

$$M_{\gamma(A)} \in \gamma^{s_*}(A), \text{ where } \gamma^{s_*}(A) = \gamma(A)_{|[0,\, s_*]};\ M_n \in \eta_p^{A^i_{k_n},\, s_*},\ \ M_{l,n}(A^i_{k_n}) \in \eta_l^{A^i_{k_n},\, s_*}.$$

By Proposition 3 the sequence of arcs $\{\eta_p^{A^i_{k_n},\, s_*}\}_{n\geq 1}$ converges in Hausdorff metric *dist* to the arc $\gamma^{s_*}(A)$. By Proposition 4 the sequence of arcs $\{\eta_l^{A^i_{k_n},\, s_*}\}_{n\geq 1}$ also converges in Hausdorff metric to the arc $\gamma^{s_*}(A)$.

Prove that the sequence of points $\{M_{l,n}(A^i_{k_n})\}_{n\geq 1}$ converges (to some point $M'_{\gamma(A)} \in \gamma^{s_*}(A)$). Note that the proof of convergence of the sequence $\{M_{l,n}(A^i_{k_n})\}_{n\geq 1}$ does not depend on belonging of the curve $\gamma(A)$ to the set $G' \setminus L_H$ or L_H. Suppose that the sequence of points $\{M_{l,n}(A^i_{k_n})\}_{n\geq 1}$ diverges. Then, this sequence has at least, two different limit points $M^{(1)}_{\gamma(A)}$, $M^{(2)}_{\gamma(A)} \in \gamma^{s_*}(A)$. Suppose that

$$A \prec M^{(1)}_{\gamma(A)} \prec M^{(2)}_{\gamma(A)}, \tag{6}$$

where " $\prec$ " is the natural order on the curve $\gamma(A)$ corresponding to growth of the parameter s (the value of $s = 0$ corresponds to the point A). Select two different subsequences of points $\{M_{l,n(1)}(A^i_{k_{n(1)}})\}_{n(1)\geq 1}$ and $\{M_{l,n(2)}(A^i_{k_{n(2)}})\}_{n(2)\geq 1}$ such that

$$M^{(1)}_{\gamma(A)} = \lim_{n(1)\to+\infty} M_{l,n(1)}(A^i_{k_{n(1)}}); \qquad M^{(2)}_{\gamma(A)} = \lim_{n(2)\to+\infty} M_{l,n(2)}(A^i_{k_{n(2)}}), \tag{7}$$

where $M_{l,n(1)}(A^i_{k_{n(1)}}) \in \eta_l^{A^i_{k_{n(1)}}, s_*}$, $M_{l,n(2)}(A^i_{k_{n(2)}}) \in \eta_l^{A^i_{k_{n(2)}}, s_*}$. Since the map F preserves the oder of points, in particular, on every curve $\eta_l^{A^i_{k_{n(1)}}}$ (see Remark 1) then, by (7) the sequence of arcs $\gamma[A^i_{k_{n(1)}}, M_{l,n(1)}(A^i_{k_{n(1)}})]$ converges in Hausdorff metric *dist* to the arc $\gamma[A, M^{(1)}_{\gamma(A)}]$. Using (6) we obtain from here the contradiction with convergence of the sequence of arcs $\eta_l^{A^i_{k_{n(1)}}, s_*}$ to the arc $\gamma^{s_*}(A)$ in Hausdorff metric. Thus, the equality $M^{(1)}_{\gamma(A)} = M^{(2)}_{\gamma(A)}$ holds. Set $M'_{\gamma(A)} = M^{(1)}_{\gamma(A)} = M^{(2)}_{\gamma(A)}$. Define $\pi(M_{\gamma(A)})$ by the equalities

$$\pi(M_{\gamma(A)}) = M'_{\gamma(A)} = \lim_{n\to+\infty} M_{l,n}(A^i_{k(n)}) = \lim_{n\to+\infty} \pi^{L_H}(M_n). \tag{8}$$

Thus, the equivalence class of points of a curve $\gamma(A) \in G' \setminus L_H$ coincides with $\gamma(A)$. Construction of the equivalence relation is finished.

Described construction of the equivalence relation (see items $1-4$) shows that the quotient space with respect to π coincides with the curves that form the set $L_{H,l} \bigcup (G' \setminus L_H)$. Moreover, every curve of the set $L_{H,l}$ consists of double points.

The following Lemma 1 proves corectness of the above construction of the equivalence relation π in the points of curves of the set $G' \setminus L_H$.

Lemma 1. *The value $\pi(M_{\gamma(A)})$ in any point $M_{\gamma(A)}$ of a curve $\gamma(A) \in G' \setminus L_H$ does not depend on the choice of a sequence of points satisfying (5).*

Proof. Let a curve $\gamma(A)$ and a point $M_{\gamma(A)}$ be as in the above item 4. Let different sequences of points M_n and M'_n $(n \geq 1)$ on curves of the set G' converge to $M_{\gamma(A)}$. Using Proposition 2 we can suppose (without loss of generality) that points M_n and M'_n belong to curves of the set L_H.

Form a new sequence $\{\overline{M}_n\}_{n\geq 1}$ with the set of values obtained as the union of the sets of values of the above sequences $\{M_n\}_{n\geq 1}$ and $\{M'_n\}_{n\geq 1}$. Then, the sequence $\{\overline{M}_n\}_{n\geq 1}$ converges to $M_{\gamma(A)}$. Therefore, by equalities (8) we have:

$$\pi(M_{\gamma(A)}) = \lim_{n\to+\infty} \pi^{L_H}(\overline{M}_n).$$

The choice of the sequence $\{\overline{M}_n\}_{n\geq 1}$ and the last equality imply

$$\lim_{n\to+\infty} \pi^{L_H}(M_n) = \lim_{n\to+\infty} \pi^{L_H}(M'_n) = \pi(M_{\gamma(A)}).$$

Lemma 1 is proved.

IOP Conf. Series: Journal of Physics: Conf. Series **990** (2018) 012003 doi:10.1088/1742-6596/990/1/012003

Lemma 2. *Equivalence relation π is continuous.*

Proof. 1. Continuity of the equivalence relation π in points of curves of the set $G' \setminus L_H$ follows from formulas (8) and Lemma 1.

2. Note that the equivalence relation π is continuous in points of curves of the set L_H. In fact, let $\gamma(A)$ here be a curve of the set $\{\eta_l^{A_2}, \eta_r^{A_2}; \eta_l^{A_k^i}, \eta_r^{A_k^i}\}$ $(k \geq 1, 1 \leq i \leq 2^k - 1)$. Let $M_{\gamma(A)}$ be a point of this curve, and $\{M_n\}_{n\geq 1}$ be a convergent sequence of points on curves of the set G'. By Proposition 2 we assume (without loss of generality) that points M_n belong to curves of the set L_H. Let the point sequence $\{M_{l,n}(A_{k_n}^i)\}_{n\geq 1}$ be chosen analogously to the point sequence from item 4 of construction of the equivalence relation π. Then, this sequence converges to some point $M'_{\gamma(A)} \in \gamma(A)$.

Consider, for determination, the case $\gamma(A) = \eta_r^{A_2}$. So, $M_{\gamma(A)} \in \eta_r^{A_2}$. To prove the continuity of π (in a point $M_{\eta_r^{A_2}}$) we must prove the equality

$$M'_{\eta_r^{A_2}} = M_{\eta_r^{A_2}}. \tag{9}$$

In fact, let $M'_{\eta_l^{A_2}}$ on the curve $\eta_l^{A_2}$ be the equivalent point for $M'_{\eta_r^{A_2}} \in \eta_r^{A_2}$. Then,

$$s([A_2, M'_{\eta_r^{A_2}}]) = s([A_2, M'_{\eta_l^{A_2}}]).$$

Therefore, if we carry the point $M'_{\eta_l^{A_2}}$ on the curve $\eta_r^{A_2}$ preserving the lenght of the arc $[A_2, M'_{\eta_l^{A_2}}]$ then, we obtain the point $M'_{\eta_r^{A_2}}$. It means that (9) is equivalent to the equality

$$M'(\eta_l^{A_2}) = \pi(M(\eta_r^{A_2})). \tag{10}$$

Correctness of the equalities (9) – (10) follows immediately from Proposition 4 and the inequalities

$$||DF_{|\eta_r^{A_2}}||_{C^0} \geq K; \; ||DF_{|\eta_l^{A_2}}|| \geq K$$

for some $K > 1$. Here $|| \cdot ||_{C^0}$ is the C^0-norm of the differential of a map:

$$||DF_{|\eta_p^{A_2}}||_{C^0} = \max\{ \sup_{x\in D(\eta_p^{A_2}),\, y\in E(\eta_p^{A_2})} |y + x(\eta_p^{A_2}(x))'| + |x|, \quad \sup_{x\in\in D(\eta_p^{A_2})} \{|2(x-2)|\}\}, \tag{11}$$

$y = \eta_p^{A_2}(x)$ is a function with the graph $\eta_p^{A_2}$, $D(\eta_p^{A_2})$ is the domain of definition, and $E(\eta_p^{A_2})$ is the set of values of $\eta_p^{A_2}$ $((p = l \text{ or } r))$. Remind that

$$D(\eta_l^{A_2}) = (2, 3], \; D(\eta_r^{A_2}) = [3, +\infty), \; E(\eta_l^{A_2}) = [1, +\infty), \; E(\eta_r^{A_2}) = (0, 1].$$

Moreover, derivative $(\eta_l^{A_2}(x))'$ increases from $-\infty$ to -1, when x changes on $(2, 3]$; and derivative $(\eta_r^{A_2}(x))'$ increases from -1 to 0 when x changes on $[3, +\infty)$.

As it follows from (11), one can assume $K = 2$ (see the above). Lemma 2 is proved.

By Lemmas 1, 2 the trace map F induces the quotient map F^* defined on the set of points of curves that form the set $L_{H,l} \bigcup (G' \setminus L_H)$. Moreover, F^* satisfies the equality

$$\pi \circ F = F^* \circ \pi. \tag{12}$$

By the equality (12) second coordinate function of the map F^* is double valued and upper semicontinuous (but not lower semicontinuous) in every point (x, y), where $(x, y) \in \eta_l^{A_2}$, $(x, y) \neq (3, 1)$; and second coordinate function of the map $(F^*)^k$ $(k \geq 2)$ is double valued and upper semicontinuous (but not lower semicontinuous) in every point (x, y), where $(x, y) \in \eta_l^{A_q^i}$, $(x, y) \neq A_q^i$ $(1 \leq q \leq k,\ 1 \leq i \leq 2^q - 1)$. By the above construction of the equivalence relation π every set $L_{H,l}$ and $G' \setminus L_H$ is invariant under F^*. Using invariance of the closed interval $x + y = 4$, $x \in [0, 4]$ under F we obtain from here that F^* is semiconjugate to the map $f(x) = x(4 - x)$ of the closed intarval $[0, 4]$.

3. Reducing to a Skew Product

We begin this part of the work from the following words by Birkhoff: "If we try to formulate the exact definition of integrability, then we see that many definitions are possible, and every of them is of some theoretical interest" [12].

Definition 1 of integrability proposed here, develops the idea by Grigorchuk of reduction of an integrable map in the plane to a skew product of maps of an interval (see [13]) and generalizes the definition of integrability from the paper [5] (see also [14]) on the case of upper semicontinuous multivalued functions defined on noncompact sets. In comparison with the definition by Grigorchuk, our definition is based on the semiconjugacy property of an integrable map in a plane with a map of an interval.

Definition 1. Let $\Pi \subset \mathbf{R}^2$ be (open or closed; bounded or not) domain in the plain $\mathbf{R}^2$. We say that a map $G : \Pi \to \Pi$ (single-valued or multi-valued) is *integrable* if there exists a self-map ψ of some interval J of the real line $\mathbf{R}^1$ so that ψ is semiconjugate with G by means a continuous surjection $\widetilde{H} : \Pi \to J$, so that

$$\widetilde{H} \circ G = \psi \circ \widetilde{H}. \tag{13}$$

The criterion for integrability of a continuous map G defined on a compact curvilinear trapezoid Π is given in [5]. In this paper we give necessary conditions for integrability in the unbounded domain.

Discuss Definition 1 for $\Pi = G_{\Delta_2}$, $J = \{(x,\, y) : x + y = 4,\, x \in [0,\, 4]\}$. Assume that $\widetilde{H}$ is one-to-one with respect to x. Then, we have:
(1.1) for every $C \geq 4$ the section of Π by a straight line $x + y = C$ is a non-degenerate compact interval corresponding $x \in [0,\, C]$;
(1.2) for all points $x',\, x'' \in [0,\, 4]$, where $x' \neq x''$, the equality $\widetilde{H}^{-1}(x',\, 4-x') \bigcap \widetilde{H}^{-1}(x'',\, 4-x'') = \emptyset$ is valid;
(1.3) the equality $\Pi = \bigcup\limits_{x' \in [0,\, 4]} \widetilde{H}^{-1}(x',\, 4-x')$ holds.

Since $\widetilde{H}$ is continuous and injective with respect to x, then for every $C > 4$ the map $\widetilde{H}$ is bijection of the closed interval $J_C = \{(x,\, y) : x + y = C,\, x \in [0,\, C]\}$ on $\widetilde{H}(J_C))$. Moreover, $\widetilde{H}(J_C)$ is a nondegenerate closed interval in J. Property (1.3) and injectivity of $\widetilde{H}$ with respect to x imply the equality $\widetilde{H}(J_C) = J$. Thus, for every $x' \in [0,\, 4]$ the set $\widetilde{H}^{-1}(x',\, 4-x')$ is the graph (denote it by $\gamma_{x'}$) of a continuous function $x = \gamma_{x'}(C)$. It means that $\widetilde{H}^{-1}(x',\, 4-x')$ is a connected set. Moreover, the unique curve $\gamma_{x'}$ passes through every point $(x',\, C-x') \in \Pi$, and $\gamma_{x'}$ intersects J_C in a unique point for every $C > 4$. Hence, by properties (1.2), (1.3) the one-parameter family of graphs $\{\gamma_{x'}\}_{x' \in [0,\, 4]}$ of the above functions defines the continuous foliation in Π (see, e.g., [15, ch. 4, § 16]). By (13) and G-invariance property of the closed interval J the foliation $\{\gamma_{x'}\}_{x' \in [0,\, 4]}$ is invariant. It means that inclusion $G(\gamma_{x'}) \subseteq \gamma_{\psi(x')}$ is valid for every $x' \in [0,\, 4]$.

Theorem 1. *Let $\Pi = G_{\Delta_2}$, and $G : \Pi \to \Pi$ be a continuous single-valued map or a map with continuous single-valued first coordinate function and double-valued upper semicontinuous second coordinate function. Moreover, in the last case there is a nonempty set of pairwise disjoint curves $\{\gamma_{x'_1}, \gamma_{x'_2}, \ldots, \gamma_{x'_r}\}$ $(1 \leq r < +\infty)$, where second coordinate function of the map G is double-valued and discontinuous in every point which does not coincide with $(x'_i,\, 4-x'_i)$ $(1 \leq i \leq r)$ (in all other points this function is single-valued and continuous).*

Let, also, the section of Π by a straight line $x + y = 4$ be G-invariant, and G be integrable in the sense of Definition 1 by means of a continuous surjection $\widetilde{H} : \Pi \to J$, which is one-to-one with respect to x (here J is defined as above).

Then there exists a homeomorphism that reduces G to a skew product of interval maps (i.e., a map $\widetilde{G}(u; v) = (\varphi(u); \psi(u; v)))$ defined on an unbounded with respect to v planar rectangle.

Proof. 1. Let the map $\theta : \Pi_{xOC} \to \Pi'_{uO'v}$ (here $\Pi_{xOC} = \Pi = G_{\Delta_2}$) be defined as follows:

$$\begin{cases} u = \widetilde{H}(x, C - x), \\ v = C - 4. \end{cases} \tag{14}$$

Then, θ is a continuous bijection of Π_{xOC} on $\Pi'_{uO'v}$.

Note that $\theta : \Pi_{xOy} \to \Pi'_{uO'v}$ is homeomorphism. In fact, by (14) the map θ is closed. Hence, θ is mutually continuous map. Remind that a continuous map $\theta : X \to Y$ is called mutually continuous, if θ is surjection of X on Y, and a set $A \subset Y$ is closed (open) iff its complete preimage $\theta^{-1}(A)$ is closed (open) [11, ch. 1, § XIII, XV]. Since θ is mutually continuous and bijective map then θ is homeomorphism [11, ch. 1, § XIII, XV].

By (14) a set $\theta(\gamma_{x'})$ is a ray $u = x'$, $v \geq 0$, for every $x' \in [0, 4]$. Hence,

$$\Pi'_{uO'v} = [0, 4] \times [0, +\infty).$$

Let $\widetilde{G}$ be the map in the plane of variables u and v corresponding to the map G in the plane of variables x and y. Then, $\widetilde{G} : \Pi'_{uO'v} \to \Pi'_{uO'v}$, $\widetilde{G}$ is topologically conjugate with G under homeomorphism θ. It means that the following equality holds:

$$\widetilde{G} = \theta \circ G \circ \theta^{-1}.$$

Denote by $\widetilde{g}_1$ and $\widetilde{g}_2$ first and second coordinate functions of the map $\widetilde{G}$ respectively. Since θ is homeomorphism then, $\widetilde{g}_1$ and $\widetilde{g}_2$ inherit properties of first and second coordinate functions of the map G respectively. It means, in particular, that $\widetilde{g}_1$ is continuous function.

2. Prove that $\widetilde{g}_1$ does not depend on variable v. In fact, $\widetilde{G}$ maps every vertical ray

$$\{\Pi'_{uO'v} \bigcap \{(x'; v) : v \geq 0\}$$

into the vertical ray

$$\{\Pi'_{uO'v} \bigcap \{(\psi(x'); v) : v \geq 0\},$$

where $\psi(x') = G(x', 4 - x')$.

Prove that partial derivative $\frac{\partial}{\partial v}\widetilde{g}_1(u, v)$ exists in every point $(u; v)$ of the rectangle $\Pi'_{uO'v}$ and equals 0. In fact, let $(x'; v')$, $(x'; v)$ be arbitrary points of a vertical ray $\Pi'_{uO'v} \bigcap \{(x'; v) : v \geq 0\}$. Then, we have:

$$\frac{\partial}{\partial v}\widetilde{g}_1(u, v) = \lim_{v \to v'} \frac{\widetilde{g}_1(x'; v) - \widetilde{g}_1(x', v')}{v - v'} = \lim_{v \to v'} \frac{\psi(x') - \psi(x')}{v - v'} = 0. \tag{15}$$

Since $\Pi'_{uO'v}$ is the convex connected set then, by (15) the coordinate function $\widetilde{g}_1$ of the map $\widetilde{G}$ does not depend on the variable v. Thus, $\widetilde{G}$ is the skew product in the plane.

Theorem 1 is proved.

Remark 3. We mean direct products as elements of the set of skew products dynamical systems.

Remark 4. Assumption on convexity of the unbounded domain $\Pi'_{uO'v}$ can not be omitted.

Remark 5. Note that in considered case we have:

$$\widetilde{G}([0, 4] \times \{0\}) = [0, 4] \times \{0\}.$$

It means that upper semicontinuity of the map $\widetilde{g}_2(u, v)$ in the points (x'_i, v) $(1 \leq i \leq r, v > 0)$ does not imply double connectivity of the set of values of $\widetilde{g}_2(u, v)$ in a vertical fiber $u = \psi(x'_i)$. Properties of $\widetilde{g}_2(u, v)$ do not influence on the proof of Theorem 1.

Using Definition 1 and results of the Subsection 1.2 we obtain the following result.

Theorem 2. *The quotient map F^* for the trace map given by formula (1) is integrable in the sense of Definition 1.*

By Theorems 1 and 2 the quotient map F^* for the trace map F is reducible to the upper semicontinuous skew product defined on the unbounded rectangle $[0, 4] \times [0, +\infty)$.

Acknowledgments
The work was partially supported by the grant No 1.3287.2017 of the Ministry of Science and Education of Russian Federation.

References

[1] Bellisard J 1990 *Number Theory and Phisics* **47** 140-50
[2] Avishai Y and Berend D 1992 *Phys Rev* B **45** 2717-24
[3] Avishai Y, Berend D and Tkachenko V 1997 *Int J Modern Physics* B **11** **(30)** 16 3525–42
[4] Baake M, Grimm U and Joseph D 1993 *Int J Modern Physics* B **7** 1527-50
[5] Belmesova S S and Efremova L S 2015 *On the Concept of Integrability for Discrete Dynamical Systems Investigation of Wandering Points of Some Trace Map* (Nonlinear Maps and Their Applications Springer Proc Math Statist vol 112) eds Lopez-Ruiz R, Fournier-Prunaret D, Nishio Y and Gracio M C (Springer Cham) chapter 7 pp 127-58
[6] Belmesova S S and Efremova L S 2012 *Vestnik Nizhegorodskogo Gosudarstvennogo Univ* **2(2)** 152–8 (Russian)
[7] Belmesova S S and Efremova L S 2013 *Russian Math* **57** 8 70–4
[8] Belmesova S S, Efremova L S and Fournier-Prunaret D 2014 *ESAIM: Proceedings and Surveys* **46** 98–110
[9] Damanik D and Gorodetski A 2009 *Nonlinearity* **22** 123–43
[10] Damanik D and Gorodetski A. 2009 *Elect Research Announcements in Math Sci* **16** 23–9
[11] Kuratovsky K 1968 *Topology* vol II (New York and London: Academic Press)
[12] Birkhoff G D 1927 *Dynamical systems* Amer Math Soc Colloq Publ vol 9 (New York: Amer Math Soc)
[13] Grigorchuk R I and Žuk A 2001 *Geometriae Dedicata* **87** 209–44
[14] Efremova L S 2017 *Russian Math Surveys* **72** 1 (101) 101-78
[15] Tamura I 1992 *Topology of foliations: Introduction*, Translations of Math Monographs **97** (New York: Amer Math Soc)

IOP Conf. Series: Journal of Physics: Conf. Series **990** (2018) 012004 doi:10.1088/1742-6596/990/1/012004

Harmonic mode-locking and sub-round-trip time nonlinear dynamics of electro-optically controlled solid state laser

M V Gorbunkov[1], Yu Ya Maslova[1], V A Petukhov[1], M A Semenov[1], Yu V Shabalin[1] and V G Tunkin[2]

[1] P.N. Lebedev Physical Insitute, Leninskii pr. 53, Moscow, Russia 119991

[2] International Laser Center, Lomonosov Moscow State University, Leninskie Gori 1, Moscow, Russia 119991

E-mail: jmaslova@lebedev.ru

Abstract. Harmonic mode-locking in a solid state laser due to optoelectronic control is studied numerically on the basis of two methods. The first one is detailed numeric simulation taking into account laser radiation fine time structure. It is shown that optimally chosen feedback delay leads to self-started mode-locking with generation of desired number of pulses in the laser cavity. The second method is based on discrete maps for short laser pulse energy. Both methods show that the application of combination of positive and negative feedback loops allows to reduce the period of regular nonlinear dynamics down to a fraction of a laser cavity round trip time.

1. Introduction

Despite the fact that the "laser era" lasts for more than 50 years, the problem of regular and chaotic temporal structure formation remains the matter of topical interest. Evidence of this is a great number of articles, reviews, and book publications, [1–4] to name a few. Soon after the creation of the first solid state laser, negative feedback was applied for spike elimination in a pulsed free-running laser [5]. Later it was shown that feedback control allows not only to stabilize laser output, but to realize the regimes that are based on nonlinear dynamics, which is inherent in this control method due to the presence of time delay [1,2]. In the present time, it is clear that the delayed feedback control makes output stabilization possible only in a limited gain range [2,6]. Maximum gain can be determined from the system of equations that describe the laser generation process. Further increase of gain above this value results initially in regular nonlinear dynamics (in the form of envelope modulation), and then in chaotic laser output. Problems of time delayed control arise frequently in the analysis of self-oscillating systems of other types. Generation of regular bursts with period in the range from several round trip times T_r up to several hundred T_r via nonlinear dynamics manifestation (using nonlinear dynamics methods) [7–9] was not only of theoretical relevance but also of practical interest. It is combination of positive and negative feedback loops that opens up new control possibilities. Such control allows to choose new time scale of regular dynamics: it was shown that combination of feedback loops can realize the time scale in the range of $(1–100)T_r$ [8, 10, 11]. Feedback erasure [12] allows to overcome the inertial action of feedback and to reduce the scale of dynamics. The control which combines two feedback loops and external modulation of losses allowed to look at the problem of the period doubling bifurcation in solid state lasers in a new fashion [10, 11]. Matter of current interest is the question of further diminishing of the time scale. Is it possible to observe nonlinear dynamics on the time scale less than T_r? Is the combination of inertial feedback loops sufficient for this purpose?

In this paper we consider the prospects for pushing the regular and chaotic nonlinear laser dynamics into the time scale below T_r based on the idea of combining harmonic mode locking [13, 14] with dual optoelectronic feedback [9]. Recent experiments on harmonic laser mode locking by means of optoelectronic feedback [15] play in favor of this perspective.

2. Harmonic mode locking in solid state laser with optoelectronic control

Feedback control gives the unique possibility to order laser radiation fine time structure. The corresponding idea of ordering radiation fine time structure was stated in [16]. The authors concluded that, by choosing the delay T_d of sufficiently fast negative feedback, it is possible to obtain either the regime of mode-locking with generation of one pulse per axial interval or the regime of smoothing of radiation fine time structure.

The importance of feedback delay T_d is clearly demonstrated by the following consideration.

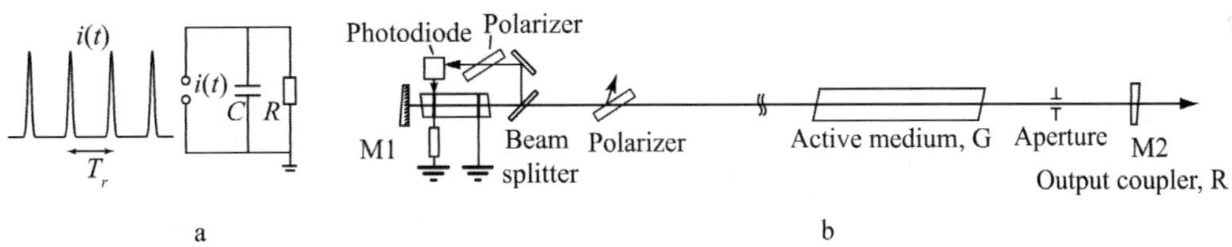

Figure 1. Delay-free-feedback controlled laser: a) *RC* circuit of the optoelectronic control scheme: $i(t)$ — photodiode current, R — discharge resistor, C — modulator capacity; b) the simplified scheme of the laser with the electro-optical modulator when delay is set to minimal technical value Δ, M1, M2 — cavity mirrors.

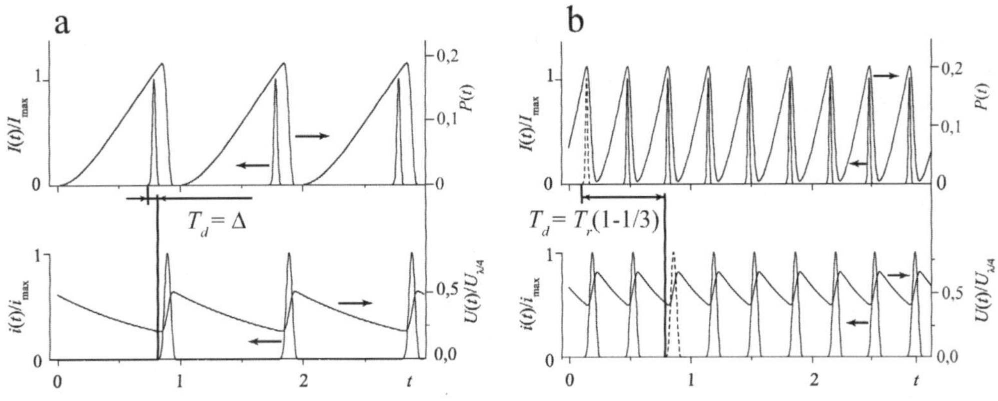

Figure 2. Temporal dependence of laser intensity $I(t)$, photocurrent $i(t)$, Pockels cell control voltage $U(t)$, and the cell transmission $P(t)$, T_d is the feedback delay, i.e. the time interval between the laser pulse arrival in the cell and the beginning of photocurrent response caused by the same pulse in the feedback control circuit. a — the case of a single pulse per axial interval, b — three pulses per axial interval. Time is normalized to T_r.

We assume the sensitivity of the negative feedback to be large enough to exclude the active medium saturation. An example of a laser controlled by the negative feedback loop based on an intracavity electro-optical modulator (Pockels cell) is presented schematically in Fig. 1. The Pockels cell is controlled by the photocurrent $i(t)$ generated in a fast photodiode by the light deflected to the feedback loop by a beam splitter. In the simplest case, electric circuit includes the Pockels cell crystal capacity C, the resistor R, and the photodiode as a current source. The time constant of this control circuit is $\tau = RC$ which characterizes the feedback inertiality. The Pockels cell transmission is

$$P(U) = \cos^2\left(\frac{\pi}{2}\cdot\frac{U+U_0}{U_{\lambda/4}}\right), \tag{1}$$

where U_0 is the static bias voltage, U is the voltage generated by laser pulse, $U_{\lambda/4}$ is the quarter-wave voltage (i.e. needed to turn off the cell). If the control is designed as negative feedback ($0 < U + U_0 < U_{\lambda/4}$), then with the increase of the laser intensity I voltage U increases and P decreases.

The choice of the feedback delay for the mode-locking regime is illustrated in Fig. 2. Suppose that a single and short enough laser pulse has been formed and circulates in a laser cavity, and the cavity round trip time is T_r. The sequence of ultrashort light pulses would generate photocurrent pulses with period T_r (we assume here that the duration of a laser pulse τ_l and the response time of the photodiode τ_{ph} are constrained by the inequality $\tau_l < \tau_{ph} \ll T_r$). In its turn, the sequence of photocurrent pulses, when fed to the RC circuit, forms on the capacitor C a sawtooth voltage with a short front (determined by the photocurrent pulse duration) and a long tail (determined by τ). The sawtooth voltage has maximum and minimum values U_{max} and U_{min}, that are related by the feedback memory

$$U_{min} = U_{max}e^{-T_r/\tau}. \tag{2}$$

The regime of mode-locking occurs in a number of regions of feedback delay time T_d. The first region corresponds to the minimal technical delay of the control action Δ (Fig. 2a), see [15] for details. In this case, photodiode current generated by the control pulse, i.e. a part of laser pulse deflected by the beam splitter, charges cell capacitor in time delay Δ after the reflection of the laser pulse by mirror M1 and its passage through Pockels cell. The rest of the same laser pulse amplified in active medium returns to the cell at the moment of its nearly maximum discharge (nearly maximum transmission). Therefore the transmission of the cell in the moment of the laser pulse passage is determined by this pulse energy and by the energy of preceding pulses taking into account the discharge of capacity in the control circuit. In the case of mode-locking with one pulse per axial interval, $[T_r$–$2T_r]$ is a suitable range of the discharge time τ for self-sustained short pulse generation [17, 18]. The increase of feedback delay by one or more T_r gives the possibility to tune the delay so that each pulse passes the cell exactly at its maximum transmission. This allows one to obtain the shortest laser pulse duration.

It is possible to determine the delay required to generate $k \geq 2$ pulses per T_r. An example with three pulses is presented in Fig. 2b. Consider the case of mode-locking with generation of a train with pulse separation interval T_r/k. We consider the case when the pulse is controlled by its part reflected into feedback loop. It passes through the modulator at the maximum discharge of its capacity. This is achieved after time interval T_r/k. Thus the delay of feedback should be equal to $(1 - 1/k)T_r$ [15]. The increase of the delay by mT_r, $m = 0, 1, 2...$, conserves the conditions for k pulses generation per T_r. In general, ordered fine time structures of a certain kind arise periodically with the increase of delay exactly by T_r. It should be noted that for short pulse generation the optimal value of discharge time τ decreases with the increase of k. The maximum number of short pulses per T_r is limited by the photocurrent response time of the photodiode used in the feedback loop.

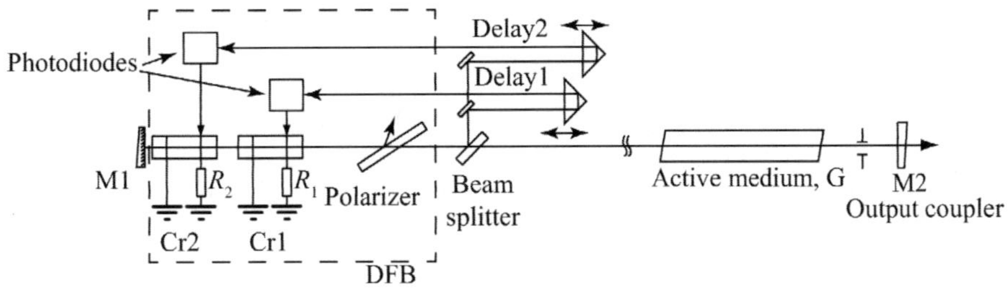

Figure 3. A simplified scheme of the laser controlled by two feedback loops based on two-sectional electro-optical modulator (crystals Cr1 and Cr2). M1, M2 — cavity mirrors, M1 is totally reflecting, M2 is partially reflecting, DFB — dual-feedback optoelectronic control system.

Finally, to find the condition for smoothing fine time structure, one can use the approach similar to that in the case of a single pulse generation. It is clear that if the feedback delay is chosen in such a way that a speculative pulse passes through the cell in the moment of minimum transmission, then feedback control results in the suppression of the pulse and thus in the smoothing of laser intensity. The best smoothing is realized when time delay is smaller than T_r approximately by $2\tau_{ph}$.

3. Numerical model of the feedback control by negative and positive loops

Let us trace the field evolution in a laser cavity one cavity-round-trip forward taking into account the action of intracavity elements in the laser with combination of two feedback loops (see Fig. 3). The dual-feedback control can be organized on the basis of two-sectional electro-optic modulator, e.g. in [7] we used temperature-compensated lithium tantalate modulator. Assume that at a laser round-trip with number K, $K = 0, 1, 2, ...K_{max}$, the laser field amplitude is expressed as $A_K(t)$, $t \in [K \cdot T_r, (K+1) \cdot T_r]$. Below each function is calculated on the same interval. The number of round-trip is incremented after the passage of intracavity polarizer. The alteration of $A_K(t)$ in the laser cavity is attributed to several factors described in Subsections 3.1–3.2.

3.1. Amplification in active medium

To describe the laser radiation passage through active medium at round-trip with number K we follow the approach of [19] which is valid for large gain values. The relation between incoming radiation amplitude $A_K^{in}(t)$ and outcoming radiation amplitude $A_K^{out}(t)$ for two-pass case is:

$$A_K^{out}(t) = \frac{\exp\left(\frac{3}{4}g_K\right)}{2\sqrt{\pi}\tau_a\sqrt{g_K}} \int_{-\infty}^{\infty} A_K^{in}(t') \exp\left(\frac{t-t'}{2\tau_a} - \frac{(t-t')^2}{4\tau_a^2 g_K}\right) dt'. \quad (3)$$

Here $\tau_a = \frac{1}{\Delta\nu_l}$ is the lifetime of upper laser level, $\Delta\nu_l$ is active medium linewidth, the form of Lorentz line $\Gamma(\nu_0 - \nu)$ is

$$\Gamma(\nu_0 - \nu) = 2\tau_a \frac{1}{[1 + (\nu_0 - \nu)^2 4\pi^2 \tau_a^2]}. \quad (4)$$

At each round-trip, the spontaneous emission noise $N(t)$ is added to the laser radiation field. In calculation it is convenient to set the field amplitude $N(t)$ of spontaneous emission of an active medium in time representation as a sum of n_{max} harmonics with arbitrary phases ϕ_n. We take $n_{max} = 2n_g$, with the number of involved harmonics n_g determined by the gain line width [20]: $n_g = \Delta\nu_l/\Delta\nu_r = 2L/(c\tau_a)$, where ν_r is the interharmonics spectral interval, c is the speed of light, L is the cavity length.

$$N_K(t) = N_s \sum_{n=0}^{n_{max}} \exp(-n^2/n_g^2) \exp\left(i\phi_n + i\frac{2\pi nt}{T_r}\right), \quad t \in [K \cdot T_r, (K+1) \cdot T_r], \tag{5}$$

where N_s is spontaneous emission noise amplitude, the first factor in the sum is mode amplitude according to the distribution of inside the gain linewidth, while the second factor corresponds to phase evolution of each longitudinal mode. Phase ϕ_n is randomly chosen from the interval $[0, 2\pi]$.

3.2. Losses and shaping in electrooptic modulator

Let the beam splitter and the output coupler reflect the fractions B and R of incident intensity correspondingly. Total passive losses equal $L = 1 - (1-B)^2\text{R}$. Linear decrease of amplitude $A_K(t)$ due to passive losses is described by a factor $P_L = \sqrt{1-L} = (1-B) \cdot \sqrt{\text{R}}$.

The laser radiation is temporarily shaped in the electro-optic modulator. Dependence of Pockels cell transmission $P(t)$ on feedback control voltage $U_1(t)$ and $U_2(t)$ is given by

$$P(t) = P(U_1(t), U_2(t)) = \cos^2\left(\frac{\pi}{2}\frac{U_0 + U_1(t) + U_2(t)}{U_{\lambda/4}}\right). \tag{6}$$

Control voltages $U_1(t)$ and $U_2(t)$ at the capacitors C_1 and C_2 are described by the following equations:

$$\frac{dU_1(t)}{dt} + \frac{U_1(t)}{\tau_1} = \frac{i_{d1}(t)}{C_1}, \qquad \frac{dU_2(t)}{dt} + \frac{U_2(t)}{\tau_2} = \frac{i_{d2}(t)}{C_2}, \tag{7}$$

where $\tau_1 = R_1C_1$, $\tau_2 = R_2C_2$ are the time constants of discharge circuits. Photocurrent $i_{d1,2}$ in each control circuit is calculated as the convolution of the laser intensity with the photocurrent response function taking into account feedback loop sensitivities and optical delays:

$$i_{d1,2} = k_{1,2}\int_0^{+\infty} I(t-t'-T_r-T_{d1,2})(1-P(t-t'-T_r-T_{d1,2}))Ph_{1,2}(t')dt', \tag{8}$$

where $Ph_{1,2}(t)$ are time response functions of photodiodes in feedback circuits. The simulation provides either the triangular or the Gaussian and rectangular photocurrent response function of a photodiode. In each case the response time of photodiode τ_{ph} is full width at half maximum of the response function. Parameters k_1 and k_2 correspond to feedback sensitivity, T_{d1} and T_{d2} are optical delays in feedback circuits. Relative sensitivity of feedback circuits is $\alpha = k_2/k_1$. Laser radiation intensity is

$$I_K(t) = A_K(t)A_K^*(t), \tag{9}$$

where A_K^* is complex conjugate of A_K.

3.3. Final form for amplitude transformation

By collecting all terms that describe energy alteration from Subsections 3.1–3.2 we write final form of pass-to-pass amplitude A transformation:

$$A_{K+1}(t) = P(t)P_L\frac{\exp(g_K)}{2\sqrt{\pi}\tau_a\sqrt{g_K}}\int_{-\infty}^{\infty} A_K(t')\exp\left(\frac{t-t'-\tau_a g_K}{2\tau_a\sqrt{g_K}}\right)^2 dt' + N_K(t). \tag{10}$$

To investigate the dynamics we calculate total energy of laser radiation at each round trip E_K:

$$E_K(t) = \int_{KT_r}^{(K+1)T_r} I(t)dt. \tag{11}$$

For convenience, we operate with gain G_K in percent of generation threshold value which is connected to g_K via the formula:

$$G_K = e^{g_K}. \tag{12}$$

As an output result we obtain arrays of total radiation energy $\{E\}$, gain $\{G\}$, intensity of a maximum spike $\{Im\}$, its location on the round-trip $\{Tm\}$ and duration $\{\tau_p\}$. For each K we obtain fine-time-structure variables, namely $I(t)$, $U_{1,2}(t)$, $i_{1,2}(t)$, $P(t)$ for $t \in [K \cdot T_r, (K+1) \cdot T_r)$ with resolution of 0.5-1 ps. Calculation of each parameter set takes from several minutes up to 24 hours with Intel PC 2.4 GHZ, 3.25 GB RAM.

The approach allows us to investigate the laser generation dynamics from self start of mode-locking regime and remains valid for pump power levels sufficient for nonlinear dynamics manifestation.

4. Nonlinear dynamics of electro-optically controlled solid state laser

Let us discuss the graphical representation of the nonlinear laser dynamics in detail. In papers written previously by other authors, the regimes with characteristic time much longer than T_r were considered [1, 2]. Our goal is to investigate the possibility of scale reduction by use of a powerful concept of feedback loops combination.

4.1. Single feedback controlled dynamics

The fastest way to test chosen combination of feedback loops is to calculate and plot $E(t)$ (total energy at each round-trip time interval) dynamics diagram. It gives the information analogous to well-known bifurcation diagrams that show dependence of steady-state amplitudes versus control parameter values. The $E(t)$ diagram is calculated with slow variation of gain in time (see Fig. 4). We fix maximum gain and find the feedback parameters that require fine tuning: relative sensitivity and delays. Then we calculate $E(t)$ while G is linearly increased from round-trip to round-trip. The gain rise is slow enough (several hundred percents per 5000–50000T_r) for the plot to show most important features of the dynamical systems. Different regimes can be distinguished in $E(t)$ dynamics diagrams:

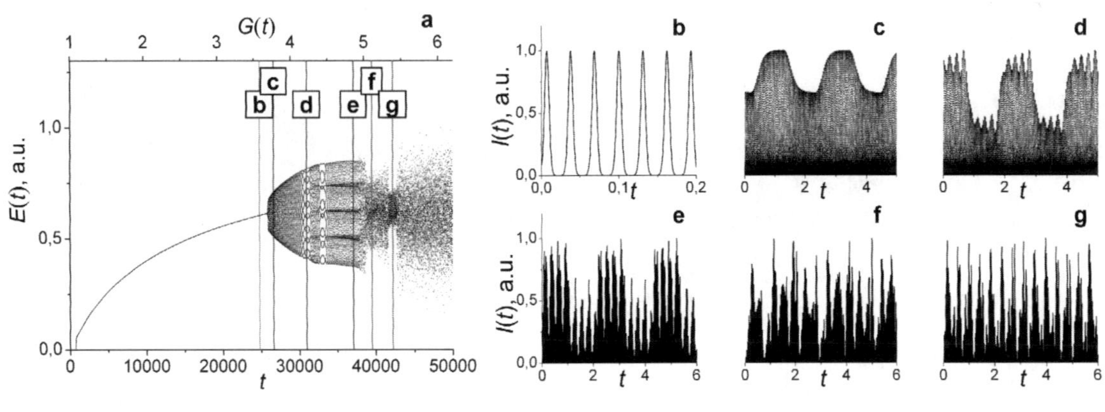

Figure 4. Dynamics of single feedback controlled laser in the case of $k = 32$ short pulses per $T_r = 10$ ns. **a**: total laser radiation energy $E(t)$ at each T_r versus time t. Gain G is linearly increased with time and shown at the upper x-axis in units of the threshold gain for reference; **b**–**g**: normalized laser radiation intensity fine time structure $I(t)$ versus time t. Time t is in units of T_r.

- when the gain is low enough, the laser operates in regular mode-locking regime and there is one branch in the diagram,
- when gain reaches bifurcation point, periodic bursting appears as several branches when period is a multiple of T_r,
- if bursting period is not a multiple of T_r, dense point mapping is observed (gray area),
- when the gain is high, it is typical to observe chaotic regimes but gray area at the diagram cannot be distinguished from the previous case. Insets with fine time structure provide detailed information on laser dynamics.

If the nonlinear dynamics is regular, and if period is a fraction of T_r, we can distinguish parallel branches at the $E(t)$ diagram (see, e.g., Fig. 4d). The most interesting regimes are then investigated with constant gain: insets I–IV show steady-state behavior at constant gain values. Each inset corresponds to the behavior type shown at the diagram with arrow, and represents laser radiation intensity versus time t (in round-trips T_r) in steady state. The dynamics studies allow to determine the parameters necessary to generate laser radiation with given properties. In simulation the following parameters of laser were used: $T_r = 10$ ns, $P_L = 50\%$, active-medium line width 120 GHz (corresponds to Nd:YAG), and photocurrent response time for both feedback loops $\tau_{ph} = 500$ ps.

It looks reasonable to seek the nonlinear dynamics regimes at a time scale less than T_r when laser is mode-locked with generation of large number of pulses on T_r. Maximum number is determined by the value of τ_{ph} and can be much larger than T_r/τ_{ph}. Example of the case with $\tau = 0.05T_r = 500$ ps $= \tau_{ph}$, $T_d = 0.98T_r$, if the laser is mode-locked by a single negative inertial feedback loop is presented in Fig. 4. The number of short laser pulses per T_r was 32, when loss modulation due to feedback control was only 5%. Maximum gain for stabilization, reaches 620% of threshold value. Despite the dense fill-in of the envelope, the period of regular nonlinear dynamics is not less than $2T_r$.

To simplify the control scenario (neighboring pulses do not interact) in search of short-period dynamics and exclude the simultaneous charging of modulator capacity by two photocurrent pulses, we consider the case when triangular photocurrent pulses do not intersect at a time line. Since the photocurrent response has full width of 1 ns at zero level, the number of short

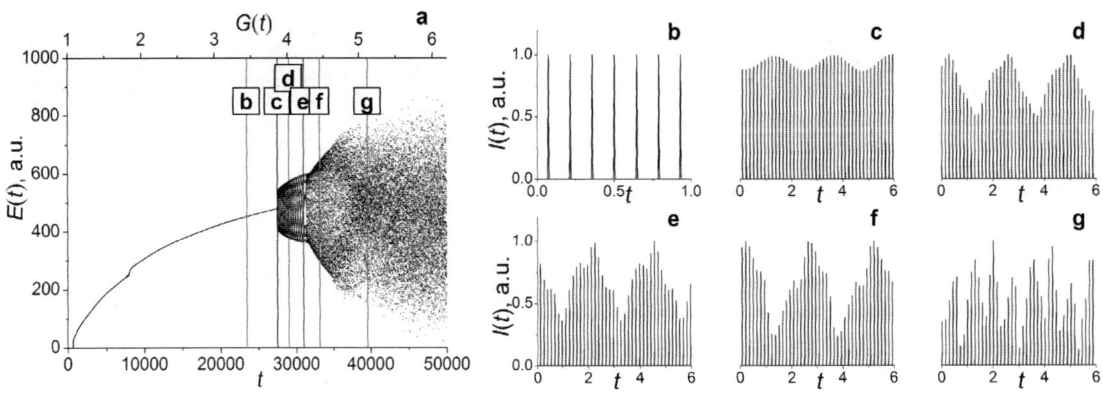

Figure 5. Dynamics of single feedback controlled laser in the case of $k = 7$ short pulses per $T_r = 10$ ns. **a**: total laser radiation energy $E(t)$ at each T_r versus time t. Gain G is linearly increased with time and shown at the upper x-axis in units of the threshold gain for reference; **b–g**: normalized laser radiation intensity fine time structure $I(t)$ versus time t. Time t is in units of T_r.

laser pulses $k = 7$ (Fig. 6) is a good choice with the photocurrent gap of few hundred ps if $T_r = 10$ ns. It is worth to note that at higher gain values the short pulse development takes less time. All regimes below are characterized by wide stability regions which we consider to be an important advantage of proposed systems. The scenario of nonlinear dynamics development for $k = 7$ presented in Fig. 5 is similar to that in the case of $k = 32$. Our investigation shows that in a single feedback controlled laser the dynamics with period less than T_r is not observed but chaotic regimes have high frequency: the characteristic time is interpulse interval T_r/k. In the following Section we show how the dynamics changes if we add the second loop in feedback control system.

4.2. Dual feedback controlled dynamics

In this Subsection we investigate the possibility to observe dynamics with periods less than T_r under dual feedback control. In our previous works, see e.g. [12], we used the method of memory erasure to switch off the action of negative inertial feedback after one T_r. This allowed to realize the dynamics of logistic map at a time scale of T_r when the relative feedback delay was set to T_r. In that study when relative delay was $2T_r$ we obtained pulsations with period $3T_r$ and $4T_r$.

Here we can generalize the memory erasure technique for the use in the case of arbitrary choice of feedback delays provided that the delay of negative feedback is shorter. If the relative sensitivity of positive feedback is chosen according to the formula

$$\alpha = e^{-\delta T_d/\tau}, \tag{13}$$

where δT_d is the relative delay of positive feedback, $\tau = \tau_1 = \tau_2$, then the action of negative feedback is switched off after a time δT_d. Note that $\alpha < 1$ has a certain physical meaning: only negative feedback can play the stabilization role, therefore its contribution should be larger.

We continue the consideration of 7th harmonics mode locking. If the relative feedback delay is set to an interpulse interval, it results in dynamics of 7 independent laser pulses (Fig. 7). Each pulse is controlled by itself on the previous round-trip. Thus the observed dynamics is close to the dynamics of independent logistic maps. We are able to trace the period doubling cascade, but the period 3 is not observable if we switch on the noise. In this case periods less than T_r are not observed but chaotic dynamics has a scale of $T_r/7$.

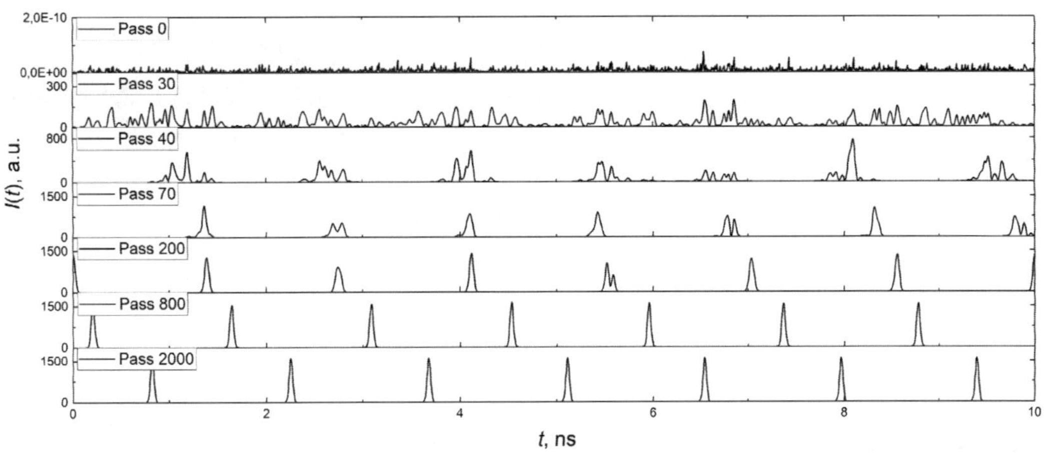

Figure 6. Development of harmonic mode-locking regime under single feedback control, $k = 7$ short pulses per $T_r = 10$ ns, $G = 3$. Laser radiation fine time structure: intensity $I(t)$ over time t on the time scale of one round-trip time.

Memory erasure technique allows to operate with feedback depth and to organize the control scenario when each pulse is controlled by a given number of pulses by increasing δT_d. If the depth is not greater than two interpulse intervals, each laser pulse is controlled by itself and its neighbor on the previous round-trip. To accomplish this we used the following parameters: negative feedback delay time $0.887T_r$ (which corresponds to the generation of 7 pulses per T_r), positive feedback delay $1.162T_r$ (relative delay $\delta T_d \sim 2T_r/7$), $\tau = 0.143T_r \sim T_r/7$ for both feedback loops, relative sensitivity of positive feedback α was set to 0.135 which corresponds to memory erasure. The behavior is presented in Fig. 9. We obtained high frequency train of short laser pulse microgroups with period $3T_r/7$.

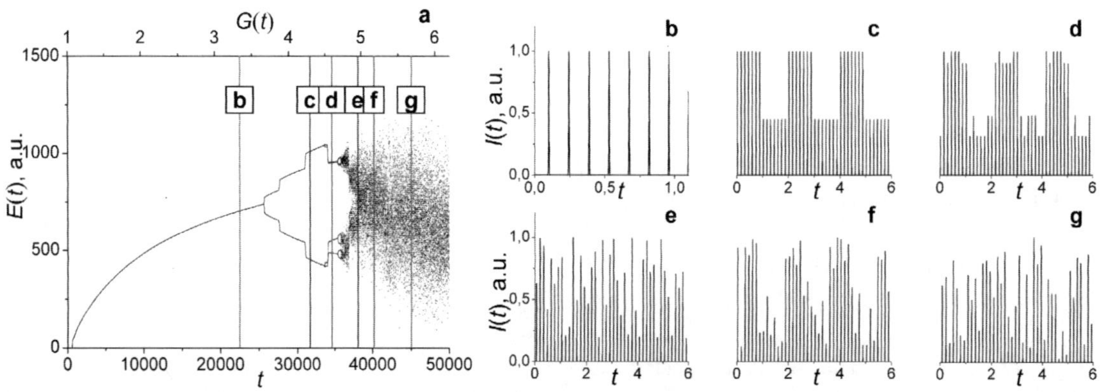

Figure 7. Dynamics of dual feedback controlled laser in the case of $k = 7$ short pulses per $T_r = 10$ ns. **a**: total laser radiation energy $E(t)$ at each T_r versus time t. Gain G is linearly increased with time and shown at the upper x-axis in units of the threshold gain for reference; **b**–**g**: normalized laser radiation intensity fine time structure $I(t)$ versus time t. Time t is in units of T_r.

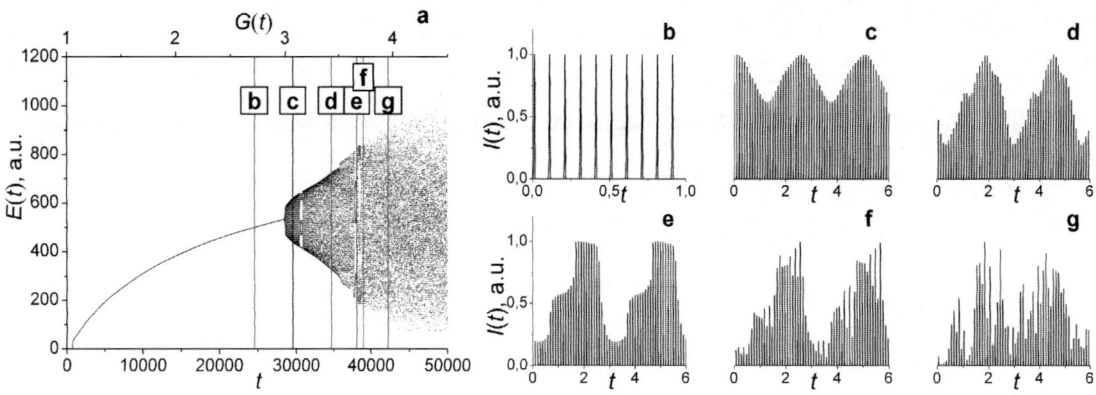

Figure 8. Dynamics of dual feedback controlled laser in the case of $k = 10$ short pulses per $T_r = 10$ ns at $T_{d1} = 0.908T_r$, $T_{d2} = 0.113T_r$, $\alpha = 0.135$, $\tau_1 = \tau_2 \sim T_r/9$. **a**: total laser radiation energy $E(t)$ at each T_r versus time t. Gain G is linearly increased with time and shown at the upper x-axis in units of the threshold gain for reference; **b**–**g**: normalized laser radiation intensity fine time structure $I(t)$ versus time t. Time t is in units of T_r. Three-stage structure with a stage width less than T_r is labelled with **e**.

More exotic regular nonlinear dynamics was obtained when positive feedback, in contrast to erasure regime, acts earlier than negative one. The dynamics in the form of three-stage structure was observed in the gain range of 3.78–3.83 (Fig. 8, III). This was demonstrated in a mode-locking regime with generation of $k = 10$ pulses per T_r, when $\delta T_d = -0.785T_r$.

The set of parameter such as harmonics number, feedback combination, relative sensitivity of loops, their inertiality form a multidimensional space which is not easy to explore due to slow procedure of numeric simulation. For better coverage of the variety of dynamical regimes it is worth to apply a simpler approach based on discrete maps.

5. Discrete maps

The discrete map approach was introduced in [21] for a picosecond laser with single non-inertial negative feedback. In the case of small changes of pulse duration the energy of $(n+1)$th pulse in the cavity was described by a recurrence relation

$$E_{n+1} = R\alpha_n^2\beta_n T_n E_n, \tag{14}$$

where R is the loss coefficient, α_n is the coefficient of amplification by active medium, β_n is the transmission coefficient of saturable absorber (dependent on pulse energy E_n), T_n is the transmission coefficient of elecro-optic modulator. In the simplest case of delay-free negative feedback

$$T_n = T_0 - \mu E_n, \tag{15}$$

where T_0 is the initial transmission of the modulator, μ is the feedback coefficient. By substitution of $x_n = \mu E_n/T_0$, authors of [21] obtained the map

$$x_{n+1} = r_n x_n(1 - x_n), \tag{16}$$

where $r_n = R\alpha_n^2\beta_n T_0$. If the variation of active medium amplification coefficient is not taken into account, than the obtained relation is reduced to the well-known logistic map.

Later the method of discrete maps was applied for the case of combination of feedback loops in our works [9, 17]. Such an approach opened up the following possibilities:

1) To allow the laser output being non-dependent of pump power and total intracavity losses variation for both continuous and pulsed lasers [17]. This is obtained due to a) the giant stability zone which is intrinsic in short-delay control and b) short steady-state build-up time which is provided by the round-trip-time delayed positive feedback.

2) In burst mode, T_r-shorter (compared to negative) positive feedback delayed control [7] allows the period increase up to several hundred round-trip time by variation of relative feedback sensitivity. Hence, the period pushed the burst repetition frequency to sub-MHz range.

3) In a dual-feedback-controlled mode-locked solid state laser energy of output pulse demonstrates the behavior typical for the logistic map: period doubling cascade, periodicity windows, intermittency, and deterministic chaos on the time scale of the cavity round trip time [12]. To achieve this, it is necessary that inertial delay-free negative feedback is combined with T_r-delayed positive feedback, which sensitivity is chosen to erase the residual action of inertial negative feedback. In this case the control parameter is pump power. This approach makes the laser to be a spectacular device for nonlinear dynamics demonstration.

4) In a dual-feedback-controlled mode-locked solid state laser burst mode with minimal period $3T_r$ can be realized. To achieve this, it is necessary that inertial delay-free negative feedback is combined with $2T_r$-delayed positive feedback, which sensitivity is chosen to erase the residual action of inertial negative feedback [11]. In this case the burst repetition frequency is equal to several tens of MHz.

Matter of current interest is to extend the map-based approach to the case of harmonically mode-locked laser. For the case of one inertial feedback and a single laser pulse in the cavity (see, e.g. [9]), we obtained a map known also as logistic map with memory

$$x_{n+1} = rx_n \left(1 - \sum_{m=0}^{\infty} x_{n-m}\gamma^m\right), \tag{17}$$

where $\gamma = e^{-T_r/\tau}$. For harmonic mode-locking with $k > 1$ we can either write down a system of k maps or compactify the system of maps into a single map with discrete time count T_r/k using

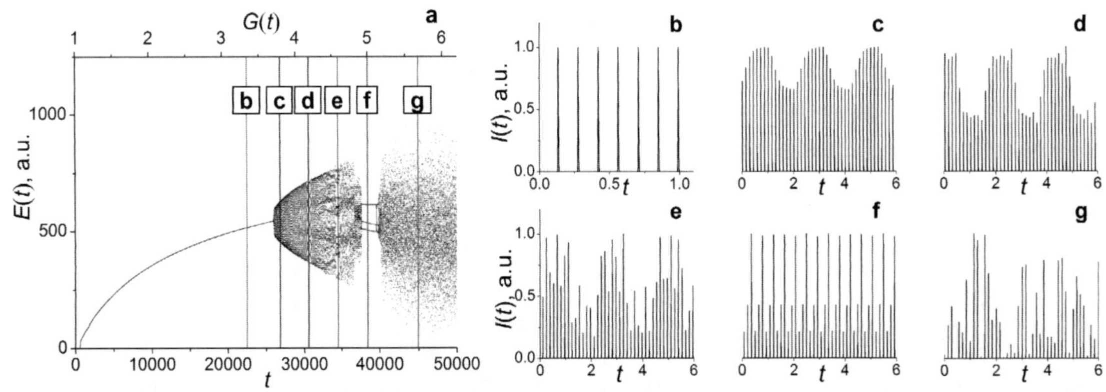

Figure 9. Dynamics of dual feedback controlled laser in the case of $k = 7$ short pulses per $T_r = 10$ ns and memory erasure with relative delay of $2T_r/7$. **a**: total laser radiation energy $E(t)$ at each T_r versus time t. Gain G is linearly increased with time and shown at the upper x-axis in units of the threshold gain for reference; **b**–**g**: normalized laser radiation intensity fine time structure $I(t)$ versus time t. Time t is in units of T_r. Dynamics with period $3T_r/7$ is labelled with **f**.

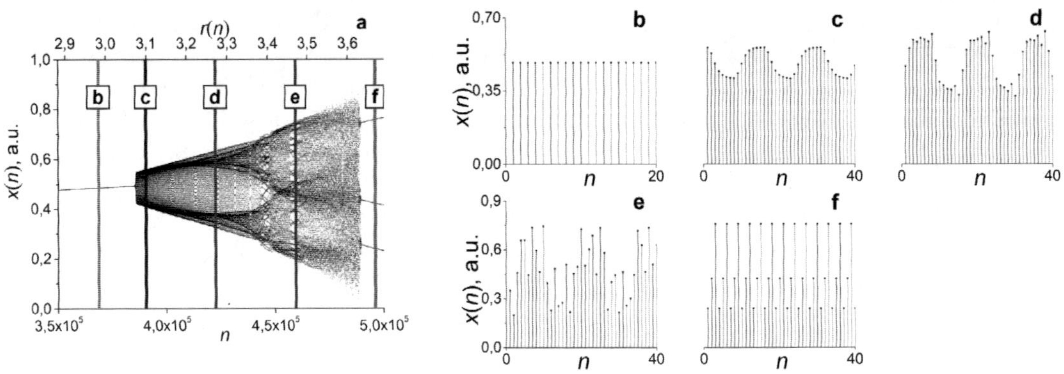

Figure 10. Dynamics of map (21). The representation is analogous to that of Fig. 9. Inset with period $3T_r/7$ is labelled with **f**. $x(n)$ is the normalized energy of a short laser pulse, n is the discrete time count in units of $T_r/7$, $k = 7$.

the cyclicality of equations. Thus the map is

$$x_{n+1} = rx_{n-(k-1)}\left(1 - \sum_{i=0} x_{n-(k-1)-k\cdot M-i}\cdot\gamma^i\right), \quad n = 0, 1, 2..., \tag{18}$$

where n is the number of laser pulse, x_n is the pulse energy, r is the total gain, $0 \leq \gamma < 1$ describes feedback inertiality. In this case all laser pulses in the cavity interact with each other due to feedback inertiality. This case is the discrete representation of laser dynamics shown in Figs. 5 and 4 in Subsection 4.1.

Non-inertial feedback case is the system of k independent logistic maps

$$x_{n+1} = rx_{n-(k-1)}\left(1 - x_{n-(k-1)}\right), \quad n = 0, 1, 2... \tag{19}$$

that describe the situation when laser pulses do not interact (analogue of the case in Fig. 7 in Subsection 4.2). The combination of negative and delayed positive feedback loops gives

$$x_{n+1} = rx_{n-(k-1)}\left(1 - \sum_{i=0} x_{n-(k-1)-k\cdot M-i}\cdot\gamma^i + \alpha\sum_{i=p} x_{n-(k-1)-k\cdot M-i}\cdot\gamma^{i-p}\right), \quad n = 0, 1, 2..., \tag{20}$$

where p is the delay and α is the relative feedback sensitivity of positive feedback loop. By choosing the value of α it is possible to cancel the members of the first sum and thus to switch off the feedback action.

Bursting with period $3T_r/7$ (Fig. 10) was observed in map

$$x_{n+1} = rx_{n-6}(1 - x_{n-6} - 0.367x_{n-7}). \tag{21}$$

The map describes the case when a pulse interacts only with its neighbour in the laser cavity. This example illustrates that the introduction of limited interpulse interaction drastically changes the generation regime in the laser system resulting in regular collective dynamics of output pulses. The results of map simulation agree with the numeric simulation taking into account the evolution of fine time structure of laser radiation (compare Figs. 9 and 10).

To summarize in this Section, discrete maps are a simple and fast instrument to describe the features of laser dynamics not only over the stability region border but far in the nonlinear dynamics region, even predicting the appearance of regular subroundtrip dynamics in a laser. On top of that, due to their simplicity discrete maps are a promising opportunity to build a bridge from a laser to other dynamical systems.

6. Conclusion

This study shows a high potential of combining the dual feedback control with self-starting harmonic mode-locking. The proposed approach to laser control gives new possibility for ordering of output time structure and may be useful for a wide community of scientists and engineers that work in the areas of nonlinear dynamics, solid state lasers and optoelectronics.

We have shown that in the case of self-started harmonic mode-locking a laser controlled by optoelectronic feedback loop is able to generate $k > T_r/\tau_{ph}$ short pulses per cavity round-trip time. Such control allowed to go into the collective dynamics of laser pulses coupled by inertial feedback action. The proposed approach is distinguished by high output radiation stability in the gain range of several hundred percent over the threshold. Single inertial optoelectronic feedback control allows one to initiate regular and chaotic nonlinear laser dynamics at a time scale of less than a laser cavity round-trip time T_r.

Regular sequence of pulses with sub-round-trip period is realized when positive feedback delay is added to erase the negative feedback action but the interaction remains at least between two pulses of k. If the pulse interaction is switched off, we are able to trace period doubling cascade of k independent logistic maps and chaotic dynamics has a scale of T_r/k. Such set of k self-starting chaotic generators in a single laser cavity is a promising object for high-frequency chaotic encoding. The detailed numerical simulation proved that all presented regimes can be realized in a solid state laser with two electro-optic feedback loops.

Acknowledgments

The work was supported by the Program of Presidium RAS 1.25P "Fundamental and applied photonics problems and new optical material physics".

References

[1] Arecchi F, Gadomski W and Meucchi R 1986 Generation of chaotic dynamics by feedback on a laser *Phys. Rev.* A **34** 1617–20
[2] Samson A, Kotomtseva L and Loiko N 1990 *Autooscillations in lasers* (Minsk: Navuka i technika)
[3] Erneux T and Glorieux P 2010 *Laser Dynamics* (Cambridge: Cambridge University Press)
[4] Grigorieva E V and Kaschenko S A 2016 *Asymptotic Representation of Relaxation Oscillations in Lasers* (Springer)
[5] Statz H and de Mars G 1965 Problem of spike elimination in lasers *J. Appl. Phys.* **36** no. 5 1510.
[6] Marshall F and Roberts D 1962 Use of electro-optical shutters to stabilize ruby laser operation *Proc. of the IRE* October 2108
[7] Gorbunkov M V, Maslova Yu Ya, Petukhov V A, Semenov M A, Shabalin Yu V and Vinogradov A V 2009 Submicrosecond regular and chaotic nonlinear dynamics in a pulsed picosecond Nd:YAG laser with millisecond pumping *Applied Optics* **48** no. 12 2267–74
[8] Gorbunkov M V, Maslova Yu Ya and Vinogradov A V 2009 Optical unit of laser-electron X-ray generator designed for medical applications *NIM* A **608** S32–5
[9] Gorbunkov M V, Maslova Yu Ya, Shabalin Yu V and Tunkin V G 2016 Dual feedback control in solid state lasers: discrete maps and experiments *J. Phys.: Conf. Series* **692** 012009-1–10
[10] Gorbunkov M V, Maslova Yu Ya, Petukhov V A, Semenov M A and Shabalin Yu V 2009 Generation of a regular sequence of short-pulse microtrains with a discretely varied repetition period *Bull. of the Lebedev Physics Institute* **36** no. 9 270–6.
[11] Gorbunkov M V, Maslova Yu Ya, Petukhov V A, Semenov M A, Shabalin Yu V and Vinogradov A V 2014 Round-trip-time nonlinear dynamics of electro-optically-controlled solid state lasers *J. of Russian Laser Research* **35** no. 5 492–500
[12] Gorbunkov M V, Maslova Yu Ya, Chaban O I and Shabalin Yu V 2009 Period doubling cascade and deterministic chaos in a laser self-mode-locked by the combination of inertial negative and positive feedbacks *Bull. of the Lebedev Physics Institute* **36** no. 5 150–6
[13] Hirano J and Kimura T 1969 Multiple mode locking of lasers *IEEE J. Quantum Electron.* **5** no. 5 219–25
[14] Becker M F, Kuizenga D J and Siegman A E 1972 Harmonic mode locking of the Nd:YAG laser *IEEE J. Quantum Electron.* **8** no. 8 687–93
[15] Gorbunkov M V, Maslova Yu Ya, Tunkin V G, Shabalin Yu V and Yakovlev D V 2016 Application of optoelectronic negative feedback to ordering of the temporal structure of the diode-pumped Nd:YLF laser radiation *Bull. of the Lebedev Physics Institute* **43** no. 7 217–22
[16] Makukha V K, Smirnov V S and Semibalamut V M 1977 Generation of ultrashort pulses in a negative-feedback laser, *Soviet J. of Quantum Electronics* **7** 573–5
[17] Gorbunkov M V and Shabalin, Yu V 2002 Two-loop feedback controlled laser: New possibilities for ultrashort pulses generation and high-level stabilization *Proc. SPIE* **4751** 463–70
[18] Vorchik D B and Gorbunkov M V 1997 Self mode locking in nd glass laser by fast negative feedback with solid state optoelectronic system *Bull. of the Lebedev Physics Institute* **12** 70–6
[19] Herrmann J and Wilhelmi B 1987 *Lasers for Ultrashort Light Pulses* (Amsterdam: North-Holland)
[20] Kryukov P G and Letokhov V S 1970 Propagation of a light pulse in a resonantly amplifying (absorbing) medium *Sov. Phys. Usp.* **12** 641–72
[21] Bayanov I M, Gordienko V M, Zvereva M G, Magnitskii S A and Tarasevich A P 1989 Highly stable picosecond YAG:Nd^{3+} laser with a negative feedback *Sov. J. Quantum Electron.* **19** 994–6

IOP Conf. Series: Journal of Physics: Conf. Series **990** (2018) 012005 doi:10.1088/1742-6596/990/1/012005

Strong generalized synchronization with a particular relationship R between the coupled systems

Clara Grácio[1], Sara Fernandes[2] and Luís Mário Lopes[3]

[1]Department of Mathematics, ECT - Universidade de Évora, CIMA-UE
Rua Romão Ramalho, 59, 7000-585 Évora, Portugal
[2]Department of Mathematics, ECT - Universidade de Évora, CIMA-UE
Rua Romão Ramalho, 59, 7000-585 Évora, Portugal
[3]Department of Mathematics, ISEL
Rua Conselheiro Emídio Navarro, 1, 1959-007 Lisboa, Portugal

E-mail: [1]`mgracio@uevora.pt`, [2]`saf@uevora.pt`, [3]`llopes@adm.isel.pt`

Abstract. The question of the chaotic synchronization of two coupled dynamical systems is an issue that interests researchers in many fields, from biology to psychology, through economics, chemistry, physics, and many others. The different forms of couplings and the different types of synchronization, give rise to many problems, most of them little studied. In this paper we deal with general couplings of two dynamical systems and we study strong generalized synchronization with a particular relationship R between them. Our results include the definition of a window in the domain of the coupling strength, where there is an exponentially stable solution, and the explicit determination of this window. In the case of unidirectional or symmetric couplings, this window is presented in terms of the maximum Lyapunov exponent of the systems. Examples of applications to chaotic systems of dimension one and two are presented.

1. Introduction

Chaotic synchronization is a very important phenomenon in many fields involving mathematical, physical, sociological, physiological, biological or other systems [1, 2, 3, 4]. Several different types of synchronization have been studied: complete synchronization [5, 6], phase synchronization [1, 7], lag synchronization [8, 9, 10], generalized synchronization [11, 12], projective synchronization [13, 14], etc. The complete synchronization is the simplest type of synchronization and a lot of work have been produced about it. Much less study have been devoted to one of the most interesting types of synchronous chaotic behavior, the generalized synchronization. When there is generalized synchronization in a coupling of two dynamical systems x_t and y_t, it is established a relationship $y_t = R(x_t)$ between the system states after the transient is finished. Depending on the properties of the relationship R, the synchronization is said to be either strong or weak. Namely, strong synchronization corresponds to a smooth relationship R, while weak synchronization corresponds to a fractal one [17]. Several methods have been suggested to detect generalized synchronization: the method of the mutual false nearest neighbors [11, 15], the auxiliary system method [16], using the conditional Lyapunov exponents [17], the modified system approach [18, 19], the phase tube approach [20, 21]. Nevertheless, most of the studies done focuses on continuous-time systems. Few of them are related to discrete-time systems

[22, 23, 24, 25, 21] and even fewer to bidirectional couplings [20, 26, 21] since generalized synchronization is traditionally introduced for two unidirectionally coupled systems [11, 17, 27].

In this paper we discuss the design of a c-family of couplings in order to achieve generalized synchronization with a desired relationship R between the coupled systems. Similar discussions have already been done for unidirectional couplings or for couplings of continuous-time systems [28, 29, 30], but now we are considering those less studied situations: bidirectional couplings of discrete-time systems. We consider conditions for which a c-family of couplings achieves the referred synchronization and for which we will be able to calculate the R-synchronization window. In fact, only some values of the coupling strength will provide exponentially stable synchronized solutions, defining the referred window.

2. A c-family of couplings that admit generalized synchronization with a particular relationship R

We consider the general coupling of two discrete n-dimensional chaotic dynamical systems x_t and y_t that is widely used in this context [28, 29, 30, 26], given by

$$\begin{cases} x_{t+1} = f(x_t) + c \cdot [F_1(x_t) + F_2(y_t)] \\ y_{t+1} = g(y_t) + c \cdot [G_1(x_t) + G_2(y_t)] \end{cases}, \tag{1}$$

where c is the coupling strength, with values in $[0,1]$ and f, g, F_1, F_2, G_1 and G_2 are appropriate functions.

The ability of the systems to synchronize depends, not only on the functions F_1, F_2, G_1 and G_2, but also on the coupling strength c. In order to consider couplings where the values of the coupling strength c for which the coupling achieves generalized synchronization with a particular relationship R is not a discrete set, we want to consider couplings that admit this synchronization for all values of c (even if it is not an exponentially stable one). That will allow us to define "synchronization windows" in a similar way that was often done for complete synchronization (either in couplings or in networks [1]). So, we consider the following c-family of couplings that provide a similar frame of analysis for the generalized synchronization with a particular relationship R:

$$\begin{cases} x_{t+1} = f(x_t) + c \cdot [-F_2(R(x_t) + F_2(y_t)] \\ y_{t+1} = g(y_t) + c \cdot [-G_2(R(x_t)) + G_2(y_t)] \end{cases}, \tag{2}$$

with $R \circ f = g \circ R$. In fact, $(x_t, y_t) = (s_t, R(s_t))$ with $s_{t+1} = f(s_t)$ is a solution of (2) for all values of c, since (2) reduces to

$$\begin{cases} s_{t+1} = f(s_t) \\ R(s_{t+1}) = g(R(s_t)) \end{cases}$$

and both equations are verified because $R \circ f = g \circ R$ and $s_{t+1} = f(s_t)$.

We note that the complete synchronization corresponds to $R(u) = u$ and the lag synchronization corresponds to $R(u) = f^{(\Delta t)}(u)$ [17, 19, 31]. Further, if R is a diffeomorphism, then f and g are topologically conjugate maps by the topological conjugacy R. So, a coupling of type (2) of two dynamical systems described by maps that are topologically conjugate by a topological conjugacy R, admits generalized synchronization with a relationship R between the coupled systems.

3. R-synchronization window

Even if a coupling admits generalized synchronization with a particular relationship R, only some values of the coupling strength (or even none) correspond to a coupling that admits a function s_t such that $(x_t, y_t) = (s_t, R(s_t))$ is an exponentially stable solution. So, we consider the following definition.

Definition 1. *For a c-family of couplings that admit generalized synchronization with a particular relationship R, we call R-synchronization window, RSW, the set of values of the coupling strength c for which there is a function s_t such that $(x_t,\ y_t) = (s_t,\ R(s_t))$ is an exponentially stable solution of (1).*

The following proposition establishes conditions for a value of c to belong to the RSW.

Proposition 1. *The R-synchronization window, RSW, of the coupling (2) satisfies*

$$\{c \in [0,1] : \mu_{rs} < 0\} \subset RSW \subset \{c \in [0,1] : \mu_{rs} \leq 0\}$$

where

$$\mu_{rs} = \max_{u_0} \lim_{T \to +\infty} \frac{1}{T} \ln \left| DM^T(s_0) \cdot \frac{u_0}{\|u_0\|} \right|,$$
$$M = g \circ R + c \cdot (G_2 \circ R - R \circ f \cdot F_2 \circ R),$$
$$DM^T(s_0) = DM(s_{T-1}) \cdot DM(s_{T-2}) \cdot \ldots \cdot DM(s_0)$$

and $s_{t+1} = f(s_t)$.

Proof. Considering $u_t = y_t - R(x_t)$, or equivalently, $y_t = R(x_t) + u_t$,

$$\begin{aligned} u_{t+1} = y_{t+1} - R(x_{t+1}) = \\ = g(R(x_t) + u_t) + c \cdot [-G_2(R(x_t)) + G_2(R(x_t) + u_t)] - \\ -R(f(x_t) + c \cdot [-F_2(R(x_t)) + F_2(R(x_t) + u_t)]) \end{aligned}$$

and, near the synchronized solution, that corresponds to $u_t = 0$ and $x_{t+1} = f(x_t)$, we have

$$\begin{aligned} u_{t+1} \simeq g(R(x_t)) + Dg(R(x_t)) \cdot u_t + c \cdot DG_2(R(x_t)) \cdot u_t - \\ -R(f(x_t)) - c \cdot DR(f(x_t)) \cdot DF_2(R(x_t)) \cdot u_t \end{aligned}$$

Since $R \circ f = g \circ R$, we obtain

$$u_{t+1} \simeq Dg(R(x_t)) \cdot u_t + c \cdot DG_2(R(x_t)) \cdot u_t - c \cdot DR(f(x_t)) \cdot DF_2(R(x_t)) \cdot u_t$$

Then, the linearization of the u_t evolution is given by

$$u_{t+1} = DM(s_t) \cdot u_t \tag{3}$$

with

$$DM(s_t) = Dg(R(s_t)) + c \cdot [DG_2(R(s_t)) - DR(f(s_t)) \cdot DF_2(R(s_t))] \tag{4}$$

and $s_{t+1} = f(s_t)$, and if, for a particular value of c, this system is exponentially stable, then the synchronized state $(x_t,\ y_t) = (s_t,\ R(s_t))$ of the coupling (2) is also exponentially stable, i.e. c belongs to its RSW.

Further,

$$u_T = DM(s_{T-1}) \cdot DM(s_{T-2}) \cdot \ldots \cdot DM(s_0) \cdot u_0 = DM^T(s_0) \cdot u_0,$$

and if $\lim_{T \to +\infty} \frac{1}{T} \ln |DM^T(s_0) \cdot u_0|$ is negative for all u_0 then (3) is exponentially stable. So we conclude that $\{c \in [0,1] : \mu_{rs} < 0\} \subset RSW$.

On the contrary, if μ_{rs} is positive, then (3) is unstable and we conclude that $RSW \subset \{c \in [0,1] : \mu_{rs} \leq 0\}$. □

IOP Conf. Series: Journal of Physics: Conf. Series **990** (2018) 012005 doi:10.1088/1742-6596/990/1/012005

We note that this proposition is valid for dynamical systems of any dimension. In the next section, we use it to confirm numerical examples of dimensions one and two.

For some couplings it is possible to explicit the RSW as a function of the maximal Lyapunov exponent of the coupled dynamical systems. In fact, if R is a diffeomorphism and $G_2 = -g$ we have for unidirectional couplings (i.e. couplings with $F_2 = 0$) or for the ones that $F_2 = R^{-1} \circ g$ (we will call them "symmetric couplings") the following propositions.

Remark 1. *If $G_2 = -g$ and $F_2 = R^{-1} \circ g$, the coupling is symmetric since (2) reduces to*

$$\begin{cases} x_{t+1} = (1-c) \cdot f(x_t) + c \cdot \left(f \circ R^{-1}\right)(y_t) \\ y_{t+1} = c \cdot (R \circ f)(x_t) + (1-c) \cdot \left(R \circ f \circ R^{-1}\right)(y_t) \end{cases},$$

Proposition 2. *For an unidirectional coupling (2) with $G_2 = -g$,*

$$\left]1 - e^{-\mu_0}, 1\right] \subset RSW \subset \left[1 - e^{-\mu_0}, 1\right],$$

where μ_0 is the maximal Lyapunov exponent of $s_{t+1} = f(s_t)$.

Proof. For such a coupling, since $G_2 = -g$ and $F_2 = 0$, (4) reduces to

$$DM(s_t) = Dg(R(s_t)) - c \cdot Dg(R(s_t)) = (1-c) \cdot Dg(\widetilde{s}_t),$$

with $\widetilde{s}_{t+1} = R(s_t)$ satisfying $\widetilde{s}_{t+1} = g(\widetilde{s}_t)$. In fact, when $s_{t+1} = f(s_t)$, we have $\widetilde{s}_{t+1} = R(s_{t+1}) = R(f(s_t)) = g(R(s_t)) = g(\widetilde{s}_t)$.

Using $Dg^T(\widetilde{s}_0)$ to stand for $Dg(\widetilde{s}_{T-1}) \cdot Dg(\widetilde{s}_{T-2}) \cdot ... \cdot Dg(\widetilde{s}_0)$, we obtain

$$\mu_{rs} = \max_{u_0} \lim_{T \to +\infty} \frac{1}{T} \ln \left| (1-c)^T \cdot Dg^T(\widetilde{s}_0) \cdot \frac{u_0}{\|u_0\|} \right| = \ln|1-c| + \widetilde{\mu}_0,$$

where $\widetilde{\mu}_0$ is the maximal Lyapunov exponent of $\widetilde{s}_{t+1} = g(\widetilde{s}_t)$. Since f and g are diffeomorphically conjugate, $\widetilde{\mu}_0 = \mu_0$ [32] and we obtain

$$\mu_{rs} < 0 \Leftrightarrow 1 - e^{-\mu_0} < c < 1 + e^{-\mu_0}$$

So, we conclude that $]1 - e^{-\mu_0}, 1] \subset RSW \subset [1 - e^{-\mu_0}, 1]$. □

Proposition 3. *For a symmetric coupling (2) with $G_2 = -g$,*

$$\left]\frac{1 - e^{-\mu_0}}{2}, \frac{1 + e^{-\mu_0}}{2}\right[\subset RSW \subset \left[\frac{1 - e^{-\mu_0}}{2}, \frac{1 + e^{-\mu_0}}{2}\right],$$

where μ_0 is the maximal Lyapunov exponent of $s_{t+1} = f(s_t)$.

Proof. For such a coupling, since $G_2 = -g$ and $F_2 = R^{-1} \circ g$, (4) reduces to

$$\begin{aligned} DM(s_t) &= Dg(R(s_t)) + c \cdot \left[-Dg(R(s_t)) - DR(f(s_t)) \cdot D\left(R^{-1} \circ g\right)(R(s_t))\right] \\ &= Dg(R(s_t)) + c \cdot \left[-Dg(R(s_t)) - DR(f(s_t)) \cdot (DR)^{-1}\left(R^{-1}(g(R(s_t)))\right) \cdot Dg(R(s_t))\right] \end{aligned}$$

and, since $R \circ f = g \circ R$, we obtain,

$$\begin{aligned} DM(s_t) &= Dg(R(s_t)) + c \cdot [-Dg(R(s_t)) - Dg(R(s_t))] \\ &= (1-2c) \cdot Dg(\widetilde{s}_t) \end{aligned}$$

with $\widetilde{s}_{t+1} = R(s_t)$, as in the previous proposition.

Also, in a similar way to the proof of the previous proposition, that leads to

$$\mu_{rs} < 0 \Leftrightarrow \frac{1 - e^{-\mu_0}}{2} < c < \frac{1 + e^{-\mu_0}}{2}$$

So, we conclude that $\left]\frac{1-e^{-\mu_0}}{2}, \frac{1+e^{-\mu_0}}{2}\right[\subset RSW \subset \left[\frac{1-e^{-\mu_0}}{2}, \frac{1+e^{-\mu_0}}{2}\right]$. □

We note that in both of the previous propositions RSW only depends on f, which means that using an unidirectional or a symmetric coupling in order to obtain an exponentially stable R-synchronization, the values of the coupling strength c that must be used are independent of R. In both cases the width of the RSW is $e^{-\mu_0}$ but for a symmetric coupling an exponentially stable R-synchronization is obtained for smaller values of the coupling strength.

4. Examples of the R-synchronization windows

In this section, in order to consider the general scope of the results obtained in the previous one, we illustrate them in situations that do not reduce neither to the complete synchronization nor to the lag synchronization, since those particularizations of the general problem have already been considered before [5, 8, 6, 9, 10], i.e. we consider $R(u)$ different from u and from $f^{(\Delta t)}(u)$. We consider couplings of one dimensional and two dimensional dynamical systems, using emblematic maps for both situations: for the couplings of one dimensional dynamical systems we consider that f is the logistic map and $R(u) = u^2$ and for the couplings of two dimensional dynamical systems we consider that f is the Hénon map and $R(u_1, u_2) = (u_2, u_1)$.

We begin by the examples of unidirectional couplings, i.e. couplings with $G_2 = -g$ and $F_2 = 0$. In figure 1 we show the computed results for both the one dimensional and two dimensional situations. In order to the RSW appear clearly, we show the graphs of the difference of the iterates y_t and $R(x_t)$ (for the two dimensional situation we show the first component of this difference), namely $y_t - x_t^2$ for the one dimensional situation and $(y_t)_1 - (x_t)_2$ for the two dimensional situation. The RSW corresponds to the values of c for which the ordinates of all the iterates are zero (after the transients died out). The RSW shown in the figures are confirmed by proposition 2, since the maximal Lyapunov exponents for the logistic and Hénon maps ($\ln 2$ and 0.419, respectively) provide $]1 - e^{-\mu_0}, 1]$ equal to $\left]\frac{1}{2}, 1\right]$ and $]0.342, 1]$, respectively. In the same

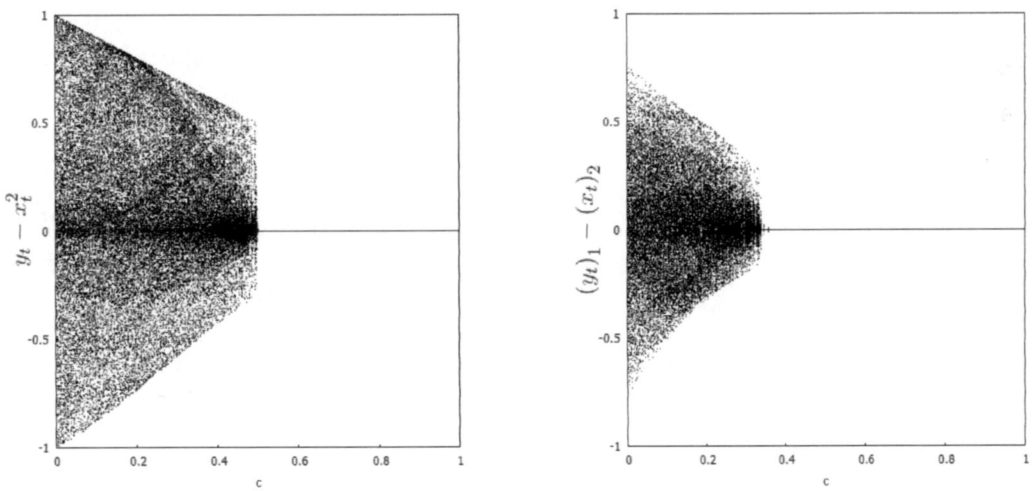

Figure 1. Graphs of the post-transient iterates $y_t - R(x_t)$ as a function of c for an unidirectional coupling with $G_2 = -g$. At the left, we consider that f is the logistic map and $R(u) = u^2$. At the right, we consider that f is the Hénon map and $R(u_1, u_2) = (u_2, u_1)$.

way, we show in figure 2 the computed results for both the one dimensional and two dimensional situations corresponding to symmetric couplings, i.e. couplings with $G_2 = -g$ and $F_2 = R^{-1} \circ g$.

Again, the RSW shown in the graphs are confirmed by the proposition 3, since $\left]\frac{1-e^{-\mu_0}}{2}, \frac{1+e^{-\mu_0}}{2}\right[$ for the one and two dimensional situations considered are $\left]\frac{1}{4}, \frac{3}{4}\right[$ and $]0.171, 0.829[$, respectively.

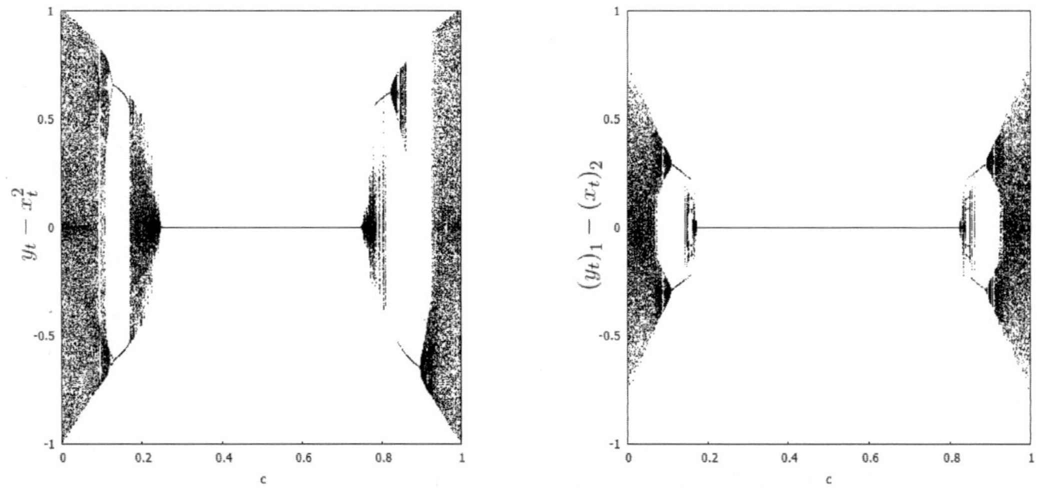

Figure 2. Graphs of the post-transient iterates $y_t - R(x_t)$ as a function of c for a symmetric coupling. At the left, we consider that f is the logistic map and $R(u) = u^2$. At the right, we consider that f is the Hénon map and $R(u_1, u_2) = (u_2, u_1)$.

We consider also two examples of couplings for which the RSW can not be calculated neither by proposition 2 nor by proposition 3. Namely, we consider examples of unidirectional couplings (i.e. couplings with $F_2 = 0$) with $G_2(u) = -g^2(u)$ for the one dimensional situation and $G_2(u_1, u_2) = \left(-g_1^2(u_1,u_2), -g_2^2(u_1,u_2)\right)$ for the two dimensional situation. We show in figure 3 the computed results and we verify that a non-empty RSW exists. In order to show that this window is confirmed by proposition 1 we show in figure 4 the value of μ_{rs}. In fact the values for which μ_{rs} are negative in the figure 4 are the ones for which the ordinates of all the post-transient iterates of $y_t - R(x_t)$ are zero.

5. Conclusions

If some conditions are satisfied, when we couple two discrete chaotic dynamical systems, they achieve generalized synchronization with a particular relationship R between the coupled systems. In fact, if the two systems are topologically conjugate by a topological conjugacy R, a coupling of the type (2) admits that synchronization and there is a synchronized exponentially stable solution for the values of the coupling strength c that belong to the R-synchronization window RSW.

So, if we are dealing with two topologically conjugate dynamical systems, we are able to synchronize them in an exponentially stable way. We just have to couple them with a coupling of type (2) and use a value of c in RSW. Or, the other way round, if we want that a dynamical system R-synchronizes with another, we choose that other system topologically conjugate to the first one by a topological conjugacy R and then we couple them in the referred way.

For couplings of type (2) an analytical expression of the RSW can be calculated. In some situations, as in the unidirectional and symmetric cases considered in propositions 2 and 3, that

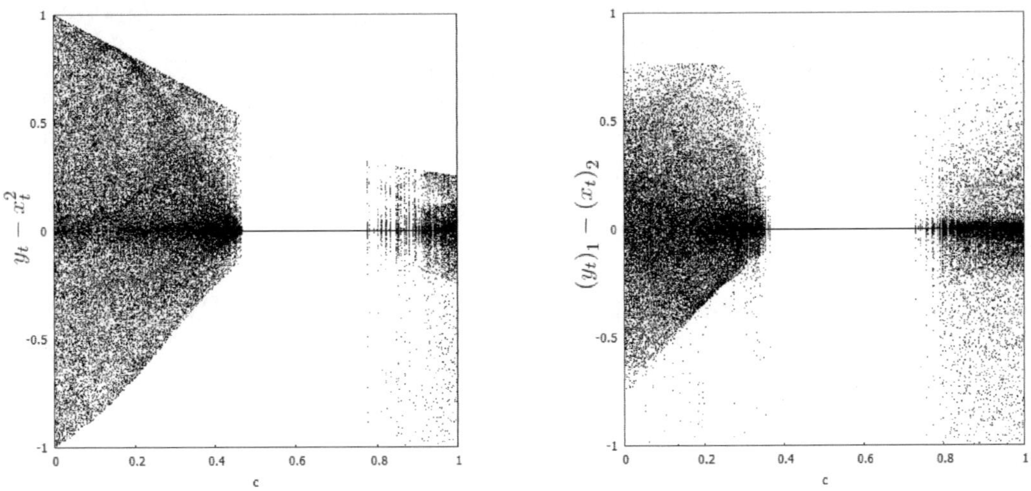

Figure 3. Graphs of the post-transient iterates $y_t - R(x_t)$ as a function of c for an unidirectional coupling. At the left, we consider that f is the logistic map, $R(u) = u^2$ and $G_2(u) = -g^2(u)$. At the right, we consider that f is the Hénon map, $R(u_1, u_2) = (u_2, u_1)$ and $G_2(u_1, u_2) = \left(-g_1^2(u_1, u_2), -g_2^2(u_1, u_2)\right)$.

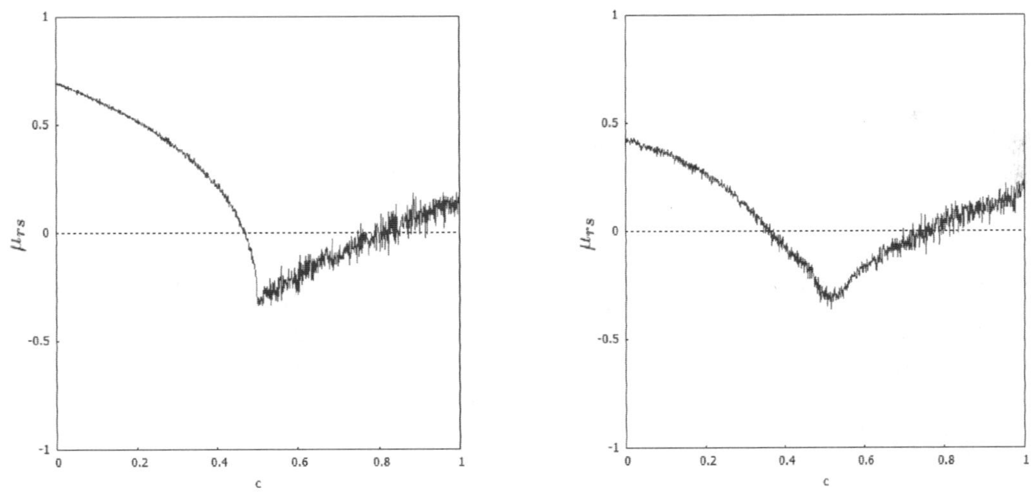

Figure 4. Graphs of μ_{rs} as a function of c for an unidirectional coupling. At the left, we consider that f is the logistic map, $R(u) = u^2$ and $G_2(u) = -g^2(u)$. At the right, we consider that f is the Hénon map, $R(u_1, u_2) = (u_2, u_1)$ and $G_2(u_1, u_2) = \left(-g_1^2(u_1, u_2), -g_2^2(u_1, u_2)\right)$.

expression can be expressed in an easy way as a function of the Lyapunov exponent of the free dynamical systems. The width of the RSW of that unidirectional coupling is the same as the one of the symmetric coupling, but if we choose a symmetric coupling instead of an unidirectional one, the systems synchronize for smaller values of the coupling strength. It is also relevant to note that those windows are independent of R.

For all the examples that we considered, even the ones that RSW is not expressed as a function of the Lyapunov exponent of the free dynamical systems, the numerical approach used provides the RSW analytically calculated, revealing that the basin of attraction of the exponentially stable synchronized solution is sufficiently large in order to avoid that the RSW is masked by trajectories corresponding to random initial values not belonging to the basin of attraction.

Acknowledgments

Research partially sponsored by national funds through the Fundação Nacional para a Ciência e Tecnologia, Portugal-FCT, under the project UID/MAT/04674/2013 (CIMA)

References

[1] Pikovsky A, Rosenblum M and Kurths J 2001 *Synchronization, a universal concept in nonlinear sciences* (England: Cambridge University Press)
[2] Glass L 2001 *Nature* **410** 277–84
[3] Boccaletti S, Kurths J, Osipov G V, Valladares D L and Zhou C S 2002 *Phys. Reports* **366** 1–101
[4] Koronovskii A A, Moskalenko O I and Hramov A E 2009 *Physics-Uspekhi* **52** 1213–38
[5] Pecora L M and Carroll T L 1990 *Phys. Rev. Lett.* **64** 821–4
[6] Luo C and Wang X 2013 *Nonlinear Dyn.* **71** 241–257
[7] Mahmoud G M and Mahmoud E E 2010 *Nonlinear Dyn.* **61** 141–152.
[8] Rosenblum M G, Pikovsky A S and Kurths J 1997 *Phys. Rev. Lett.* **78** 4193–6
[9] Lopes L, Fernandes S and Grácio C 2014 *Nonlinear Dyn.* **79** 1615–24
[10] Zhang F 2015 *Entropy* **17** 4974–85
[11] Rulkov N F, Sushchik M M, Tsimring L S and Abarbanel H D I 1995 *Phys. Rev. E* **51** 980–94
[12] Zhang G, Liu Z R and Ma Z J 2007 *Chaos, Solitons & Fractals* **32** 773–779
[13] Grassi G and Miller D A 2007 *Int. J. Bifurcation Chaos* **17** 1337-42
[14] Nian F, Wan X and Zheng P 2013 *Int. J. Mod. Phys. B* **27** 2252–79
[15] Pecora L M, Carroll T L and Heagy J F 1995 *Phys. Rev. E* **52** 3420–39
[16] Abarbanel H D I, Rulkov N F and Sushchik M M 1996 *Phys. Rev. E* **53** 4528–35.
[17] Pyragas K 1996 *Phys. Rev. E* **54** R4508–11
[18] Hramov A E and Koronovskii A A 2005 *Phys. Rev. E* **71** 067201
[19] Koronovskii A A, Moskalenko O I and Hramov A E 2006 *Technical Physics* **51** 143–150
[20] Koronovskii A A, Moskalenko O I and Hramov A E 2011 *Phys. Rev. E* **84** 037201
[21] Koronovskii A A, Moskalenko O I, Shurygina S A and Hramov A E 2013 *Chaos, Solitons & Fractals* **46** 12-18
[22] Rulkov N F, Afraimovich V S, Lewis C T, Chazottes J R and Cordonet A 2001 *Phys. Rev. E* **64** 016217
[23] Afraimovich V S, Chazottes J R and Cordonet A 2001 *Discrete Contin. Dyn. Syst. B* **1** 21–442
[24] So P, Barreto E, Josic K, Sander E and Schiff S J 2002 *Phys. Rev. E* **65** 046225
[25] Grassi G 2012 *Chin. Phys. B* **21** 050505
[26] Yuan Z, Xu Z and Guo L 2012 *Commun. Nonlinear Sci. Numer. Simulat.* **17** 992-1002
[27] Hramov A E, Koronovskii A A and Moskalenko O I 2005 *Europhys. Lett.* **72** 901–7.
[28] Yang X S and Chen G 2002 *Chaos, Solitons & Fractals* **13**1303–8
[29] Lu J and Xi Y 2003 *Chaos, Solitons & Fractals* **17** 825–31
[30] Jie S, Bollt E M and Nishikawa T 2009 *SIAM Journal on Applied Dynamical Systems* **8** 202–221
[31] Lopes L, Fernandes S and Grácio C 2016 *Journal of Physics: Conference Series* **692** 012008
[32] Bollt E M and Santitissadeekorn N 2013 *Applied and computational measurable dynamics* (Society for Industrial and Applied Mathematics)

IOP Conf. Series: Journal of Physics: Conf. Series **990** (2018) 012006 doi:10.1088/1742-6596/990/1/012006

On strong sensitive points of continuous maps on dendrites

Elena Makhrova

Institute of Information Technology, Mathematics and Mechanics, Lobachevsky State University of Nizhny Novgorod, Nizhny Novgorod, Russia

E-mail: elena_makhrova@inbox.ru

Abstract. Let $f : I \to I$ be a continuous map of a closed interval I. It is known that the positiveness of topological entropy of f, the existence of a horseshoe and the existence of a strong sensitive point from $\overline{Per(f)} \setminus Per(f)$ are mutually equivalent. The aim of the paper is to investigate some relations between the mentioned properties for continuous maps of dendrites (locally connected continua without subsets homeomorphic to the circle).

1. Introduction

For continuous maps on a closed interval the next theorem is true:

Theorem 1 [1], [2] *Let $f : I \to I$ be a continuous map of a closed interval I. Then the next statements are equivalent:*
(i) topological entropy of f is positive;
(ii) f^m has a horseshoe for some natural number $m \geq 1$;
(iii) f has a strong sensitive point $x \in \overline{Per(f)} \setminus Per(f)$.

It is known that implication $(ii) \Rightarrow (i)$ is true for any continuous map $f : X \to X$ of a compact metric set X. In [3] it is shown that implications $(i) \Rightarrow (ii)$ and $(i) \Rightarrow (iii)$ are not true for continuous maps on dendrites.

Example 2 [3, Example 6] *There are a dendrite X and a continuous map $f : X \to X$ such that*
(2.1) topological entropy of f is positive;
(2.2) f^m doesn't have a horseshoe for any natural number $m \geq 1$;
(2.3) there doesn't exist a strong sensitive point from $\overline{Per(f)} \setminus Per(f)$.

In the article we investigate some relations between $(i) - (iii)$ of theorem 1 for continuous maps on dendrites. We show that (iii) does not imply (i) and (ii) for such maps (section 3).

Example 3 *There are a dendrite X and a continuous map $f : X \to X$ such that*
(3.1) topological entropy of f is zero;
(3.2) f^m doesn't have a horseshoe for any natural number $m \geq 1$;
(3.3) f has a strong sensitive point $x \in \overline{Per(f)} \setminus Per(f)$.

We study the structure of dendrites on which the implication $(ii) \Rightarrow (iii)$ is true for any continuous maps. We prove the next theorem (section 4).

Theorem 4 *Let X be a dendrite with a countable set of end points, $f : X \to X$ be a continuous map and f has a horseshoe. Then the set $\overline{Per(f)} \setminus Per(f)$ contains a strong sensitive point.*

2. Preliminaries

2.1. We use **N** to denote the set of natural numbers. The symbol $Card\,(A)$ stands for the cardinality of A, $diam\,(A)$ means the diameter of A.

By *continuum* we mean a compact connected metric space. A *dendrite* is a locally connected continuum without subsets homeomorphic to a circle. An *arc* is a set which is homeomorphic to the closed interval $[0, 1]$. It is well known that for every two points in a dendrite X there is a unique arc containing these points (see [4, §51]).

Let points $x, y \in X$. We denote by $[x, y]$ an arc joining x and y and containing these points; $(x, y) = [x, y] \setminus \{x, y\}$, $(x, y] = [x, y] \setminus \{x\}$ and $[x, y) = [x, y] \setminus \{y\}$.

Let X be a dendrite, a point $p \in X$. The set $X\setminus\{p\}$ consists of one or more connected sets. Each such set is called a *component of a point p*.

Definition 5 *A point $p \in X$ is called to be*
- an end point of X if a point p has one component;
- a branch point of X if a point p has at least three components.

Denote by $E(X)$ the set of end points of X, by $B(X)$ the set of branch points of X.

It is known that the set $B(X)$ and the number of components of any point $p \in X$ are at most countable (see [4, §51]). If a number of components of a point p is finite and equals n then we say that an *order of a point p in X* is equal to n; written $ord\,p = n$. If a number of components of a point p is infinite then we say $ord\,p = \omega$.

Definition 6 *Let $f : X \to X$ be a continuous map, a point $p \in X$. p is called to be*
- a fixed point of f, if $f(p) = p$;
- a periodic point of f, if there is a natural number $n \geq 1$ such that $f^n(p) = p$;
- a recurrent point of f, if for any neighborhood $U(p)$ of a point p there is a natural number $n \geq 1$ such that $f^n(p) \in U(p)$.
- a strong sensitive, if there is $\varepsilon > 0$ such that for any neighborhood $U(p)$ of a point p there exists a natural number $N \geq 1$ so that $diam(f^n(U(p))) > \varepsilon$ for any $n > N$.

We denote by $Fix(f)$, $Per(f)$, $Rec(f)$ the set of fixed points, periodic points, recurrent points of f respectively.

Definition 7 *Let $f : X \to X$ be a continuous map of a dendrite X. We say that f has a horseshoe (A, B) if A, B are non-empty closed subcontinua such that $A \cap B = \emptyset$ and $f(A) \cap f(B) \supset A \cup B$.*

If A, B lay on the one arc then a horseshoe (A, B) is called an arc horseshoe.

2.2. We need the next auxiliary results.

The next theorem follows from [5, Theorem 1] and its proof.

Theorem 8 [5] *Let X be a dendrite with a countable set of end point, $f : X \to X$ be a continuous map and there is a subset $A \subset X$ such that $A \subset f(A)$. Then there exists a periodic point α of f such that $\alpha, f(\alpha) \in A$.*

Theorem 9 [6] *Let X be a dendrite with a countable set of end points and a continuous map $f : X \to X$ has a horseshoe. Then there is a natural number $m \geq 1$ such that f^m has an arc horseshoe.*

Lemma 10 [7] *Let $\{Y_n\}_{n\geq 1}$ be a sequence of connected subsets of a dendrite X. If $Y_i \cap Y_j = \emptyset$ for all $i \neq j$, then $\lim\limits_{n\to\infty} diam\,(Y_n) = 0$.*

3. Example 3

Here we construct a dendrite X and a continuous map $f : X \to X$ such that
(3.1) topological entropy of f is zero;
(3.2) f^m doesn't have a horseshoe for any natural number $m \geq 1$;
(3.3) f has a strong sensitive point $x \in \overline{Per(f)} \setminus Per(f)$.

This example was constructed jointly with Mikushin D.

Let X be Gehman dendrite (see [8, Example 10.39]). Gehman dendrite X can be characterized as the dendrite whose the set of end points is homeomorphic to the Cantor set, and whose branch points are of order 3 only. Note that the infinite binary tree is another name of this dendrite.

We denote by o the vertex of a dendrite X, $a_{i_1 \dots i_n}$; the branch point of X so as on the figure 1.

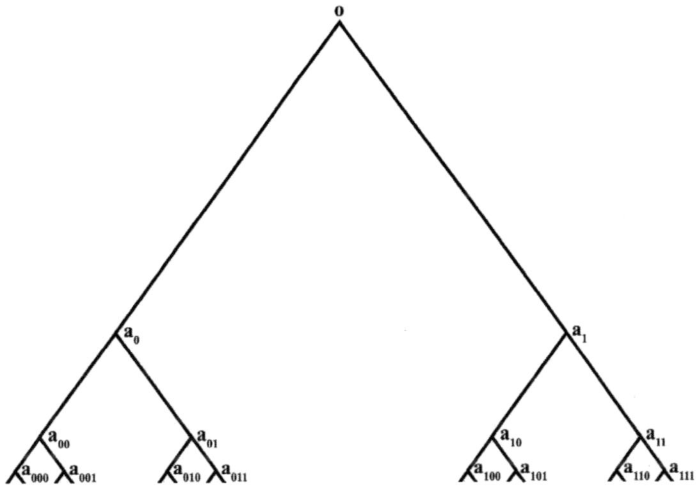

Figure 1. The subset of Gehman dendrite

We need two auxiliary maps. We use the first map $\varphi : X \to X$ from [9, Theorem C]. Recall that a continuous map φ is constructed so that
∘ $\varphi(o) = o$;
∘ any point $x \in [a_i, o)$ is a periodic point of period 2 for all $i \in \{0, 1\}$;
∘ any point $x \in [a_{i_1 \dots i_n}, a_{i_1 \dots i_{n-1}})$ is a periodic point of period 2^n for all $n \geq 2$ and all numbers $i_1, \dots, i_n \in \{0, 1\}$;
∘ any end point of a dendrite X is a recurrent nonperiodic point of φ.

Thus, we have
$(B1)$ $Per(\varphi) = X \setminus E(X)$;
$(B2)$ $\varphi(E(X)) = E(X)$ and $E(X) = Rec(\varphi) \setminus Per(\varphi)$;
$(B3)$ topological entropy of φ is equal to 0.

Now we construct a second map $\psi : X \to X$. Denote by b_i the middle of the arc $[o, a_i]$ for $i \in \{0, 1\}$. Denote by $b_{i_1 \dots i_n}$ the middle of the arc $[a_{i_1 \dots i_n}, a_{i_1 \dots i_{n-1}}]$ for all $n \geq 2$ and any $i_1, \dots, i_n \in \{0, 1\}$.

Define a continuous map ψ by the following way:
∘ $\psi([b_0, b_1]) = o$;
∘ $\psi : [a_i, b_i] \to [a_i, o]$ is a linear homeomorphism so that $\psi(a_i) = a_i$, $\psi(b_i) = o$, for all $i \in \{0, 1\}$;
∘ $\psi : [a_{i_1 \dots i_{n-1}}, b_{i_1 \dots i_n}] \to [a_{i_1 \dots i_{n-1}}, b_{i_1 \dots i_{n-1}}]$ is a linear homeomorphism so that $\psi(a_{i_1 \dots i_{n-1}}) = a_{i_1 \dots i_{n-1}}$, $\psi(b_{i_1 \dots i_n}) = b_{i_1 \dots i_{n-1}}$ for all $n \geq 2$ and any $i_1, \dots, i_n \in \{0, 1\}$;
∘ $\psi : [b_{i_1 \dots i_n}, a_{i_1 \dots i_n}] \to [b_{i_1 \dots i_{n-1}}, a_{i_1 \dots i_n}]$ is a linear homeomorphism so that $\psi(b_{i_1 \dots i_n}) = b_{i_1 \dots i_{n-1}}$,

$\psi(a_{i_1\dots i_n}) = a_{i_1\dots i_n}$ for all $n \geq 2$ and any $i_1, \dots, i_n \in \{0,1\}$;
$\circ$ $\psi(x) = x$ for any point $x \in E(X)$.

Note that a map ψ has the following properties:
$(C1)$ $Per(\psi) = Fix(\psi) = B(X) \cup E(X) \cup \{o\}$;
$(C2)$ for any point $x \in X \setminus Fix(\psi)$ there is $N \in \mathbf{N}$ so that $\psi^N(x) = o$;
$(C3)$ for any point $y \in E(X)$ and any its neighborhood $U(y)$ there exists $N \in \mathbf{N}$ such that $\psi^n(U(y)) \cap \{o\} \neq \emptyset$ for any $n \geq N$;
$(C4)$ topological entropy of ψ is equal to 0.

We set $f = \psi \circ \varphi : X \to X$. Then f is a continuous map and by $(B1)$, $(B2)$ and $(C1)$, $Per(f) = B(X) \cup \{o\}$. Hence,

$$\overline{Per(f)} \backslash Per(f) = E(X).$$

By $(B1)$ and $(C2)$, for any point $x \in X \setminus (B(X) \cup E(X) \cup \{o\})$ there is $i \in \mathbf{N}$ such that $f^i(x) = o$. So for any point $y \in E(X)$ and any neighborhood $U(y)$ of a point y there exists $N \in \mathbf{N}$ so that $f^n(U(y)) \cap \{o\} \neq \emptyset$ for any $n \geq N$. By $(B2)$ and $(C1)$, $f(E(X)) = E(X)$. Hence, $f^n(U(y)) \cap E(X) \neq \emptyset$ for any $n \geq 1$. So any point $x \in E(X)$ is a strong sensitive point. Moreover, f^m doesn't have a horseshoe for any $m \geq 1$.

According to $(B3)$ and $(C4)$, topological entropy of f equals 0.

Thus, conditions (3.1) – (3.3) of example 3 hold.

4. Proof of theorem 4

Let X be a dendrite, points y, $z \in X$. We denote by $X_y z$ $(X_z(y))$ the component of a point z (y) containing y (z); $X_{(y,z)} = X_y(z) \cap X_z(y)$; $X_{[y,z]} = \overline{X}_{(y,z)}$. Note that $X_{[y,z]} = X_{(y,z)} \cup \{y, z\}$.

Since a dendrite X has a countable set of end points and f has a horseshoe then by theorem 9, there is a natural number $m \geq 1$ such that f^m has an arc horseshoe; denote it by $([a,b],[c,d])$. Without of generality we can assume that points $b, c \in (a,d)$. We set $g = f^m$, then

$$g([a,b]) \cap g([c,d]) \supset [a,b] \cup [c,d]. \tag{1}$$

By (1), $g^{-1}(c) \cap [a,b] \neq \emptyset$, $g^{-1}(d) \cap [a,b] \neq \emptyset$. Since g is a continuous map then the sets $g^{-1}(c) \cap [a,b]$, $g^{-1}(d) \cap [a,b]$ are closed. Hence, there are points $c_1 \in g^{-1}(c) \cap [a,b]$, $d_1 \in g^{-1}(d) \cap [a,b]$ so that $(c_1, d_1) \cap (g^{-1}(c) \cup g^{-1}(d)) = \emptyset$. It follows from here that

$$g([c_1, d_1]) \supset [c,d]. \tag{2}$$

and

$$g([c_1, d_1]) \subset X_{[c,d]}. \tag{3}$$

So by (1) and (2), we have

$$g^2([c_1, d_1]) \supset [a,b] \supset [c_1, d_1].$$

According to theorem 8, there is a point $\alpha_1 \in Per(g^2) \cap [c_1, d_1]$ so that $g^2(\alpha_1) \in [c_1, d_1]$. By (3), $g(\alpha_1) \in X_{[c,d]}$.

Again by (1), $g([a,b]) \supset [c_1, d_1]$, so

$$g^{-1}(c_1) \cap [a,b] \neq \emptyset,\ g^{-1}(d_1) \cap [a,b] \neq \emptyset.$$

By continuity of g, there are points $c_2^{(1)} \in g^{-1}(c_1) \cap [a,b]$, $d_2^{(1)} \in g^{-1}(d_1) \cap [a,b]$ so that $(c_2^{(1)}, d_2^{(1)}) \cap (g^{-1}(c_1) \cup g^{-1}(d_1)) = \emptyset$. So

$$g([c_2^{(1)}, d_2^{(1)}]) \supset [c_1, d_1], \tag{4}$$

and

$$g([c_2^{(1)}, d_2^{(1)}]) \subset X_{[c_1,d_1]}. \tag{5}$$

Since $X_{[c,d]} \cap X_{[c_1,d_1]} = \emptyset$ then by (3) and (5), $[c_1, d_1] \cap [c_2^{(1)}, d_2^{(1)}] = \emptyset$.

By (2) and (4), $g^2([c_2^{(1)}, d_2^{(1)}]) \supset [c, d]$. So $g^{-2}(c) \cap [c_2^{(1)}, d_2^{(1)}] \neq \emptyset$, $g^{-2}(d) \cap [c_2^{(1)}, d_2^{(1)}] \neq \emptyset$. Now we choose an arc $[c_2, d_2] \subset [c_2^{(1)}, d_2^{(1)}]$, so that

$$(c_2, d_2) \cap (g^{-2}(c) \cap g^{-2}(d)) = \emptyset.$$

Then

$$g^2([c_2, d_2]) \subset X_{[c,d]}, \tag{6}$$

and

$$g^2([c_2, d_2]) \supset [c, d]. \tag{7}$$

According to (5),

$$g([c_2, d_2]) \subset X_{[c_1,d_1]}. \tag{8}$$

By (7) and (1), we have $g^3([c_2, d_2]) \supset [c_2, d_2]$. According to theorem 8, there is a point $\alpha_2 \in Per(g^3) \cap [c_2, d_2]$ so that $g^3(\alpha_2) \in [c_2, d_2]$. By (8), $g(\alpha_2) \in X_{[c_1,d_1]}$; by (6), $g^2(\alpha_2) \in X_{[c,d]}$.

Describe the construction of an arc $[c_3, d_3]$. Firstly we choose an arc $[c_3^{(1)}, d_3^{(1)}] \subset [a, b]$ so that $c_3^{(1)} \in g^{-1}(c_2) \cap [a, b]$, $d_3^{(1)} \in g^{-1}(d_2) \cap [a, b]$ and $(c_3^{(1)}, d_3^{(1)}) \cap (g^{-1}(c_2) \cup g^{-1}(d_2)) = \emptyset$. Then $g([c_3^{(1)}, d_3^{(1)}]) \subset X_{[c_2,d_2]}$. It follows from here, (3) and (5) that $[c_3^{(1)}, d_3^{(1)}] \cap [c_i, d_i] = \emptyset$ for $i \in \{1, 2\}$.

Since $g^2([c_3^{(1)}, d_3^{(1)}]) \supset [c_1, d_1]$, then we choose an arc $[c_3^{(2)}, d_3^{(2)}] \subset [c_3^{(1)}, d_3^{(1)}]$ so that $c_3^{(2)} \in g^{-2}(c_1) \cap [c_3^{(1)}, d_3^{(1)}]$, $d_3^{(2)} \in g^{-2}(d_2) \cap [c_3^{(1)}, d_3^{(1)}]$ and $(c_3^{(2)}, d_3^{(2)}) \cap (g^{-2}(c_2) \cup g^{-2}(d_2)) = \emptyset$. Then $g^2([c_3^{(2)}, d_3^{(2)}]) \supset [c_1, d_1]$ and $g^2([c_3^{(2)}, d_3^{(2)}]) \subset X_{[c_1,d_1]}$.

Since $g^3([c_3^{(2)}, d_3^{(2)}]) \supset g([c_1, d_1]) \supset [c, d]$, then we choose an arc $[c_3, d_3] \subset [c_3^{(2)}, d_3^{(2)}]$ so that $c_3 \in g^{-3}(c) \cap [c_3^{(2)}, d_3^{(2)}]$, $d_3 \in g^{-3}(d) \cap [c_3^{(2)}, d_3^{(2)}]$ and $(c_3, d_3) \cap (g^{-3}(c) \cup g^{-3}(d)) = \emptyset$. Then $g^3([c_3, d_3]) \subset X_{[c,d]}$ and $g^4([c_3, d_3]) \supset [c_3, d_3]$. Applying theorem 8, we get the existence of a periodic point $\alpha_3 \in [c_3, d_3]$. Note that $g^i(\alpha_3) \subset X_{[c_{3-i},d_{3-i}]}$ for $i \in \{1, 2\}$, $g^3(\alpha_3) \in X_{[c,d]}$ and $g^4(\alpha_3) \in [c_3, d_3]$.

Continuing in this manner, we get a sequence of arcs $\{[c_i, d_i]\}_{i \geq 1} \subset [a, b]$ and a sequence of periodic points $\{\alpha_i\} \in [c_i, d_i]$ with the following properties:

(A1) $[c_i, d_i] \cap [c_j, d_j] = \emptyset$ for any $i \neq j$;

(A2) $\alpha_i, g^{i+1}(\alpha_i) \in [c_i, d_i]$ for all $i \geq 1$;

(A3) $g^j([c_i, d_i]) \subset X_{[c_{i-j},d_{i-j}]}$ and so $g^j(\alpha_i) \in X_{[c_{i-j},d_{i-j}]}$ for all $i \geq 1$ and $1 \leq j \leq i-1$;

(A4) $g^i([c_i, d_i]) \supset [c, d]$, $g^i(\alpha_i) \in X_{[c,d]}$ for all $i \geq 1$.

By (A1), $X_{[c_i,d_i]} \cap X_{[c_j,d_j]} = \emptyset$ for $i \neq j$. Applying lemma 10, we get that

$$\lim_{i\to\infty} diam\, X_{[c_i,d_i]} = 0. \tag{9}$$

Let β be one of the limit points of a sequence $\{g^i(\alpha_i)\}_{i \geq 1}$, that is there is a subsequence $\{g^{i_j}(\alpha_{i_j})\}_{j \geq 1}$ so that

$$\beta = \lim_{j\to\infty} g^{i_j}(\alpha_{i_j}).$$

By condition (A4), $\beta \in X_{[c,d]}$. By continuity of g and condition (A2),

$$g(\beta) = \lim_{j\to\infty} g^{i_j+1}(\alpha_{i_j}) \in [a, b].$$

IOP Conf. Series: Journal of Physics: Conf. Series **990** (2018) 012006 doi:10.1088/1742-6596/990/1/012006

We show that $g^k(\beta) \in [a, b]$ for any natural number $k \geq 2$. By continuity of g,

$$g^k(\beta) = \lim_{j \to \infty} g^{k+i_j}(\alpha_{i_j}). \tag{10}$$

Let J be a natural number so that $i_J > k$. Then by (A3),

$$g^{k+i_j}(\alpha_{i_j}) = g^{k-1}(g^{(1+i_j)}(\alpha_{i_j})) \in X_{[c_{i_j-(k-1)}, d_{i_j-(k-1)}]}.$$

It follows from here, (9), (10) and condition $X_{[c_i,d_i]} \cap [a, b] \neq \emptyset$ that

$$g^k(\beta) \in [a, b],\ k \geq 1. \tag{11}$$

Now we show that β is a strong sensitive point.

Let $U(\beta)$ be any neighborhood of a point β. Let $\varepsilon^* = diam([b, c])$. There is a natural number N such that $[c_N, d_N] \subset f(U(\beta))$. According to (A4) $g^N([c_N, d_N]) \supset [c, d]$. Hence, by (1) and (11), $diam(g^n(U(\beta))) \geq diam(g^n([c_N, d_N])) > \varepsilon^*$ for any $n \geq N$. So

$$diam(f^{mn}(U(\beta))) > \varepsilon^* \text{ for all } n > N. \tag{12}$$

Now we take

$$\varepsilon = \min[\varepsilon^*, diam(f(U(\beta))), \ldots, diam(f^{m-1}(U(\beta)))].$$

Then by (12),

$$diam(f^n(U(\beta))) > \varepsilon \text{ for all } n > N.$$

Theorem 4 is proved.

References

[1] Misiureicz M 1979 *Bull. Acad. PO. Sci., Ser. Sci. Math.* **27** 167

[2] Blokh A M 1980 *Uspekhi Mat. Nauk* **37** 189

[3] Kočan Zd, Kornecká-Kurková V and Málek M 2010 *Ergod. Th. & Dynam. Sys.* **22** 253

[4] Kuratovski K 1968 *Topology, vol.2* (New York: Academic Press)

[5] Makhrova E N 2011 *Russian Mathematics (Izvestiya VUZ. Matematika)* **55** 33

[6] Makhrova E N 2015 *Russsian Mathematics (Izvestiya VUZ. Matematika)* **59** 52

[7] Mai J and Shi E 2009 *Int. J. Bifurcation and Chaos* **19** 1391

[8] Nadler S 1992 *Continuum theory* (Marcel Dekker, New York)

[9] Efremova L S and Makhrova E N 2001 *Sb. Math.* **192** 807

Constructions of conformal measures for generalized interval exchange transformations

Mikhail Malkin, Klim Safonov

Lobachevsky State University of Nizhny Novgorod 603950, Nizhny Novgorod, RUSSIA, Gagarin avenue, 23

E-mail: `malkin@unn.ru, safonov.klim@yandex.ru`

Abstract. For onedimensional piecewise monotone discontionuous maps without periodic points, the 1-conformal measures are constructed and, as a corollary, semiconjugacy to piecewise linear models of constant (in absolute value) slope is obtained. It turns out that for generalized interval exchange transformations, the constructed semiconjugacy is, in fact, conjugacy .

1. Introduction

In this paper we construct piecewise linear molels of constant (in absolute value) slope for generalized interval exchange transformations, and also for more general class of onedimensional discontinuous maps without periodic points. In [1], a similar problem was considered by J. Milnor and W. Thurston for continuous piecewise monotone maps with positive topological entropy, and in [6], [7] by M. Malkin for discontinuous maps of Lorenz type. It was proved in those papers that a map f under consideration with positive topological entropy ($h_{top}(f) > 0$) is semicongugate to a piecewise linear map (a model map) of constant slope $e^{h_{top}(f)}$; besides, sufficient conditions were given which provide, in fact, conjugacy to the model map. Note that the Milnor and Thurston kneading technique for constructing such a (semi)conjugacy is based on the properties that the kneading series of f in the case when $h_{top}(f) > 0$, converges in the (open) unit disc, and its minimal root is a positive number strictly less than one (namely, it is equal to $e^{-h_{top}(f)}$). These properties allow one to apply for constructing the piecewise linear model, the methods of analytical and meromorphic functions. On the other hand, if a given map has zero topological entropy, the mentioned analytical methods fail, and thus the problem becomes much more complicated.

Definition 1. *A map $f : I \to I$ of the unit interval I is called a piecewise monotone map (PMM), if I has a finite partition $A = \{(c_i, c_{i+1}), i = \overline{0, n-1}\}$ such that the restriction of f to each interval (c_i, c_{i+1}) of this partition is continuous and monotone.*

If among possible partitions above, we choose the minimal one (with respect to n) then $c_i (i = \overline{1, n-1})$ are the critical points or points of discontinuity for f, and the points c_0 and c_n are the boundary points of I (i.e., ∂I).

Definition 2. *An interval translation map (ITM) is a PMM $f : I \to I$ such that the restriction of f to each element (c_i, c_{i+1}) of the partition is isometry, i.e., it is either of the form $x \mapsto x + \beta_i$ or $x \mapsto -x + \beta_i$ for some β_i.*

We will show in this paper that ITMs surves as model maps for wide classes of discontinuous maps with zero entropy. Such a problem was considered previously in [2] by using countable topological Markov chains (the so called "Hofbauer towers"); in that paper, Buzzi and Hubert obtained certain conditions for conjugacy to piecewise linear models and gave estimates on the number of ergodic measures. We will construct (semi)conjugacy by using different tools; namely we generalize ideas of Milnor and Thurston to show that the kneading technique can be applied also for maps with zero entropy without periodic points. Moreover we will show that if a map is the disjoint union of minimal PMMs, then the constructed semiconjugacy is in fact, a conjugacy. In particular, this result can be applied to generalized interval exchange transformations and thus, it generalizes Buzzi's theorem which states that a minimal PMM is conjugate to its piecewise linear model.

From now on we consider a PMM f satisfying the following assumptions.

- The set of preimages $D = \cup_{i=1}^{n-1} \cup_{j=0}^{\infty} f^{-j}(c_i)$ is dense on I.
- f has no periodic points including discontinous ones, i.e.,

$$f^k(c_i+) \neq c_i, f^k(c_i-) \neq c_i \text{ for all } k \in \mathbf{N}$$

The assumptions above imply absense of wandering intervals, i.e., intervals $J \in I$ whose iterates $J, f(J), f^2(J) \dots$ are dijoint. The latter property is natural because ITMs have no wandering intervals. If an ITM had a periodic orbit then it would have a whole periodic interval That is why when constructing piecewise linear models for maps with zero topological entropy, it is natural to exclude maps with periodic points.

Actually to construct a semiconjugacy between a PMM $f : I \to I$ and its piecewise linear model of constant slope $s = s(f)$, it is enough to show existence of a $\frac{1}{s}$-conformal measure (see [1]).

Definition 3. *Given a map $T : X \to X$ and a nonnegative function $f : X \to \mathbf{R}$, a measure μ on X is called f-conformal if the following equality holds*

$$\mu(T(A)) = \int_A f d\mu,$$

for all measurable sets $A \subset X$ at which the restriction $T : A \to T(A)$ is one-to-one.

The structure of the paper is as follows. Section 2 is devoted to construction of special 1-conformal measures for PMMs withous periodic points (more general constructions of f-conformal measures can be found in [8]). In Section 3 we consider one-to-one interval maps, i.e., the generalized interval exchange transformations. By using technique of [4] and [5] we show that if such a map has neithe periodic points nor wandering intrvals then it is conjugate to its piecewise linear model constructed in Section 2.

2. Construction of 1-conformal measures

Given a PMM $f : I \to I$ let us denote by $C = \{c_i, i = 1, \dots, n-1\}$ the set of points of discontinuity, and by

$$D_k = \{x \in I | f^k(x) \in I\}, D = \cup_k D_k$$

the sets of preimages of discontinuity points.

Following [1], introduce for subinterval $J = (a, b)$, the formal power series

$$L(J, t) = \sum_{k=0}^{\infty} l_k(J) t^k, \text{ where } l_k(J) = card(D_k \cap J).$$

IOP Conf. Series: Journal of Physics: Conf. Series **990** (2018) 012007 doi:10.1088/1742-6596/990/1/012007

Since we consider maps without periodic points, their topological entropy equals zero. According [3], the topological entropy for piecewise monotone one-dimensional map f can be calculated as follows.

$$h_{top}(f) = \lim_{k\to\infty} \frac{\ln\, l_k(I)}{k} \tag{1}$$

For any piecewise monotone map, the limit above exists and belongs to $[0, \log n]$, where n is the (minimal) number of its monotonicity intervals. It follows from the Cauchy-Hadamard formula that the radius of convergence of the series $L(I,t)$ is equal to

$$r(L(I,t)) = \frac{1}{\lim_{n\to\infty} \sqrt[n]{cardD_n}} \tag{2}$$

Thus, the equalities (1) and (2) imply $r(L(I,t)) = 1$. Since $0 \le l_k(J) \le l_k(I)$, the power series $L(J,t)$ define analytical functions in the open unit disc. Note that if a series $L(J,t)$ has infinitely many nonzero coefficients than its radius of convergence is precisely 1 and the analytical function $L(J,t)$ has singularity at $t = 1$; otherwise, $L(J,t)$ is a polynomial.

Now for any $t \in (0,1)$ we construct the probability measure

$$\mu_t = \frac{1}{L(I,t)} \sum_{k=0}^{\infty} t^k \sum_{x\in D_k} \delta_x,$$

where δ_x is the Dirac (atomic) measure with support x.

Note that for any interval J its measure $\mu_t(J) = \dfrac{L(J,t)}{L(I,t)}$ is well defined for all $t \in (0,1)$, and $\mu_t(I) = 1$. Let us consider a limit point (in the weak topology) μ for some sequence $\mu_{t_k} \to \mu$ as $t_k \to 1-0$. We will show that μ is 1-conformal.

Propostion 1. *μ is a probability 1-conformal measure.*

Proof. First we show that μ is nonatomic, i.e., $\mu(\{x\}) = 0)$ for any $x \in I$. Indeed, consider the set of continous functions $f_\varepsilon : I \to \mathbf{R}$ which are equal to 1 in the neighborhood $(x-\varepsilon, x+\varepsilon)$ and are zero outside the interval $(x-2\varepsilon, x+2\varepsilon)$. Then

$$\mu(\{x\}) = \lim_{\varepsilon\to 0} \int_I f_\varepsilon d\mu = \lim_{\varepsilon\to 0} \lim_{k\to\infty} \int_I f_\varepsilon d\mu_{t_k}.$$

Since f has no periodic points, it follows that $card(D_k \cap (x-\varepsilon, x+\varepsilon))$ tends to zero as $\varepsilon \to 0$ for all k, except for possibly one k (in the case when x is exactly the k-th preimage of some point of discontinuity). Hence

$$\int_I f_\varepsilon d\mu_{t_k} \to 0 \text{ as } \varepsilon \to 0,$$

and therefore $\mu(\{x\}) = 0$. Then for any subinterval J one has $\mu(J) = \lim_{k\to\infty} \mu_{t_k}(J)$ as $t_k \to 1-0$. If we take subinterval $J \subset I$ such that f restricted to J is one-to-one, then $card(D_0 \cap J) = \emptyset$. Since $card(D_i \cap J) = card(D_{i-1} \cap f(J))$, we have

$$\mu(f(J)) = \lim_{k\to\infty} \mu_{t_k}(f(J)) = \lim_{k\to\infty} \frac{1}{t_k} \mu_{t_k}(J) = \mu(J).$$

□

Hence one gets the result.

Theorem 1. *Let $f : I \to I$ be a PMM without periodic points. Then f has a probability atomless 1-conformal measure.*

Now we show how by using the measure from Theorem 1, one can construct a semiconjugacy to a piecewise linear map of unit slope. So we have to show that there exists an ITM $T : I \to I$ and a monotone nondecreasing surjective continuous map $h : I \to I$ such that

$$T \circ h = h \circ f.$$

Consider the function $h(x) = \mu[0, x)$. Since μ is atomless, h is a continuous map of I onto inself. We define the map T as follows

$$T(x) = h(f(c_i+)) - h(c_i) + \varepsilon_i x, \text{ for } x \in (h(c_i), h(c_{i+1})),$$

where ε_i is either 1 or -1, depending on whether f is increasing or decreasing on (c_i, c_{i+1}). Now we check the identity $T \circ h = h \circ f$. Indeed, in the case when f is increasing on (c_i, c_{i+1}), one has

$$h(f(x)) = \mu[0, f(x)) = \mu[0, f(c_i+)) + \mu[f(c_i+), f(x)) =$$
$$= \mu[0, f(c_i+)) + \mu[c_i, x) = \mu[0, f(c_i+)) - \mu[0, c_i) + \mu[0, x) = T(h(x))$$

In the decreasing case, the identity is checked similarly. Hence we get:

Theorem 2. *Let f be a piecewise monotone map without periodic points. Then f is semiconjugate to some interval translation map.*

3. Generelized interval exchange transformations

In this section we consider one-to-one discontinuous piecewise monotone maps, we call them *generelized interval exchange transformations (GIET)*. As before denote by $\mathcal{A}$ the partition of I into monotonicity intervals of a GIET $f : I \to I$. Let $E = \{c_i^{\pm}\}$ be the set of onesided boundary points of $\mathcal{A}$. To any discontinuity point c_i one associates two elements $c_i^{\pm}$ of E, we also include in E two boundary points $\{c_0^+, c_n^-\} = \partial I$. Consider also subpartitions of $\mathcal{A}$

$$\mathcal{A}^{k+1} = \mathcal{A} \vee f^{-1}(\mathcal{A}) \vee \ldots \vee f^{-k}(\mathcal{A})$$

Since the set of preimages of discontinuity points is dense, it follows that for any sequence of cylinders $A_k \in \mathcal{A}^{k+1}$, their intersection is either empty or a single point. For GIETs we will prove a result which can be regarded as generalization of the well known Poincaré theorem on conjugacy of a transitive circle homeomorphism of degree 1 with a rigid (irrational) rotation.

Theorem 3. *Let $f : I \to I$ be a GIET without periodic points and wandering intervals. Then there exists a linear interval exchange transformation $T : I \to I$ such that f and T are topologically conjugate.*

To prove this theorem we need some terminology and results from [4] and [5].

Definition 4. *Hofbauer diagram is the graph (D) with vertices*

$$\{f^k(A) : k \geq 0 \text{ and } A \in \mathcal{A}^{k+1}\}$$

and edges

$$D \to E \iff \exists A \in \mathcal{A} \text{ such that } E = f(D) \cap A.$$

For $e \in E$ let us define the set $D(e,k) = f^k(A_{k+1})$, where A_{k+1} is the only element from $\mathcal{A}^{k+1}$ for which e is the boundary point. It turns out that the set of vertices of the graph (D) can be represented as $\cup_E \cup_N D(e,k)$. The union of vertices $\cup_N D(e,n)$ is called the *branch associated with the boundary point* $e \in E$.

Set $H(e,k) = \{D(e,l) | l \geq k\}$. On E one can construct the graph $\mathcal{E}$ by the following rule: there is an arc $e \to f$ whenever for any natural k there is an arc (D) from some element $H(e,k)$ to some element $H(f,k)$. Subgraph$\mathcal{N} \subset \mathcal{E}$ is called irreducible if for any to vertices $e, f \in \mathcal{N}$ there is a path in $\mathcal{E}$ connecting these two vertices, and the graph $\mathcal{N}$ is maximal (with respect to inclusion).

It turns out that for one-to-one maps, the graph $\mathcal{E}$ has the following property: if there is a path from $e \in \mathcal{E}$ to $f \in \mathcal{E}$, then there is also a return path from f to e. The latter property implies that $\mathcal{E}$ can be devided into irreducible parts each of which is a single connected component. We now prove how this property will imply theorem 3. Remark that in [2] a similar result was proved for the case when the whole graph $\mathcal{E}$ represents a single irreducible component.

Let $f : I \to I$ be a GIET having neither periodic points nor wandering intervals. It follows from theorem 2 that f is semiconjugate to some IET $T : I \to I$ by a continuous surjective map $h(x) = \mu(0,x)$. We need to prove that $h : I \to I$ is one-to-one. To this end we have to check that $\mu(D(e,k)) > 0$ for all vertices $D(e,k) \in \mathcal{D}$.

Suppose by the contrary that $\mu(D(e,k)) = 0$ for some $e \in E$. Then $\mathcal{N}(e) \subset \mathcal{E}$ is an irreducible component which contains e. Since for any vertex $D \in \mathcal{D}$ one has $\mu(D) = \mu(f(D))$, it follows that all the vertices which are accessible from $D(e,k)$ have measure zero. Hence, for any $h \in \mathcal{N}(e)$ there is a natural number k_i such that

$$\mu(D(h,k)) = 0 \text{ for } k > k_i.$$

Thus, the measure of the set $H = \cup_{h \in N(e)} H(h,k_i)$ is 0. On the other hand, there are no edges from $\mathcal{N}(e)$ to other vertices of $\mathcal{E}$. Therefore, H contains all (except for finitely many) preimages of points $h \in N(e)$. This implies that there is a natural number p such that for bigger numbers k, the coefficients of the series $L(H,t)$ satisfy the inequality $l_k(H) \geq 1$.

Note that the series $L(I,t)$ for one-to-one maps is of the form

$$L(I,t) = (n-1)(1 + t + t^2 + \ldots).$$

So we get the contradiction:

$$\mu(H) = \lim_{k\to\infty} \frac{L(H,t_k)}{L(I,t_k)} > \lim_{k\to\infty} \frac{t_k^p(1 + t_k + t_k^2 + \ldots)}{(n-1)(1 + t_k + t_k^2 + \ldots)} = \frac{1}{n-1} > 0,$$

which proves our result.

Acknowledgments

The research was partially supported by RFBR (grant 16-01-00364), and by Education and Science Ministry of Russia (project 1.3287.2017).

References

[1] Milnor J and Thurston W 1977 On iterated maps of the interval *preprint*.

[2] Buzzi J and Hubert P 2004 Piecewise monotone maps without periodic points: rigity, measures and complexity *Ergod. Th.& Dynam. Sys.* **24** 383-405.

[3] Young L S 1981 On the prevalence of horseshoes. *Trans. AMS* **263** 75-88.

[4] Hofbauer F 1986 Piecewise invertible dynamical systems. *Prob.Th. Relat. Fields* **72** 359-86.

[5] Hofbauer F and Raith P 1989 Topologically transitive subsets of piecewise monotonic maps which contains no periodic points. *Monat. Math* **107** 217-9.

IOP Conf. Series: Journal of Physics: Conf. Series **990** (2018) 012007 doi:10.1088/1742-6596/990/1/012007

[6] Malkin M 1989 On continuity of entropy of discontinuous mappings of the interval *Selecta Mathematica Sovietica* 131-9
[7] Malkin M. 1987 Piecewise linear models for maps of Lorenz type. *Methods of the qualitative theory of differential equations. Gorky State University Edition. Editor E.A. Leontovich-Andronova* 148-61.
[8] Denker M and Urbanski M 1991 On the existence of conformal meausure. *Trans. AMS* **328** 563-87.

IOP Conf. Series: Journal of Physics: Conf. Series **990** (2018) 012008 doi:10.1088/1742-6596/990/1/012008

Experimental research of iterated dynamics for the complex exponentials with linear term

Igor V. Matyushkin[1,2], **Maria A. Zapletina**[2,3]

[1]JSC Molecular Electronics Research Institute, [2]National Research University of Electronic Technology, [3]Institute for Design Problems in Microelectronics RAS, Moscow, Russia.

E-mail: imatyushkin@niime.ru, zapletina.mariya@gmail.com

Abstract. The research of the orbit of the point zero, fixed points, Julia and Fatou sets for the iterated complex-valued exponential is carried by means of computer experiment. The object of study is three one-parameter families based on $\exp(iz)$: $f : z \to (1 + \mu)\exp(iz)$, $g : z \to (1 + \mu |z - z^*|)\exp(iz)$, $h : z \to (1 + \mu(z - z^*))\exp(iz)$. Here . For the first family 17- and 2-periodic regimes are detected when passing near the bifurcation value$\mu \approx 2.475$i, while the multiplicator equals 1. The second family shows a more interesting behavior: (i) three-valley structure of isolines of the convergence rate near fixpoint z^* at $\mu = 0 + 1 \cdot i$; (ii) saddle-node transition when the parameter moves along a straight line Re$\mu = 0$, leading to the appearance of a second fixpoint and loss of stability by the old fixpoint at Im$\mu = 2.1682$; (iii) the nontrivial nature of the orbits of points in the vicinity of the new fixpoint and the presence of false fixpoints in the portrait of the Julia set; (iv) second phase transition leading to a radical change in the form of the Julia and Fatou sets at $\mu \simeq 2.5i$. The dynamics of the third family during movement at Re$\mu = 0$ is similar to the first case, but 17th and 2nd periodic modes are replaced by 39th and 3rd modes. Transitions $17 \to 2$ and $39 \to 3$ seem to be rapid and discreet while their geometric interpretation matches the ratios 17=1+2*8, 39=13*3. At $\mu = |z^*|^{-1}$ for the h- family Julia set fills the entire complex plane.

1. Introduction

As is well known, the topology of the Mandelbrot and Julia sets of exponential maps of the type [1,2] essentially differs from the classical case of quadratic ones , showing the Cantor bouquet elements [3,4] (the Cantors set analog for two dimensions). Although it is difficult to find a physical system modeled by an iterated exponential, nevertheless, its analogue in modular arithmetic traces its links to cryptography [5]. For exponential maps a number of theoretical results was obtained, one of which [1] emphasizes the point (0,0) in particular: 1) if there is an attractor, the orbit of point 0 tends to it; 2) if the orbit of zero goes to infinity, then the whole complex plane will be the Julia set.

This paper belongs to the area of experimental mathematics [6], and the phenomenological description serves to attract attention to the previously unnoticed features of exponential maps. Earlier [7] we investigated the base map e^{iz}, easily reducible to the standard form by substitution, and:

- found a single stable fixpoint $z^* \approx 0.576 + 0.375$i, $|z^*| \approx 0.687$ and unstable multiple-periodic orbits;

- made sure that the point 0 is indeed a hidden fixpoint (a neighborhood of zero passes in a small neighborhood of z* in approximately 15-20 iterations);
- portrayed many Julia and Fatou sets, in the structures of which the attention was paid to the local increase in the convergence rate regions;
- portrayed Mandelbrot set for the family (starting point is zero).

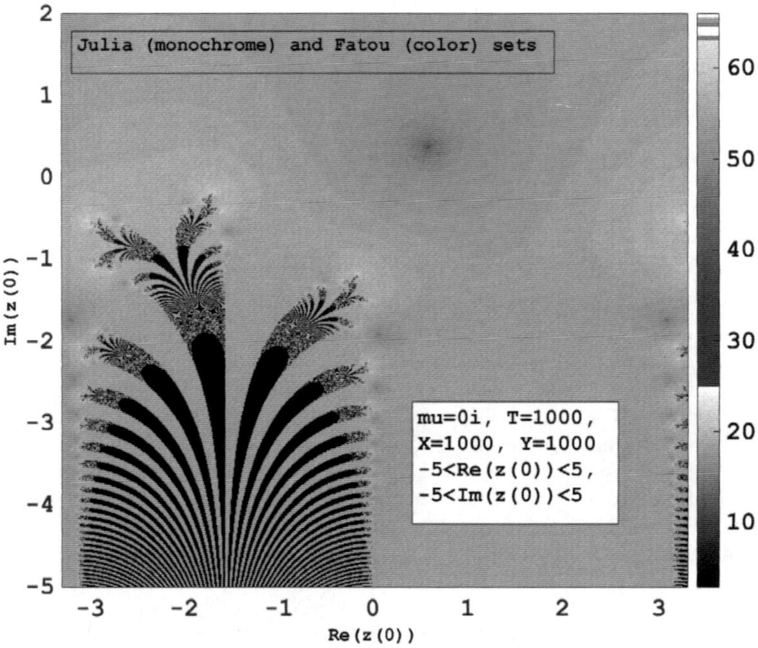

Figure 1. The topology of Cantor's bouquets on Julia and Fatou sets for the map $f(z)$, $\mu = 0$. It shows the unique attractive point z*. Here and further on the axes *abs* : $\mathrm{Re}(z^0)$, *ord* : $\mathrm{Im}(z^0)$are indicated. The image is rotated by 900 compared to the standard case, which facilitates perception.

Here we focus on parametric dependencies for three similar families: $f : z \to (1 + \mu)\exp(iz), \mu \in \mathbf{C}$, and $g : z \to (1 + \mu|z - z^*|)\exp(iz)$, $z^* : g(z^*) = z^*$, and also $h : z \to (1 + \mu(z - z^*))\exp(iz)$. As before, Fatou and Julia sets as well as the fixed points stability and orbit of zero will be the subject of our study. The second family is caused by the search for homogeneous equilibria in an ensemble of linearly coupled exponentials, i.e..:

$$z_s^{t+1} = \lambda(t)\exp\left(iz_s^t\right), \lambda(t) = \mu_0 + \mu\left|\frac{1}{n}\sum_{k=1}^{n} z_k^t - z^*(\mu_0)\right|, \quad \mu_0 \doteq 1$$

Here t is discrete time, s is a spatial index of map instance, k is a cell index of the neighborhood template of size n. It is appropriate to use the terminology of cellular automata (CA); if you allow the cell state to take value in the complex plane, then there is no difference with the coupled map lattice or ensemble [8,9].

2. Experimental method

The calculations were performed in MATLAB R2012 environment with double precision (double, 64 bits). No special measures to improve accuracy have been made; the change to algebraic computation (command *vpa*) would lead to a significant slowdown in computations by two orders of magnitude or more. As soon as the value of z(t) reached the constant Inf (1.79769e + 308), it was assumed that the orbit goes to infinity, and the calculation passed to the next point. Note that the truth of the calculations is questioned much earlier - upon reaching$\simeq 10^{15}$, because to correctly calculate the expression $\exp(iz)$ it is required to store at least a second digit after the comma of a large number (an error of 0.1 gives an error in the phase by 360). Taking into account the iterativity the requirements for accuracy is probably even tougher. We believe, however, that the quality of the dynamics does not affect the features of computer computation.

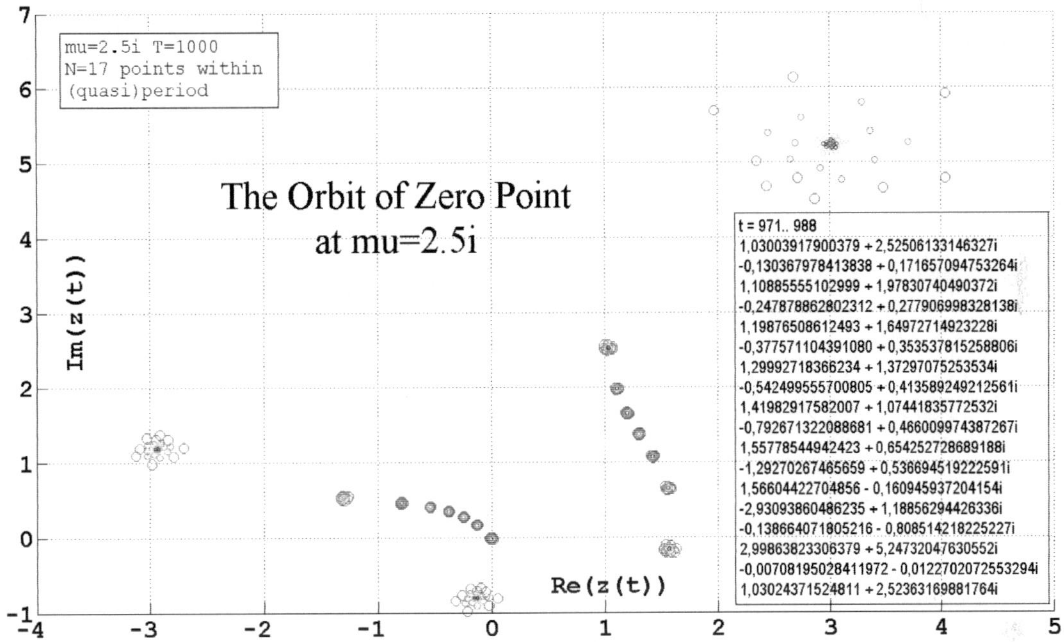

Figure 2. The orbit of the point (0,0). The entire orbit range 0..T is divided into several parts according to the colors of the rainbow, and the size of the marker-circle decreases with increasing point number t. Two almost linear sections are distinguished on the period of the orbit. The differences between z(971) and z(988) suggest that convergence has not been reached; perhaps it is a case of quasiperiodicity.

By default, the depth of the iterations (time) was T = 1000, and the spatial grid in constructing the Julia set was X Y = 1000 1000 of original size of the region. A standard "window" for the region window of $[-5;5]^2$. Sometimes smaller values were taken, primarily for CA. The orbit of the point z for the map f was assumed to be the finite sequence$\left\langle z^0 \equiv z, z^1 = f(z), \ldots, z^T = f\left(z^{T-1}\right)\right\rangle$. The Julia set was considered equivalent to the set of escape points, i.e. $J \simeq J_{esc} = \left\{z : \left(z^0 = z\right) \Rightarrow (\exists t \leq T)\left(z^t = \text{Inf}\right)\right\}$. In theory, Julia set is introduced in more complex ways, but given the limitation of the computer here we follow this simple definition. The standard imaging technique was used: the color of the points was indicated according to the speed of convergence. The Julia set is displayed in grey color scheme, a dark tone corresponds to small values of t and bright colors to a slow escaping into infinity.

For the Fatou set two cases F1 and F2 were distinguished: 1) the basin of the focus; 2) other cases: periodic, quasi-periodic or chaotic orbit. The stop condition in the first case is

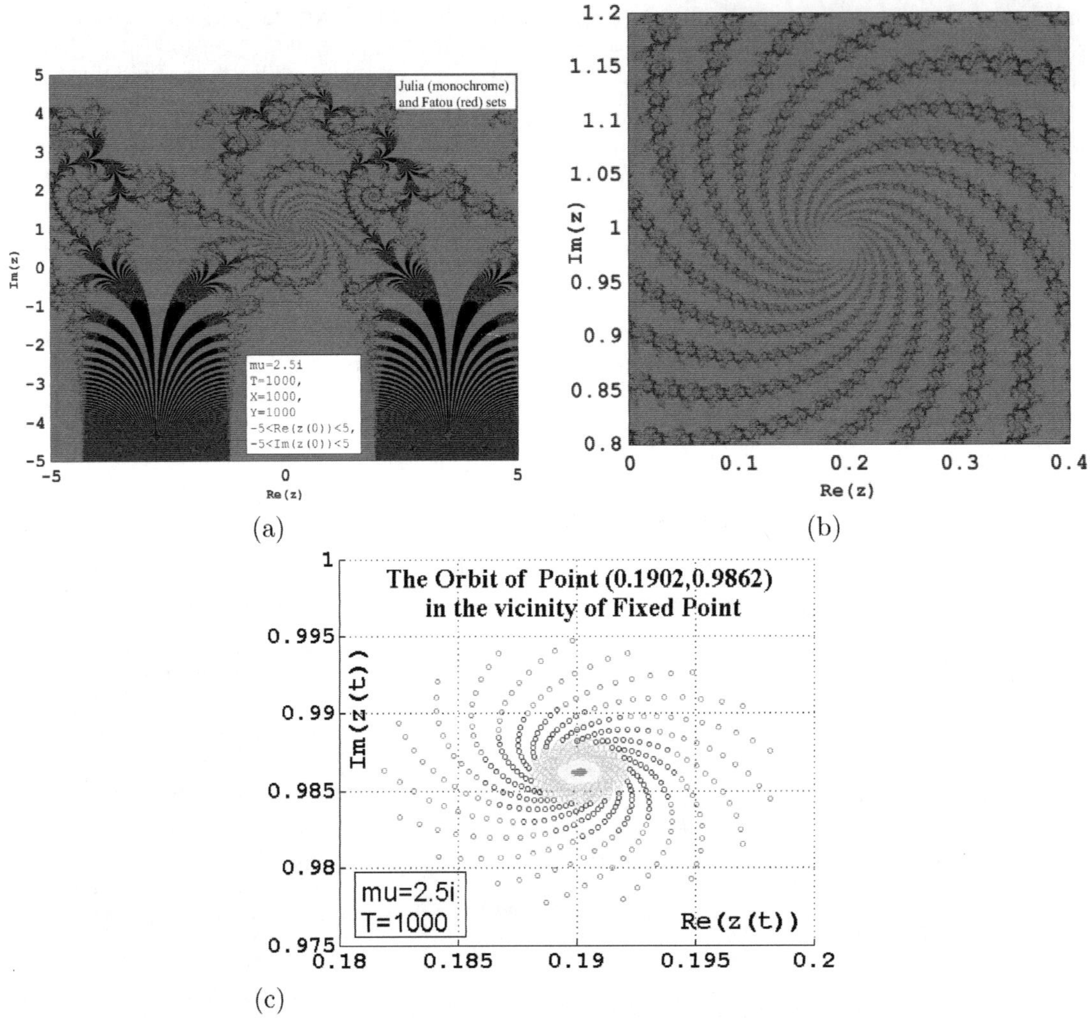

Figure 3. Julia and Fatou sets for the case $\mu = 2.5\text{i}$ in two "windows": (a) [-5..5, -5..5]; (b) [0..0.4; 0.8..1.2]. In the center of the spiral arms there is, apparently, a Siegel disk with a diameter of no more than 0.05 and fixpoint $z^{*,2.5} \approx 0.1902{+}0.9862\text{i}$. The violet color corresponds to longer times (up to T = 1000).

either$\left|z^{t+1} - z^t\right| < \varepsilon_1 = 10^{-5}$, or $\left|z^{T+1} - z^T\right| < \varepsilon_2 = 10^{-2}$, if the convergence is slow. If the last condition is not satisfied, then the point comes from F2. By default, F1-points are marked with colors from blue (fast convergence) to green, and F2-points - in yellow-red scale (here the color tone has no meaning).

While rendering the command MATLAB *pcolor* was used instead of the standard *contourf* to suppress undesirable color interpolation effects. We refer the rhomboid-like patterns in Fig.1 to the discreteness of the grid, where the color is attributed to the entire patch, based on the color of one calculated point.

3. The map $z \to (1+\mu)\exp(iz)$

At $\mu = 0$ we have basic map with simple dynamics: the points either escape to infinity, or are attracted by the point (0,0), and then by the fixpoint z*, or, passing zero, directly fall into the neighborhood of z*. For this case the Julia and Fatou sets are already known (and, obviously, 2-periodic in Re z), Fig.1. The orbit of zero represents the vertices of concentric triangles that slightly rotate around the center (that is, for three iterations the rotation is almost 2). Visualized by small areas of increased convergence, subsets of Fatou set are treated as multiplication (copies), possibly with the rotation of one wide subset near the fixpoint.

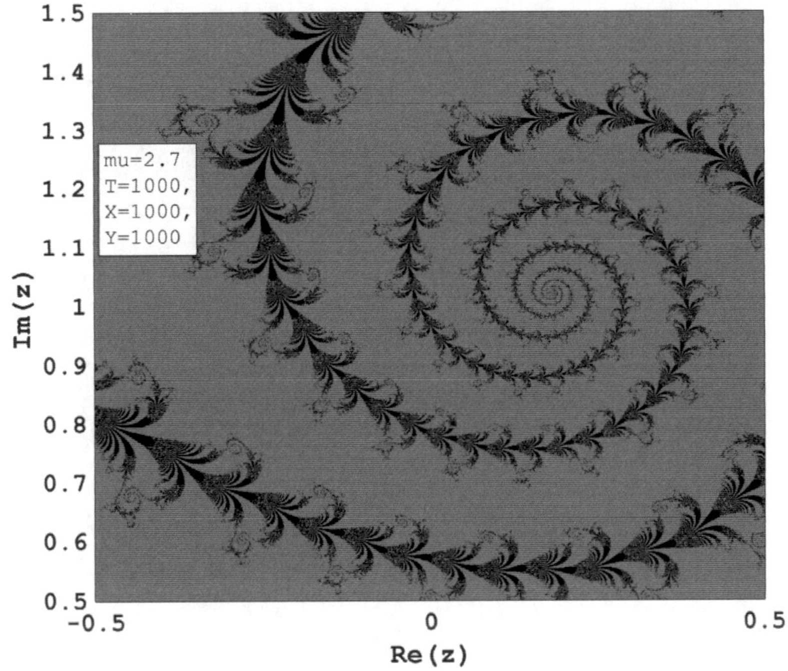

Figure 4. Julia (monochrome) and Fatou (red) sets for the case $\mu = 2.7$i in 1x1 size window near the fixpoint.

The Mandelbrot set [7] of family $f_\lambda = \lambda \exp(iz)$ shows the presence near the zero of the basin of attraction restricted by a curve resembling a cardioid rotated by 900. Areas of periodic cycles and a network of regions with a chaotic regime lie outside it. The boundaries of this basin on a straight line $\mathrm{Re}\lambda = 1$ lie within $\approx -0.9 \leq \mathrm{Im}\mu \leq 2.5$. Let's move in the direction of$\mathrm{Re}\mu = 0, \mathrm{Im}\mu > 0, \mu \nearrow$. At $\mu = 1.5$i no qualitative change in dynamics (in comparison with the base map) is observed: 1) "bushes" are shifted to the left only, keeping distance between themselves equal to ; 2) the fixpoint coordinates change; 3) convergence slows down (in the worst case: from 30 to 80 iterations). At $\mu = 2.0$i and $\mu = 2.25$i these changes increase, while the orbit of zero acquires a two-arm spiral shape, and the number of iterations to convergence increases to 192 and 384, respectively. Note that at $\mu \approx 2.475$i the multiplicator $|f'(\mu, z^*(\mu))|$becomes 1.

At $\mu = 2.5$i the convergence regime is replaced by a multiple-periodic one (Table 1). In Fig.2. the orbit of zero characterized by a cycle with period 17 is considered. The orbit of the point (0, -2) or (2, -2) is similar to it - its period is also 17 (or a multiple of 17). In this case, the Julia set has a more complex structure (Fig. 3): the sequences of Cantor's bouquets that stretch from each "bush" join each other and transform into a twisting multy-arm spiral. There is fixpoint $z^{*,2.5} \approx 0.1902{+}0.9862$i at its center (the multiplicator is 1.0045, i.e. the point is almost neutral), around which Fatou set component is located in the shape of a small disk identified by us as a Siegel disk. It is not possible to distinguish experimentally Siegel's disk from extremely slow escaping to infinity, as in the case of fixpoint $z^{*,2.5}$ or, more accurately, of a point in its small neighborhood (Fig. 3b). The spirality of the sleeves, which is also characteristic of the orbit of fixpoint vicinity, confirms it, but the distance from the center contradicts it. Visually on the outer circumference of the Siegel disk 17 spiral arms, the components of the Julia set, begin (for the near-to-disk points, escaping to infinity is slower than in the remote ones). Let us recall the special role of 17 in mathematics (the discovery of young Gauss).

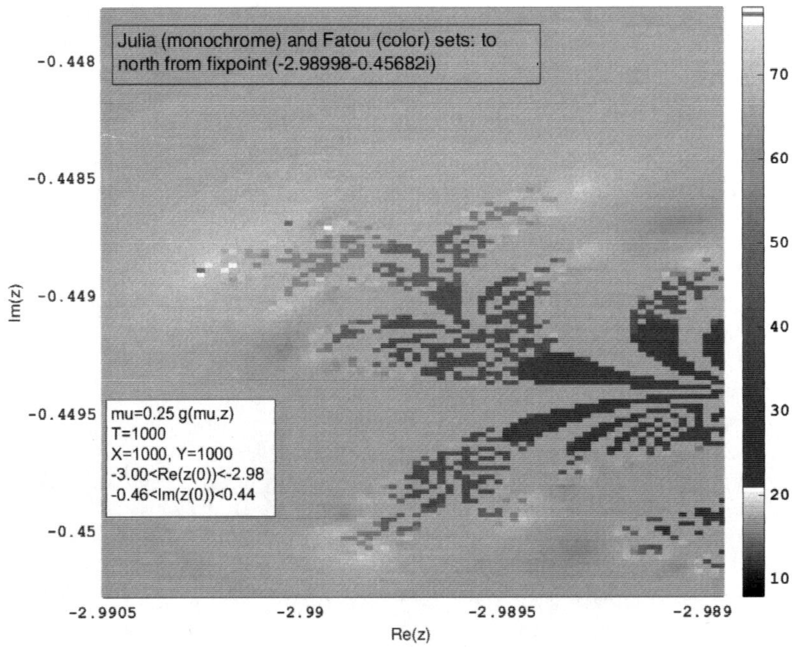

Figure 5. The neighborhood of fixpoint $z^{*1}, \mu = 0.25$ (Table 1). Such points exist that do not have time to converge neither to infinity nor to z *. This again proves Julia and Fatou sets fractality. The pixelization effect is connected with a limited number of digits used in computations, and not with rendering or grid granularity.

At $\mu = 2.7$i the shape of the Julia set changes again, the spiral becomes a two-arm (Fig. 4), and the multiplicity of the period for the points of Fatou set becomes 2 (such is the orbit of zero). As an example we give the orbit of the point 0 + 1i: during the first 60 iterations the point "sways" between the two fixpoints, gradually moving away from them, then at the 62nd iteration $|z^{62}| > 2000$, and at the 63rd iteration the point escapes to infinity.

At $\mu = 3$i the portrait of Julia and Fatou sets is simplified: for the most points, including zero, there is a two-period mode with stable fixpoints -0.262 + 0.073i and 1.621 + 2.452i. The unstable fixpoint $z^{*,3} \approx (0.1634, 1.0702)$ with the multiplicator 1.08 is fairly fast, for 200 iterations, goes into this periodic mode. At $\mu = 6$i the orbit of the formal fixpoint converges to a cycle in about 20 iterations.

4. Proposition 1:

If the map of $z \to \lambda \exp(iz)$, $\mathrm{Re}\lambda = 1$ kind has a periodic mode, then for all such points of the complex plane its multiplicity n is uniquely determined by the parameter, i.e. $n = n(\lambda) < \infty$. The periods multiplicity n is assumed to be only 2 and 17 (or multiples of 17).

5. The map $z \to (1 + \mu |z - z^*|) \exp(iz)$

The sign of the module instead of the brackets in the expression $g(z, \mu)$ makes the analysis more difficult, but simplifies the understanding. In Table 2 for selected μ values some of the equilibria closest to z* are listed; they are all unstable, except for the case $\mu = 2.5i$, z^{*3}. Let us discuss the appearance of such a fixpoint in more detail.

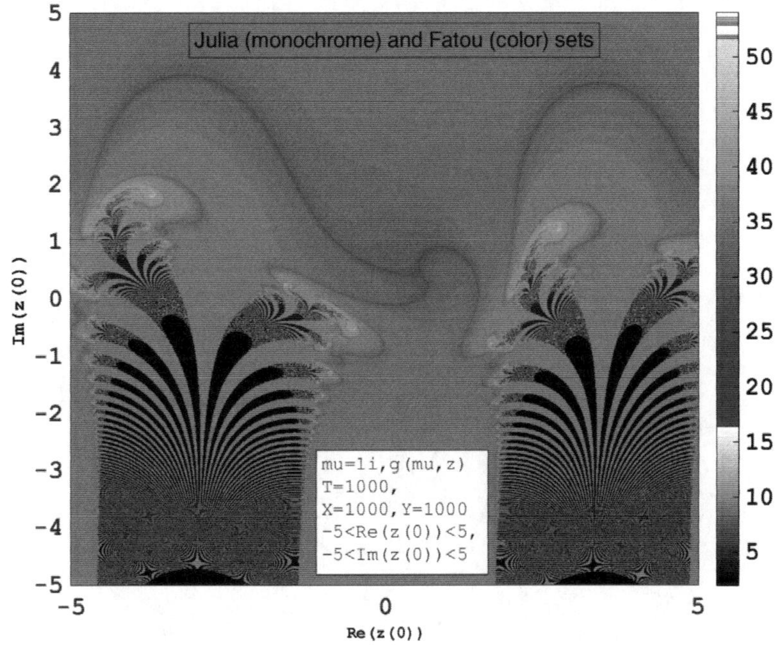

Figure 6. Julia and Fatou sets for the $\mu = 0 + 1 \cdot i$ case

Table 1. Cycle multiplicities after bifurcation

Parameter $\text{Im}\mu > \text{Im}\mu^*$	Cycle multiplicity n	Bifurcation value
2.48	49	$\text{Im}\mu^*$=2.474440
2.49	51	
2.50	17	$z^*(\mu^*)$=0.191436735086597 + 0.981615010858628i
2.51	17	

Note: Some points of the n-periodic orbit are located near each other and clusterize on the plane. For example, Fig. 2 shows three clusters: 1, 1+7, 1+7. The last two clusters exhibit the geometric resemblance

Let us suggest in case of $\mu = \mu^{double}$: $\text{Re}\mu = 0$ the simultaneous existence of two equilibria $z^*, z^* + \varepsilon$next to each other, and, relatively speaking, the root of the transcendental equation $z = g(\mu, z)$ has multiplicity 2: $z^* = g(\mu, z^*)$, $z^* + \varepsilon = g(\mu, z^* + \varepsilon)$, $\varepsilon \to 0$. If we neglect terms of second order, we get:

$$
\begin{aligned}
&z^* + \varepsilon \approx z^* \left(1 + \mu \left|\delta z\right|\right)\left(1 + \mathrm{i}\varepsilon\right) \Rightarrow \varepsilon \left(\tfrac{1}{z^*} - \mathrm{i}\right) = \mu \left|\varepsilon\right| \\
&\mu^{double} = \tfrac{\varepsilon}{|\varepsilon|}\left(\tfrac{1}{z^*} - i\right) \underset{\substack{\mathrm{Re}\mu=0,\\ \varepsilon=|\varepsilon|\exp(i\theta)}}{\Rightarrow} \quad \theta = \tfrac{\pi}{2} - angle\left(\tfrac{1}{z^*} - i\right) \qquad (1) \\
&\left|\mu^{double}\right| = 2.1682, \quad \theta = 2.5442 \approx 146^0
\end{aligned}
$$

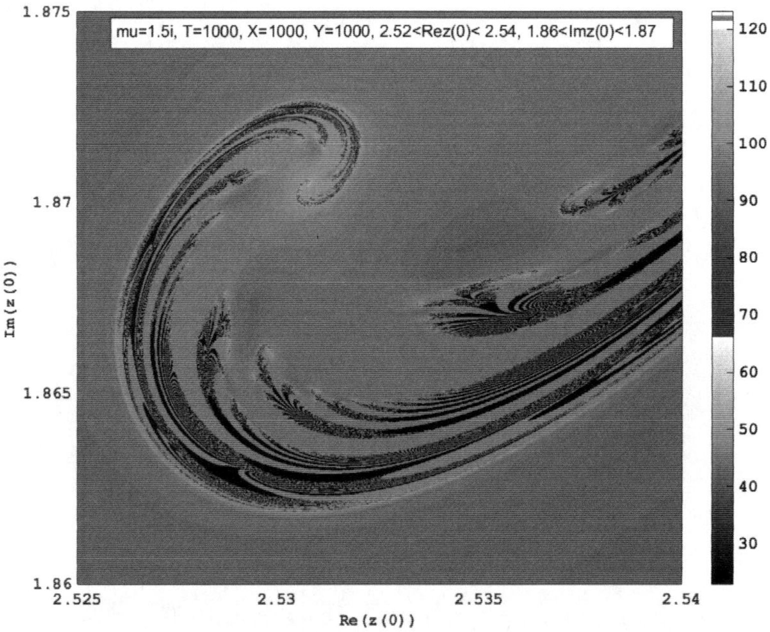

Figure 7. Julia and Fatou sets for the $\mu = 1.5\mathrm{i}$ case: fragment. The strong deformation of Cantor's bouquets is evident.

Table 2. Some fixpoints of the $g(z, \mu)$ map for different μ

μ	0 (base)	0.25	0.25i	-0.9i	2.5i
z^{*1}	-2.76636-1.08965i	-2.98998 - 0.45682i	-3.74118-0.93942i	-2.07187+0.21721i	-4.84608+1.01677i
z^{*2}	-9.18489-2.24661i	-9.319401 - 0.98569i	-10.53346-1.26968i	-7.97970-0.02618i	-11.04836+0.96522i
z^{*3}	5.98583-1.83427i	6.13132 - 0.93833i	5.16985-1.21627i	7.67724-0.16871i	0.37322+0.58097i

It is easy to see that the multiplicator $|g'(\mu, z^*)|$ is equal to 1 at $\mathrm{Im}\mu = \mu^{double}$. In comparison with the base map at $\mu = 1\mathrm{i}$ the Fatou set is more complicated: the Julia set bush is more asymmetric relative to its central axis; three curvilinear areas (valleys) gather at the point z*, where the convergence rate is increased, as well as several areas in the form of the Cantor's bouquets halo, where the convergence is locally reduced (Fig. 5, Fig. 6). This "three-valley" effect is absent for quadratic maps and is observed for the first time. At $\mu = 1.5i$ the changes are more striking, despite the presence of two convergence rate maxima instead of the "valley" (plus two more maxima to the right) inside the Fatou set, nevertheless the fixpoint z* remains unique (and the second point is a false fixpoint). Three "valleys" disappear, but one of them turns into a dumbbell shape. We can point out deformation and "vorticity" of the outer left and right bouquets of a bush (Fig. 7)

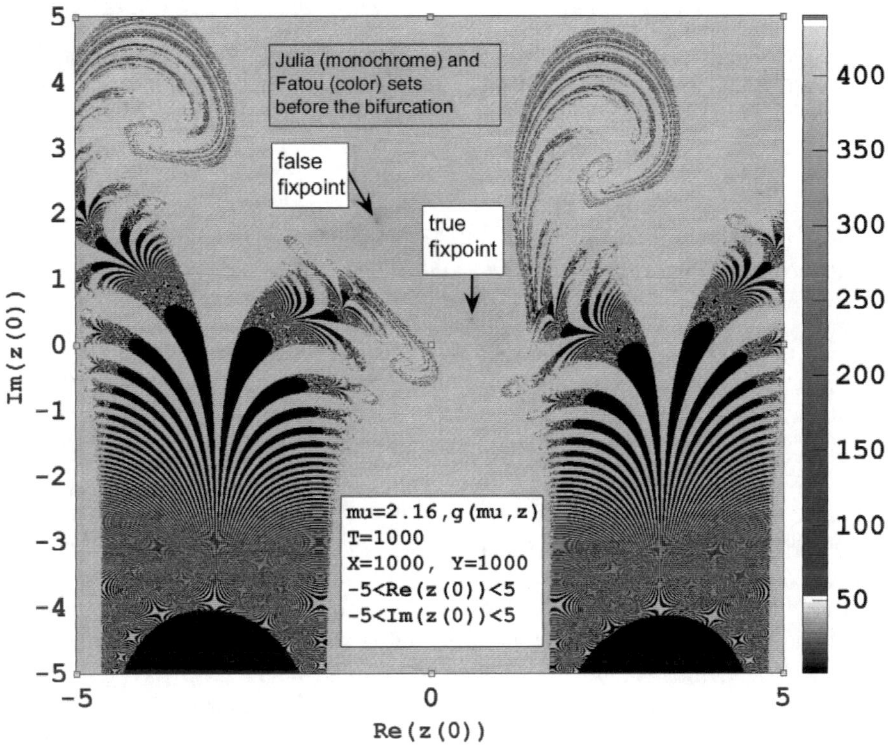

Figure 8. Julia and Fatou sets before the bifurcation at $\mu = 2.16i$.

Near the bifurcation $\mathrm{Im}\mu = 2.16 < \left|\mu^{double}\right|$ in addition to significantly slowing down of convergence (300-400 iterations) two fixpoints are clearly observed (Fig.8): true (z*) and false $z^{*,\mu} \approx -0.765 + 1.875i$, $g\left(\mu, z^{*,\mu}\right) = z^*$ false fixpoint passes in the true for the next iteration. The bifurcation itself is analogous to the saddle-node transition because of doubling and divarication the fixpoints from each other in the direction of the angle $\approx 135^0$, as was assumed in the derivation of Eq.(1).

Let us consider the neighborhood of the fixpoints immediately after the bifurcation via Fatou set (Fig.9). ecause of the second order terms the old fixpoint z * remains stable in a very small vicinity; for $z^0 = z^* + 10^{-4}(1 + \mathrm{i})$ orbit tends to z*, but the orbit of

$$z^0 = z^* + 10^{-3}(1 + \mathrm{i})$$

converges to a new fixpoint directly along a straight line connecting both fixpoints without any helical patterns. The number of iterations level-lines look quite unusual (Fig.9): for fast

IOP Conf. Series: Journal of Physics: Conf. Series **990** (2018) 012008 doi:10.1088/1742-6596/990/1/012008

convergence it has the shape of an irregular ring, and the attractor is located in the surrounding area, and not in the center.

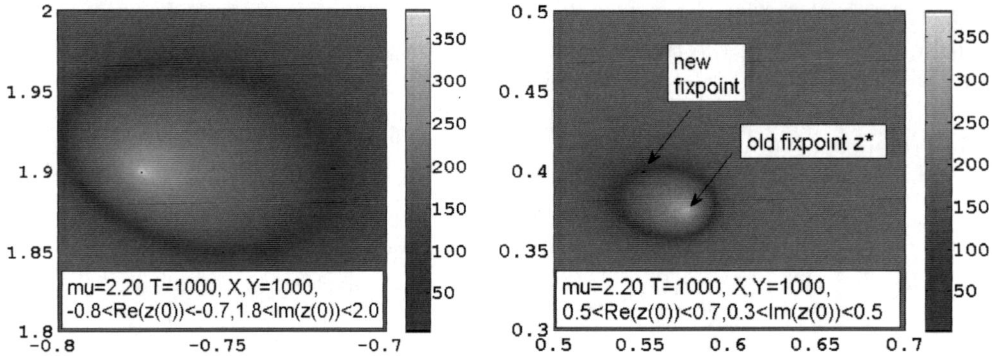

(a) (b)

Figure 9. Fatou set after bifurcation at $\mu = 2.20$inear false (a) and true (b) fixpoint. The true fixpoint has new coordinates (0.545 + 0.398i) and is near z *.

At Im$\mu = 2.5 > \left|\mu^{double}\right|$ the transition of the neighborhoods $U\left(z^{*}\right) \to U\left(z^{*3}\right)$requires about 200 iterations, the orbit of zero goes to infinity. However, the Fatou and Julia sets themselves differ substantially (Fig. 10), even in comparison with the previous case $\mu = 2.2i$. The area between the two "bushes" became predominantly the part of Julia set. The Fatou set loses its simple connectedness and resembles "tadpoles"; some components of Julia set show variations of Cantors set in the form of "lunettes". On the selected window (Fig. 10a), the areas of these sets make about 80% (Julia) and 20% (Fatou). Obviously another bifurcation has occurred, but it is difficult for us to specify a concrete value.

In particular, at Im$\mu = 2.4$ the orbit of zero has an interesting 14-periodic mode, which consists of 7-member cycles of increasing amplitude, the latter of which takes the point (-983.25+268.05i) in a vicinity of zero. Points (0-2i) and (-3+1i) also have 14-periodic mode. The same points at Im$\mu = 2.6$ gave 11-periodic mode, but whereas the penultimate point of the period showed a second order of magnitude, the latter point was 1049 to start a new period from (0,0) vicinity. This order of magnitude casts doubt on the computer calculation. At Im$\mu = 3$ the orbit of zero is 8-periodic, and in general within Fatou set periodic modes dominates over the share of convergence to single fixpoint. While inside "tadpoles" the mode returns to the valley pattern, although the valley is one (not three, as previously). At Im$\mu = 4$ the orbit of zero is 32-periodic, and the component of Fatou set with convergence to fixpoint ("tadpoles") disappears, and its place is occupied by a chaotic mix of escaping points and points with periodic orbits. At periphery of this region there is an alternation of the bands of periodicity and, probably, of the deformed Cantor bouquets for escaping points, but closer to its center we find no signs of orderliness.

6. Proposition 2

Let $g(z) = (1 + \mu\left|z - z^{*}\right|)\exp(iz)$ map be given for any μ : $Re\mu = 0$. Then: 1) if the orbit of zero goes to infinity, then there are no n-periodic ($n > 1$) regimes; 2) if the orbit of some point goes to the n-periodic regime, then the point (0,0) belongs to this period, i.e. $(\forall\varepsilon < 0)(\exists t \in \mathbf{N})\left(\left|\zeta^{t}\right| < \varepsilon\right)$.

As we can see, the Sullivans results in a somewhat weakened form are also fair for our maps. They can be generalized, probably, either to linear-exponential maps $z \to (\mu_0 + \mu z)\exp(z)$or to quasi-polynomials $P(z)\exp(z)$.

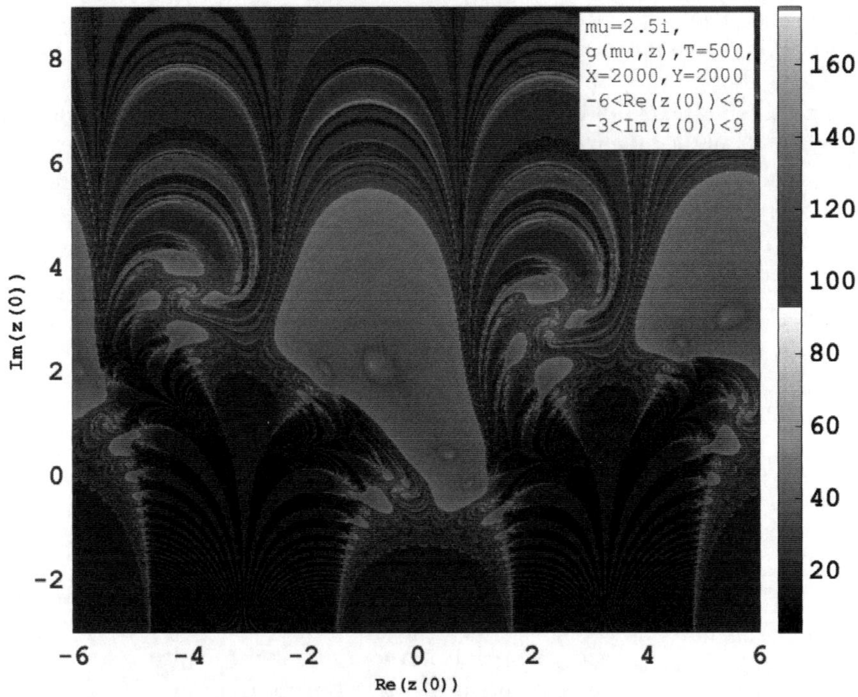

(a)

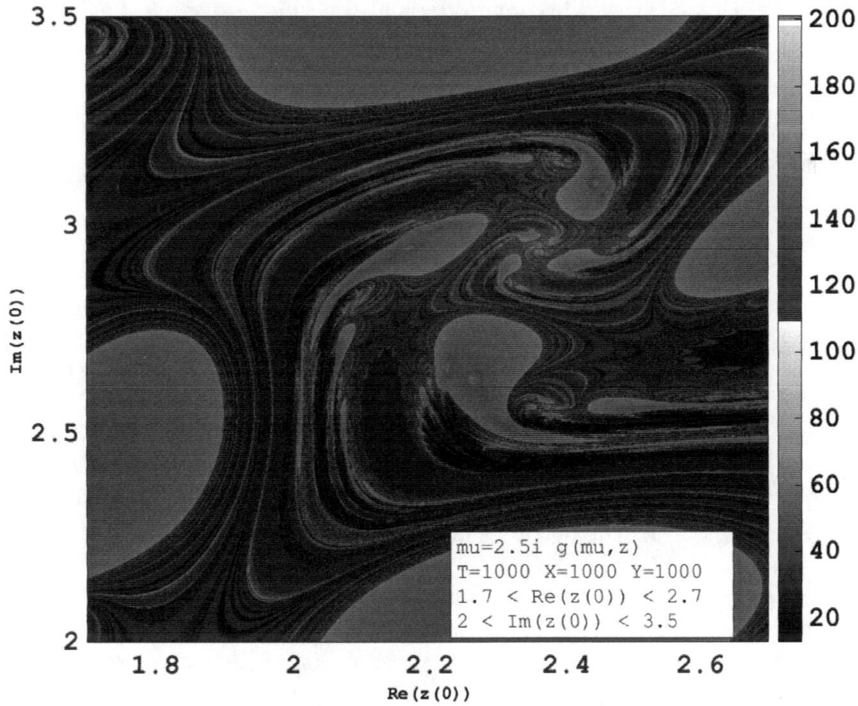

(b)

Figure 10. Julia and Fatou sets at $\mu = 2.5\mathrm{i}$: General view (a) and Detailed view (b).

IOP Conf. Series: Journal of Physics: Conf. Series **990** (2018) 012008 doi:10.1088/1742-6596/990/1/012008

7. The map $z \to (1 + \mu(z - z^*)) \exp(iz)$

The dynamics of this view is more simplified than the previous one and resembles more a classical exponent, albeit in shape f (z). But unlike the exponent of Fatou and Julia sets they lose their periodicity along the real axis, and the neighboring "bushes" seem to stretch out and begin to hang over the "bush" close to zero. Tables 3 and 4 summarize the main features of dynamics and Julia / Fatou sets when the parameter is varied, mainly along the imaginary axis, i.e. $\mathrm{Re}\mu = 0$, $\mathrm{Im}\mu > 0$. For this mapping figures 3 and 39 instead of 2 or 17 have value. In addition, examples have been identified when Fatou set disappears.

Table 3. Summary mapping table $h(z)$ (Part 1)

$Im(\mu)$	Quality description
During movement along imaginary axis	
0.1	Convergence according to the trefoil scheme (analogous to Figure 11a), a fix point is a focus (See Note)
0.25	The same
0.5	A three-period cycle (the fix points are removed from zero), each fix point is a focus (one-arm spiral), Julia set: adjacent three-arm spiral - two sleeves from one "bush", one from the other
0.75	A three-period cycle (one of the fix point is shifted by 0.1 from zero), convergence to each point is fast in type of node; touching of the three arms (Fig. 11a)
1	The same
1.5	3-periodic cycle, one fix point is in the neighborhood of zero; three "bush" converge at a single fix point (Fig. 11b)
2	3-periodic cycle, one fix point is at a distance of no more than $2 * 10^{-4}$ from zero
3	The same, with two fix points - 0 and "big" in module
6	Very rapid escape to infinity, there is no visualization of Fatou and Julia sets
9	Julia set fills the entire plane, a rapid escape to infinity through 0
12	The same
20	3-period mode on the verge of loss of accuracy, i.e. escaping from it to infinity: -27394509,8784864 + 9868306,50217099i $\to$ 0,00000000000000+ 0,00000000000000i$\Leftrightarrow$ 8,49398041474234 - 11,5282544606287i

Note: We refer the terms focus and node to visual characteristics of Julia set which is close to the standard graphic definitions of the focus and node as of the characteristics of fix point (trajectories of convergence). The difference is nevertheless substantial: the visual node can match the fixpoint-focus.

It is interesting to find such a value $\mathrm{Im}\mu = \mu^{bf}$ when the multiplicator turns to unity at the very fixpoint. Analysis of formal critical points does not provide useful information; these values are:

$$\mu = \frac{-i}{1+i(z^{cr}-z^*)} \Leftrightarrow z^{cr} = z^* + \left(i - \frac{1}{\mu}\right),$$
$$\mu^{bf} = |z^*|^{-1} - 1 = 0.454553558270876, \quad z^{cr,bf} \equiv z^{cr}\left(\mu^{bf}\right) = z^* + 3.2000i$$

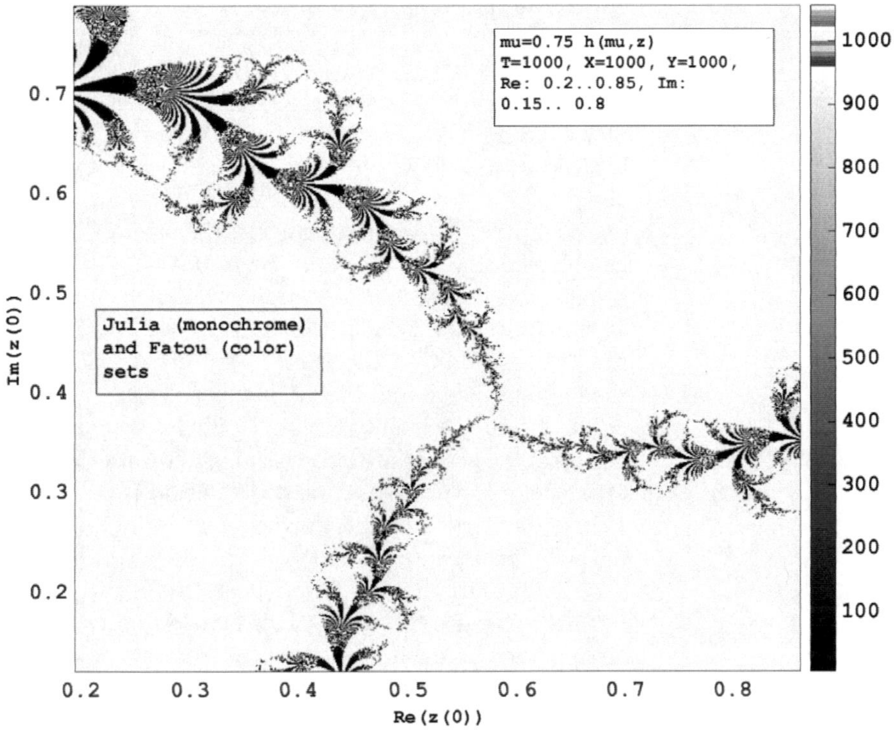

(a)

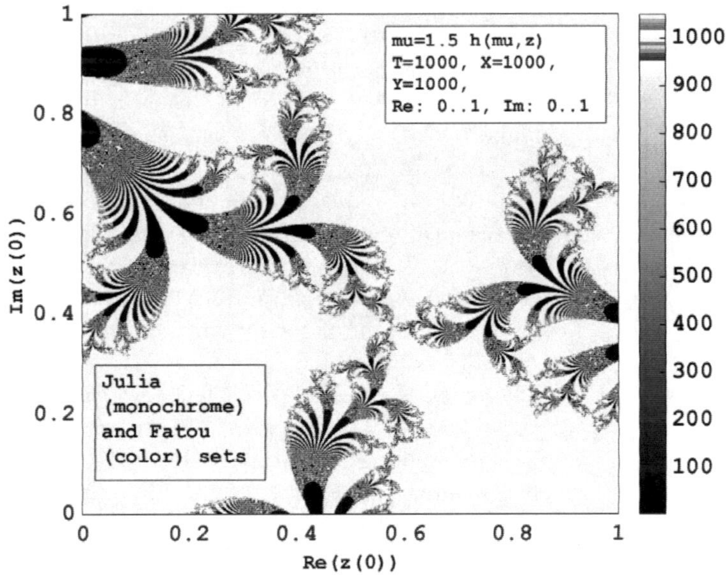

(b)

Figure 11. Julia and Fatou sets for Im$\mu = 0.75 \to 1.5$ (from (a) to (b)). The formation of the trefoil joint near the fixpoint z^*.

Table 4. Summary mapping table $h(z)$ (Part 2)

$Im(\mu)$	Quality description
At the bifurcation value	
0.33	Convergence according to the trefoil scheme
0.45	Slow convergence (more than 3800) iterations, but convergence is achieved (at 10000 iteration deepth)
0.46	39-fold cycle - two elements are deleted, the others merge into one strip, zero belongs to the cycle; the orbit of the point in the neighborhood of the fix point is $z^* + 0.05 * (1 + i)$ at first shows a quasiperiodicity with a period of 3, then infinity is broken at the 720th iteration; if the shift is by 0.01 (1 + i), then at the 1156th iteration the value of 10128 is reached as well as then going to zero and the 39-member cycle; chaos near the fix point, formed by points with periodic and runaway orbits.
μ^{bf}= 0.454553558270876	The orbit of zero is an attractor in the form of a (semi) closed tape, there are no large values, the period is not detected (Fig. 12), zero enters the attractor; Orbit z * + s (1 + i) is attractor in the form of a ring of size (2 3) * s, the period is not detected, s = 0.001, 0.01, 0.1; a similar pattern appears when shifting by 0.01, ring size (2 3) * 0.01; When s = 1, the ring resembles a tape for the orbit of zero; Orbit of the critical point: an attractor, there is no period, the neighborhood of zero enters the attractor, the part of the attractor (southern) is dense, the part contains discontinuities (northern). Julia Set: the spirals pass at some distance from fix point.
0.47	3-cycle, point 0 does not enter the cycle
0.465	39-cycle for the orbit of zero, while the fix points cluster into three self-similar groups: in each group there is a "leader" standing at a distance from the semicircle of his twelve brothers; all the fix points are concentrated on the "window" [-1.5..2; -14]; point (0,0) is next to the cycle at a distance of 0.01; 3-arm spiral on Julia set.
0.4675	3-cycle, 13 arms are stretched to each fix points, i.e. convergence according to 13-leaf scheme (Fig. 13); it is easy to see the origin of the previous 39-fold cycle, because 39 = 3 * 13.
0.4545	The orbit of zero - convergence to zero over a complex closed curve (possibly of fractional dimension) for 323422 iterations.
0.4546	The orbit of zero - the orbit returns to zero after leaving to large values of the order of 10^{12}; For the first time this occurs at the 2149th iteration, i.e. at 2149th cycle

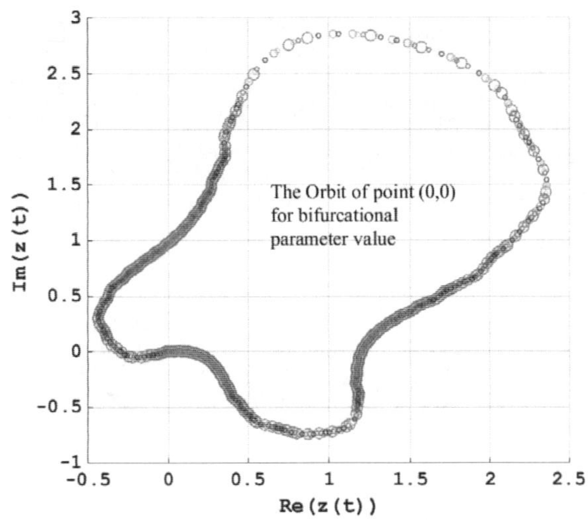

Figure 12. The zero orbit at the bifurcation value of the parameterμ^{bf}. The designations of the axes, colors and dimensions of the marker are the same as in Fig. 2. The depth of calculation is 3000 iterations. The northern part of the orbit is notable: is this portion of the orbit a dense set if the number of iterations is unlimited?

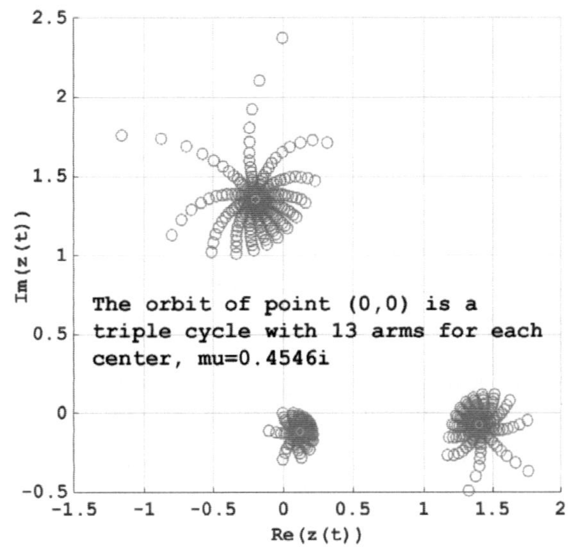

Figure 13. Degeneration of 39-member cycle into the triple cycle with 13 arms near bifurcation.

8. Conclusion

The main content of the paper is presented in the figures, which is a feature of experimental mathematics in application to nonlinear dynamics. Its goal was once again to draw attention to the iterated exponential (and its modifications) on the complex plane. We have seen at least two bifurcation transitions. For the first one, the critical value of the parameter was derived, but for the second one it was not.

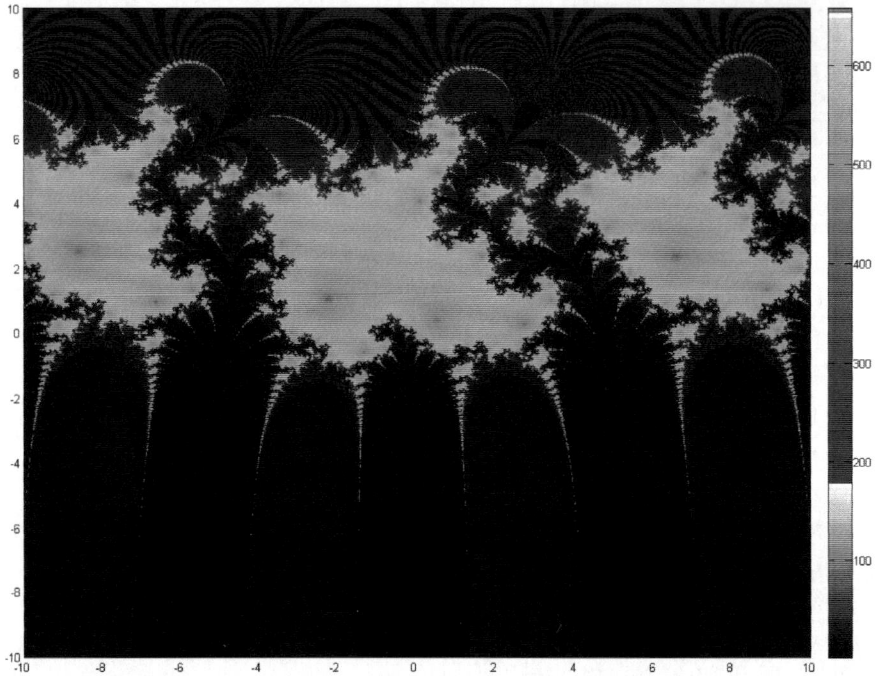

Figure 14. Julia and Fatou sets at the real $\mu = 1$: general view. Large window [-10.. 10; -10.. 10], resolution 30003000, depth of calculation - 700.

In conclusion we formulate the questions that need to be solved by theoretical mathematicians:

- Within Fatou set there are regions of local increase of the convergence rate in the form of small circles (Fig. 2) or valleys (Fig. 6). This phenomenon we call "yellow spots" is absent for quadratic mappings. Can it be given a strictly formal definition avoiding reference to quantities?
- Is there a Siegel disk for some value of the parameter in the three studied variants of linear-exponential mappings? (Figure 3b leaves us in doubt).
- For the given families Proposition 1 needs to be proved with respect to the following: multiplicity of a cycle in a periodic mode, if such cycle exists, does not depend on the choice of the starting point.
- For these families Proposition 1 needs to be proved with respect to the following: at a continuous variation of the parameter near the bifurcation the cycle multiplicity varies discretely running through some special "magic" values (for the family f: 51, 17, 2).
- For these families under what conditions does the orbit contain a cycle after the quasiperiodic section?
- For the given families Proposition 2 needs to be proved, i.e. if the orbit of the point 0 does not contain a cycle, then periodic orbits are absent for any initial point.
- What are topological properties (is a set dense everywhere?) of an attractor for h-family for a parameter in the neighborhood of bifurcation value?

Figure 15. A fragment of Julia set at the real$\mu = 1/|z^*|$. Its complex internal structure in the window can be seen [-3..0;0..3], resolution 10001000, depth of calculation - 3000.

9. References

[1] Ghys E, Goldberg L R and Sullivan D P 1985 On the measurable dynamics of $z \mapsto e^z$ *Ergodic Theory Dynam. Systems* **5** 3 329-35
[2] Devaney R L 1991 e^z - Dynamics and Bifurcation *Int.l Journal of Bifurcation and Chaos* **1** 287-308
[3] Aarts J and Oversteegen L 1993 The geometry of Julia sets *Trans. Amer. Math. Soc.* **338** 897-918
[4] Kotus J 2009 *Cantor bouquets for non-entire meromorphic functions* (Workshop on Cantor bouquets in hedgehogs and transcendental iteration, June 16th-19th 2009, Toulouse, France)
[5] Glebsky L 2016 *Proc. Of Int. Conf.-School Dynamics, Bifurcations and Chaos 2016 (DBC III), Nizhny Novgorod, Russia, July 18 22* (Nizhny Novgorod: Lobachevsky University Press) p 12
[6] Kawahira T 2016 The Riemann Hypothesis and Holomorphic Index in Complex Dynamics *Experimental Mathematics* **25** 1 37-46
[7] Matyushkin I V 2016 On some properties of an *exp(iz)* map *Russian Journal of Nonlinear Dynamics* **12** 1 3-15
[8] Kaneko K 1992 Overview of coupled map lattices *Chaos: An Interdisciplinary Journal of Nonlinear Science* **2** 3 279-82
[9] Garcia-Morales V 2016 From deterministic cellular automata to coupled map lattices *J. Phys. A: Math. Theor.* **49** 295101

Dynamics of one model of the fast kinematic dynamo

Timur Medvedev[1], Vladislav Medvedev[2] and Evgeny Zhuzhoma[2]

[1] National Research University Higher School of Economics, 136 Rodionova Street, Niznhy Novgorod, Russia

[2] National Research University Higher School of Economics, 25/12 Bolshaya Pecherckaya Street, Niznhy Novgorod, Russia

E-mail: tmedvedev@hse.ru

Abstract. We suggest a new model of the fast nondissipative kinematic dynamo which describes the phenomenon of exponential growth of the magnetic field caused by the motion of the conducting medium. This phenomenon is known to occur in the evolution of magnetic fields of astrophysical bodies. In the 1970s A.D. Sakharov and Ya.B. Zeldovich proposed a "rope" scheme of this process which in terms of the modern theory of dynamical systems can be described as Smale solenoid. The main disadvantage of this scheme is that it is non-conservative. Our model is a modification of the Sakharov-Zeldovich's model. We apply methods of the theory of dynamical systems to prove that it is free of this fault in the neighborhood of the nonwandering set.

1. Introduction

One of the fundamental problems of the natural sciences is formation and evolution of magnetic fields of astrophysical bodies. Particularly, the theory of the kinematic dynamo studies the evolution of the magnetic fields of electrically conducting fluids [1, 2, 3]. The velocity field $\vec{v}$ of an incompressible conducting medium (fluid, gas or plasma) is supposed to be given while the subject of interest being the magnetic field $\vec{H}$ stretched by the fluid flow in the presence of a low diffusion dissipating the magnetic energy. The kinematic dynamo is described by the following equations

$$\frac{\partial \vec{H}}{\partial t} = \operatorname{rot}\left[\vec{v}\vec{H}\right] + \eta\triangle\vec{H}, \quad \operatorname{div}\vec{H} = 0, \quad \operatorname{div}\vec{v} = 0,$$

here η is the magnetic diffusivity which is in inverse proportion to Magnetic Reynolds number $R_m = \frac{1}{\eta}$ (see the main notions and definitions in [4, 5, 6]). The literature on the magnetohydrodynamics often uses $\vec{B}$ to describe the magnetic field, where $\vec{B} = \mu\vec{H}$ and μ is the permeability of the medium (for us the difference between $\vec{H}$ and $\vec{B}$ is irrelevant). One of the important aspects of the kinematic dynamo is the fast kinematic dynamo when the motion of the conducting medium causes an exponential growth of the magnetic field for a small magnetic diffusion.

The discrete (in time) version of this problem studies the growth of the magnetic field at moments $t = 1, 2, \ldots$. Instead of the transport of the flow and the continuous diffusion of the magnetic fields one considers the composition of these processes. That is, for a given conservative (volume preserving) diffeomorphism $f : M \to M$ the magnetic field is considered to be first transported to the field $f_*(\vec{\boldsymbol{H}})$ and then to be dissipated as the solution of the equation $\frac{\partial f_*(\vec{\boldsymbol{H}})}{\partial t} = \eta \triangle \left(f_*(\vec{\boldsymbol{H}}) \right)$.

A kinematic dynamo is said to be dissipative ("realistic") if $\eta \to +0$ or nondissipative ("idealistic") for $\eta = 0$. For the nondissipative dynamo the magnetic field is "frozen" into the movement of the medium [7] and one usually studies the exponential growth of the energy of this field. According to [8] a fast nondissipative dynamo occurs if the diffeomophism f has a nonzero topological entropy (i.e if f chaotic enough). In the 1970s A.D. Sakharov and Ya.B. Zeldovich proposed a scheme of a so called rope dynamo the idea of which modern 3-dimensional models of the fast dynamo widely exploit [1, 2]. From the point of view of the modern theory of the dynamical systems the construction of Sakharov-Zeldovich is an Ω-stable map of the solid torus into itself suggested by S. Smale [9]. The nonwandering set of this map is a topological solenoid and an expanding attractor, therefore the nonwandering set in the Smale's construction is often called the Smale solenoid or the Smale attractor [10]. Notice that the topological entropy of this map is positive. The main disadvantage of this scheme for the theory of the kinematic dynamo is that the suggested map is not conservative (this subject is well discussed in [1], chapter V). In this paper we suggest a modification of the Sakharov-Zeldovich's scheme without this fault in the a neighborhood of the nonwandering set. Notice that the solid torus is naturally foliated by 2-disks perpendicular to its axis and that the Smale map preserves this disk structure. We modify the scheme in such a way that the solid torus maps into its neighborhood, the disk structure is preserved but the intersection of each disk with its image becomes two symmetric Smale horseshoes [11, 9]. Since there are horseshoes with an arbitrary Jacobian determinant it is possible to construct a map which is conservative in the neighborhood of the nonwandering set. This map can be extended to a diffeomorphism of the 3-sphere S^3 or of the Euclidean space $\mathbb{R}^3$, though such an extension is not guaranteed to be conservative on the entire manifold.

The example we suggest is of its own interest for the theory of the dynamical systems. To describe the nonwandering set we give the symbolic model of the restriction of the diffeomorphism to its wandering set. We show that the nonwandering set is the nontrivial 1-dimensional basic set (see basic notions of the theory of the dynamical systems in [12, 13]) and it is in the class of so called solenoidal sets [14].

2. The main construction

Consider the direct product $K \times [0;1]$, where $K = [-1;+1] \times [-1;+1]$ is the square on the plane $\mathbb{R}^2$ with the Cartesian coordinates (x, y). Let $R_t : \mathbb{R}^2 \to \mathbb{R}^2$ denote the counterclockwise rotation

$$\begin{cases} \bar{x} &= x\cos\pi t - y \sin\pi t \\ \bar{y} &= x \sin\pi t + y\cos\pi t \end{cases}$$

of the plane $\mathbb{R}^2$ through the angle πt. The set $\bigcup\limits_{0\le t\le 1} (t, R_t(K))$ is homeomorphic to $K \times [0;1]$ because R_t is a homeomorphism for each t. Since $R_1(K) = K$, the squares $K \times \{0\}$, $K \times \{1\}$ can be naturally identified by $id : K \times \{1\} \to K \times \{0\}$. Let B be the body $\bigcup\limits_{0\le t\le 1} (t, R_t(K))$ with the squares $K \times \{0\}$ and $K \times \{1\}$ identified by id:

$$\bigcup_{0\le t\le 1} (t, R_t(K)) \,/\, \left(K \times \{1\} \overset{id}{\sim} K \times \{0\}\right) \overset{\text{def}}{=} B.$$

We assume that the identification id reverses the orientation if the initial orientation of the squares $K \times \{0\}$, $K \times \{1\}$ is induced by an arbitrary orientation of the body $\bigcup_{0\le t\le 1} (t, R_t(K))$. The body B is a twisted cylinder shown in Fig. 1, where (a) shows the part of B for the values $0 \le t \le \frac{1}{2}$ while (b) shows it for $\frac{1}{2} \le t \le 1$.

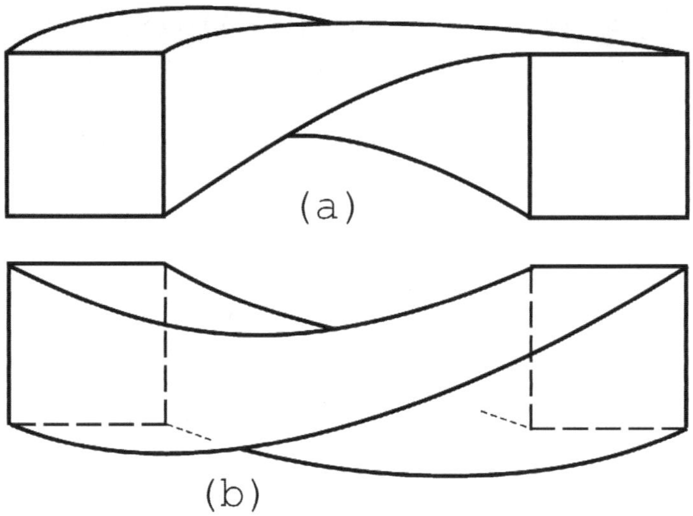

Figure 1. Two halves of the body B. The body is obtained by gluing the halves by the squares.

It is clear that the set

$$\bigcup_{0\leq t<1} (t,(0,0)) \stackrel{\text{def}}{=} S_0^1$$

is a circle on which the quotient map $[0;1] \to [0;1]/(0 \sim 1) = S^1$ induces the cyclic coordinate t mod 1. We say the circle S_0^1 to be the *axis* of the body B. We embed B into $\mathbb{R}^3$ in such a way that the axis B has no knots in $\mathbb{R}^3$ and we consider B to be identical to its embedding. First we are going to construct the desired diffeomorphism $F : B \to f(B) \subset \mathbb{R}^3$ of the body B onto its image in some neighborhood homeomorphic to the solid torus. Notice that the natural projection $K \times [0;1] \to [0;1]$ is a trivial bundle and it induces the locally trivial bundle $p_1 : B \to S_0^1$ with the fiber K. Let D_t denote the fiber over $t \in S_0^1$, $D_t = p_1^{-1}(t)$. Evidently D_t can be considered as $R_t(K)$, i.e. as the result of the rotation R_t of the square K.

To define the diffeomorphism F we need to modify the map introduced by Stephen Smale known as the Smale horseshoe [11, 9]. Recall that the classic Smale horseshoe is a diffeomorphism of some disk, containing the square $K = D_0^2$ on the plane $\mathbb{R}^2$, into itself. The diffeomorphism $w : D_0^2 = D^2 \to \mathbb{R}^2$ of this square is the composition of a contraction along the axis Ox, an expansion along the axis Oy, a bend (the direction of the bend is irrelevant) of the resulting rectangle and, finally, its translation in such a way that the intersection $D^2 \cap w(D^2)$ is the union of two disjoint strips which are symmetric with respect to the axis Oy[1] (see Fig. 2(a)). Clearly, the contraction and the expansion could be chosen in such a way that the Jacobian determinant $J(w)$ of w on D^2 equals $\frac{1}{2}$. From now on we suppose these conditions to be satisfied. Denote by $sh_0 : \mathbb{R}^2 \to \mathbb{R}^2$ the translation $(x;y) \longrightarrow (x + \frac{1}{2};y)$ along the axis Ox and let $w_0 = sh_0 \circ w : D^2 \to \mathbb{R}^2$. Let $S_0 : \mathbb{R}^2 \to \mathbb{R}^2$ denote the inversion with respect to the origin $(0;0)$, $S_0(x;y) = (-x;-y)$. Again one can pick the contraction, the expansion and the bend such that the following conditions are satisfied (see Fig. 2 (b) on the right):

(i) the intersection $D^2 \cap w_0(D^2)$ consists of two disjoint strips;

(ii) the sets $w_0(D^2)$, $S_0\left(w_0(D^2)\right)$ are disjoint,

[1] The horseshoe is sometimes defined as the diffeomorphism of the square which is afterwards extended to the entire plane. It is known [12, 11] that w can be extended to a map of the entire plane $\mathbb{R}^2$ in such a way that this map is the identity outside some neighborhood of D^2.

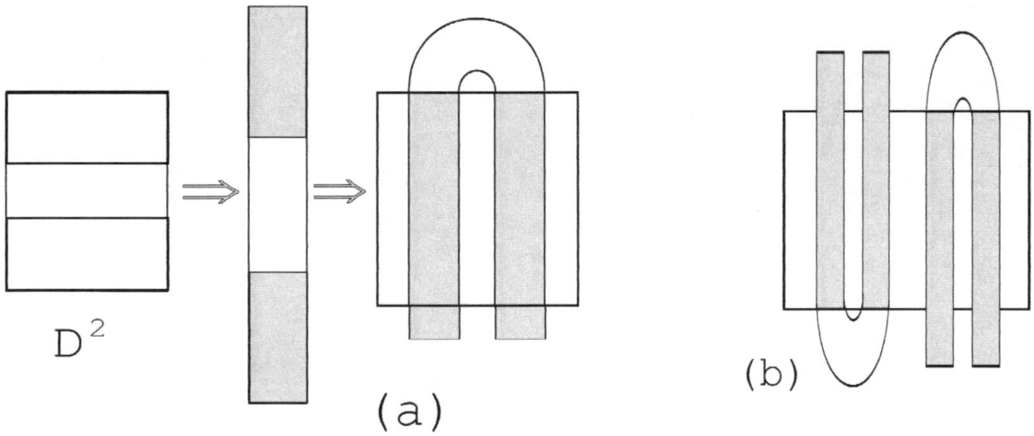

Figure 2. The Smale horseshoes $w(D^2)$ (a) and $w_0(D^2) = sh_0 \circ w(D^2)$ (b).

$$w_0(D^2) \bigcap (S_0 \circ w_0(D^2)) = \emptyset. \tag{1}$$

The first condition means that the map $sh_0 \circ w \stackrel{\text{def}}{=} w_0$ is a Smale horseshoe whose symmetry line is perpendicular to the axis Ox. The second condition means that the horseshoe $w_0(D^2)$ is disjoint from its reflection by S_0. Notice that $S_0 \circ w_0(D^2)$ is a horseshoe as well.

Consider a neighborhood $S^1 \times B^2$ of the body B homeomorphic to the solid torus; here $B^2 \subset \mathbb{R}^2$ is a disk large enough containing the square K and S^1 is a circle with cyclic coordinate t mod 1. Below we identify a neighborhood of B with $B^2 \times S^1$. Without loss of generality one assumes S^1 to coincide with the axis S_0^1 of B and that $D_t \subset \{t\} \times B^2$ for each t. Recall that the square D_t is the result of rotation $R_t(K)$ of the square K, therefore we can define a Smale horseshoe on D_t. Let

$$w_{0t} = R_t \circ w_0 \circ R_{-t} : D_t \to \{t\} \times B^2.$$

This map forms the horseshoe in the direction of the line $y = x \cdot \tan \pi t$ only when the symmetry line of the horseshoe $w_{0t}(D^2)$ is perpendicular to the line $y = x \cdot \tan \pi t$.

Let $S^1 = [0;1]/(0 \sim 1)$ be a circle with the natural parametrization $[0;1] \to [0;1]/(0 \sim 1) = S^1$. The map $E_2 : S^1 \to S^1$ of the form $t \to 2t \mod 1$ is an expanding endomorphism of the circle of degree 2 [15]. We now define the map $F : B \to S^1 \times B^2$ in the following way: for every $t \in [0;1)$ and every $z \in D_t$ let

$$(t; z) \longmapsto (E_2(t); R_t \circ w_{0t}(z)), \qquad t \in [0;1), \quad z \in D_t.$$

Notice that from the definition of F if follows that $F(D_t) \subset B_{2t \mod 1}$, (Fig. 3).

Lemma 1 *The map $F : B \to F(B) \subset S^1 \times B^2$ is a diffeomorphism onto its image.*

Proof Assume $F(t_1; z_1) \cap F(t_2; z_2) \neq \emptyset$, then $F(D_{t_1}) \cap F(D_{t_2}) \neq \emptyset$. From the definition of F it follows that $E_2(t_2) = E_2(t_1)$, i.e. $2t_1 \mod 1 = 2t_2$. Since the map w_{0t} is a diffeomorphism to its image one assumes $t_1 \neq t_2$, therefore $t_2 = t_1 + \frac{1}{2}$. Then $F(D_{t_1}) = R_{2t_1} \circ w_0 \circ R_{-t_1}(D_{t_1})$,

$$F(D_{t_2}) = F(D_{t_1+\frac{1}{2}}) = R_{2t_1+1} \circ w_0 \circ R_{-t_1-\frac{1}{2}}(D_{t_1}) = R_1 \circ R_{2t_1} \circ w_0 \circ R_{-t_1-\frac{1}{2}}(D_{t_1}).$$

Since R_1 is the rotation through π, the horseshoes $F(D_{t_1})$ and $S_0 \circ F(D_{t_1})$ must intersect and this contradicts to (1).□

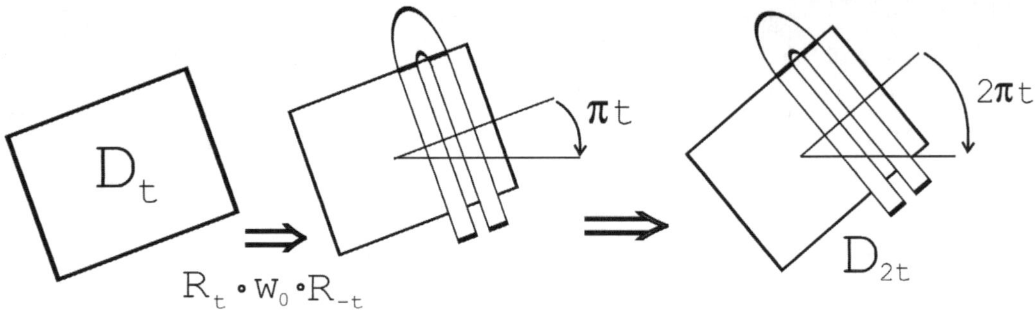

Figure 3. Construction of the map F.

Notice that since the Jacobian determinant $J(w)$ of the map w on D^2 equals $\frac{1}{2}$, the Jacobian determinant of F equals $J(F) = J(w) \cdot DE_2 = \frac{1}{2} \cdot 2 = 1$ and therefore, F is a conservative diffeomorphism to its image. The union $\mathbb{R}^3 \cup \{\infty\}$ of the Euclidean space and the point at infinity $\{\infty\}$ can be identified with the 3-sphere S^3 in the standard way.

Lemma 2 *The map $F : B \to F(B) \subset S^1 \times B^2 \subset \mathbb{R}^3$ can be extended to a diffeomorphism $f : S^3 \to S^3$ which is conservative in some neighborhood of B.*

Proof By construction the circle S_0^1 is the axis of the solid torus $S^1 \times B^2$ and the body B as well as $S^1 \times B^2$ are its tubular neighborhoods. The diffeomorphism of the square in the Smale horseshoe can be extended to a diffeomorphism of a disk large enough (see [12]), therefore F can be extended to a diffeomorphism (which we again denote by F) of the solid torus $S^1 \times B^2$ which preserves the disk structure. Without loss of generality one assumes that F is conservative in some neighborhood of B (otherwise one can consider the square K a bit larger). If follows from the construction that the curves S_0^1 and $F(S_0^1)$ are knot free in $\mathbb{R}^3$. Therefore there is a deformation of S_0^1 to $F(S_0^1)$ which can be easily extended to a deformation of their tubular neighborhoods $\varphi_0 : S^1 \times B^2 \to F(S^1 \times B^2)$. Clearly φ_0 can be made conservative. It follows from [16] that φ can be extended to the desired diffeomorphism $f : S^3 \to S^3$.□

3. The dynamics of the nonwandering set

We say the solid torus B embedded into S^3 to be *basic* and we denote it by $\mathcal{B}$. Let

$$\Omega = \bigcap_{n=-\infty}^{\infty} f^n(\mathcal{B}). \tag{2}$$

The set Ω is invariant with respect to f ([12])and it is not empty because it contains in $D_0 = \{0\} \times D^2 \subset \mathcal{B}$ the invariant nontrivial (0-dimensional) set Ω_0 of the Smale horseshoe ([12, 11, 9]). Let $Diff^1(S^3)$ denote the space of diffeomorphisms of the 3-sphere S^3 with C^1 topology.

Lemma 3 *The set Ω is hyperbolic and the restriction $f|_\Omega$ of f to Ω is of positive topological entropy. Moreover, there is a neighborhood $U(f)$ of f in the space $Diff^1(S^3)$ such that every diffeomorphism $g \in U(f)$ has a hyperbolic invariant set $\Omega_g \subset \mathcal{B}$, the diffeomorphisms $f|_\Omega$, $g|_{\Omega_g}$ are conjugate and the entropy of the restriction $g|_{\Omega_g}$ is positive.*

Proof By construction the Jacobian determinant of the map $f|_\mathcal{B} : S^1 \times D^2 \to \mathbb{R}^3$ is equal to $J(f) = \begin{pmatrix} \frac{1}{2} & 0 \\ 0 & 2 \end{pmatrix}$. Therefore f is hyperbolic (not only on Ω but on $\mathcal{B}$ as well) and it follows that the set Ω is hyperbolic. The restriction $f|_{\Omega_0} : \Omega_0 \to \Omega_0$ has a positive entropy ([12, 13]). Then it follows from [17] that the entropy of the restriction $f|_\Omega$ is positive as well. Since hyperbolic sets are stable under C^1-small perturbations, the desired neighborhood $U(f)$ exists because the entropy is invariant with respect to conjugacy.□

Now we study the dynamics of the restriction of the diffeomorphism $f : S^3 \to S^3$ to its nonwandering set belonging to the basic solid torus $\mathcal{B}$. To do this we construct a symbolic model of $f|_\Omega$ on the invariant set Ω. Let $t_0 \in S^1$, $0 \le t_0 < 1$ be fixed. Consider the intersection of Ω and the disk $D_{t_0} = \{t_0\} \times B^2 \subset S^1 \times B^2$.

Following the symbolic model of the classic Smale horseshoe we define two vertical and two horizontal (in the usual sense) strips in the square D_{t_0} in the following way. Recall that the intersection of the square D_{t_0} with its image with respect to the map of the Smale horseshoe w_{0t_0} consists of the two vertical strips,

$$w_{0t_0}(D_{t_0}) \cap D_{t_0} = R_0(t_0) \cup R_1(t_0),$$

where $R_0(t_0)$ (respectively $R_1(t_0)$) is the strip nearest to the center (respectively, the farthest). It follows from the construction that in D_{t_0} there are two disjoint horizontal (perpendicular to $R_0(t_0)$, $R_1(t_0)$) strips (we denote them by $R_0^{(-1)}(t_0)$ and $R_1^{(-1)}(t_0)$) such that

$$w_{0t_0}(R_0^{(-1)}(t_0)) = R_0(t_0) \text{ and } w_{0t_0}(R_1^{(-1)}(t_0)) = R_1(t_0). \tag{3}$$

We now show that the intersection $f^{-1}(\mathcal{B}) \cap D_{t_0} \cap f(\mathcal{B})$ consists of eight rectangles. By construction the intersection $D_{t_0} \cap f(\mathcal{B}) \cap \mathcal{B} = D_{t_0} \cap f(\mathcal{B})$ consists of four strips. Indeed, there are exactly two points $t_1' = \frac{t_0}{2}$, $t_1'' = \frac{t_0+1}{2} \in S^1$ such that $t_0 = E_2(t_1')$ and $t_0 = E_2(t_1'')$ and $D_{t_0} \cap f(D_\mu) = \emptyset$ for every $\mu \ne t_1', t_1''$, $0 \le \mu < 1$. Then

$$D_{t_0} \bigcap f(\mathcal{B}) = D_{t_0} \bigcap \left(f(D_{t_1'}) \cup f(D_{t_1''})\right) = \left(D_{t_0} \cap f(D_{t_1'})\right) \bigcup \left(D_{t_0} \cap f(D_{t_1''})\right).$$

Each intersection $D_{t_0} \cap f(D_{t_1'})$, $D_{t_0} \cap f(D_{t_1''})$ consists of two strips. Notice that $D_{t_0} \cap f(D_{t_1'}) = R_0(t_0) \cup R_1(t_0)$. Then by the definition of f

$$f(R_i(t_0)) = R_i(E_2(t_0)), \quad i = 1, 2. \tag{4}$$

Applying (3) with t_0 changed to $E_2(t_0)$ and (4) we get

$$R_0^{(-1)}(t_0) \cap R_1^{(-1)}(t_0) = f^{-1}\left(R_0(E_2(t_0)) \cup R_1(E_2(t_0))\right).$$

Therefore the intersection $f^{-1}(\mathcal{B}) \cap D_{t_0}$ consists of two horizontal strips $R_0^{(-1)}(t_0) \cap R_1^{(-1)}(t_0)$. Then the intersection $f^{-1}(\mathcal{B}) \cap D_{t_0} \cap f(\mathcal{B})$ consists of 8 rectangles which we say to be rectangles of the first degree. They are the intersection of four vertical strips and two horizontal strips, see Fig. 4 (a).

Similar to the standard process of coding for the classic Smale horseshoe we code the rectangles of first degree in the following way. Recall that $D_{t_0} \cap f(D_{t_1'})$ is two vertical strips and each strip divides the disk. We assign "0" to the strip of $D_{t_0} \cap f(D_{t_1'})$ which is the closest to the coordinate origin and we assign "+1" to the other strip. In the same way we assign "0" and "+1" to the two strips of the intersection $D_{t_0} \cap f(D_{t_1''})$. Notice that $D_{t_0} \cap f(D_{t_1'}) = R_0(t_0) \cup R_1(t_0)$ and we have assigned "0" to the strip $R_0(t_0)$ and we have assigned "+1" to the strip $R_1(t_0)$. We assign $\omega_0' = 0$ and $\omega_0'' = 1$ to the respective horizontal strips $R_0^{(-1)}$ and $R_1^{(-1)}$. Now each rectangle of the first degree has the corresponding block $[(t_0, \omega_0); (t_1, \omega_1)]$ where $t_0 = E_2(t_1)$, $\omega_0 \in \{0; 1\}$, $\omega_1 \in \{0; 1\}$. Denote by $V^{(1)}[(t_0, \omega_0); (t_1, \omega_1)]$ the rectangle with this block. Since

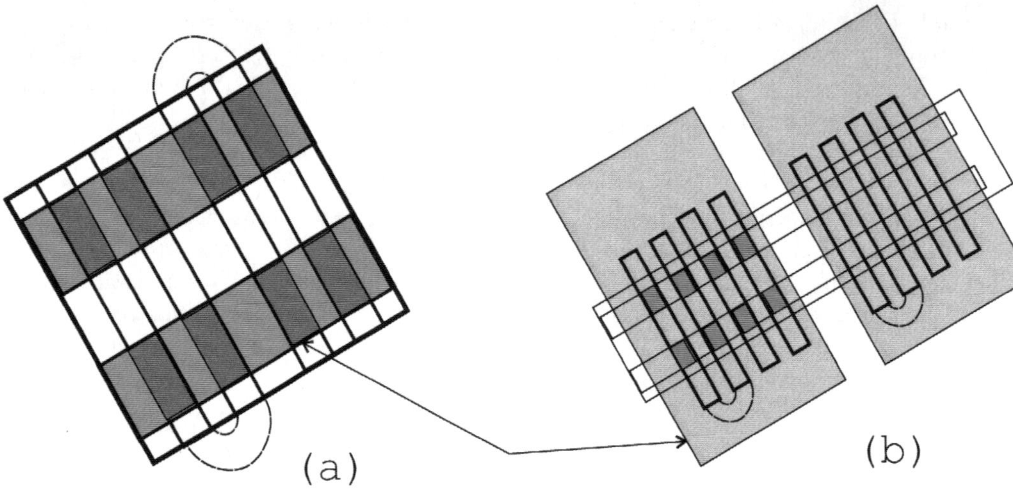

Figure 4. Rectangles of the first degree (a), rectangles of the second degree (b).

the rectangles of the first degree are pairwise disjoint it follows from (2) that an arbitrary point $P \in \Omega \cap D_{t_0}$ belongs to exactly one octagon and $P \in V^{(1)}[(t_0,\omega_0);(t_1,\omega_1)]$ has been assigned the (initial) block $[(t_0,\omega_0);(t_1,\omega_1)]$ of the first degree.

Analogously the intersection

$$f^{-2}(\mathcal{B}) \cap f^{-1}(\mathcal{B}) \cap D_{t_0} \cap f(\mathcal{B}) \cap f^2(\mathcal{B}) = f^{-1}\left[f^{-1}(\mathcal{B}) \cap \mathcal{B}\right] \cap D_{t_0} \cap f\left[\mathcal{B} \cap f(\mathcal{B})\right]$$

consists of 8^2 rectangles which we call the rectangles of the second degree, see Fig. 4 (b). It is easy to see that each rectangle of the first degree contains 8 disjoint rectangles of the second degree. Using the rectangle of the first degree $V^{(1)}[(t_0,\omega_0);(t_1,\omega_1)]$ instead of the disk we analogously assign the block $V^{(2)}[(t_{-1},\omega_{-1});(t_0,\omega_0);(t_1,\omega_1);(t_2,\omega_2)]$ to the rectangle of the second degree $V^{(1)}[(t_0,\omega_0);(t_1,\omega_1)]$, where $t_{-1} = E_2(t_0), t_1 = E_2(t_2), \omega_j \in \{0;1\}, j = -1,0,1,2$. Since a point $P \in \Omega \cap D_{t_0}$ is contained in exactly one octagon of the second degree, say $V^{(2)}[(t_{-1},\omega_{-1});(t_0,\omega_0);(t_1,\omega_1);(t_2,\omega_2)]$, the unique block of the second degree $[(t_{-1},\omega_{-1});(t_0,\omega_0);(t_1,\omega_1);(t_2,\omega_2)]$ is assigned to it. If we continue this procedure then for each point $P \in \Omega \cap D_{t_0}$ we get the bilateral sequence

$$\widehat{P} \stackrel{\text{def}}{=} [\cdots(t_{-n},\omega_{-n});\cdots;(t_{-1},\omega_{-1});\underbrace{(t_0,\omega_0)};(t_1,\omega_1);(t_2,\omega_2);\cdots;(t_n,\omega_n);\cdots],$$

where $\omega_j \in \{0;1\}$, $j \in \mathbb{Z}$, $E_2(t_{i+1}) = t_i$, $i \in \mathbb{Z}$. The underlined pair is conventionally assumed to be at the position 0.

Let $\Sigma_2(E_2)$ denote the set of all sequences of the type

$$[\cdots(t_{-n},\omega_{-n});\cdots;(t_{-1},\omega_{-1});\underbrace{(t_0,\omega_0)};(t_1,\omega_1);(t_2,\omega_2);\cdots;(t_n,\omega_n);\cdots],$$

where $\omega_j \in \{0;1\}$, $j \in \mathbb{Z}$, $E_2(t_{i+1}) = t_i$, $i \in \mathbb{Z}$. Fix a sequence $\widehat{P}^{(0)} \in \Sigma_2(E_2)$, $\widehat{P}^{(0)} = \{(t_i^{(0)},\omega_i^{(0)}\}_{i=-\infty}^{\infty}$ and fix numbers $r \in \mathbb{N}$, $\varepsilon > 0$. A (r,ε)-neighborhood $U_{r,\varepsilon}\left(\widehat{P}^{(0)}\right)$ of the sequence $\widehat{P}^{(0)}$ is the set of sequences $\widehat{P} \in \Sigma_2(E_2)$, $\widehat{P} = \{(t_i,\omega_i)\}_{i=-\infty}^{\infty}$ which satisfy

$$|t_i^{(0)} - t_i| < \varepsilon \text{ for all } -r \le i \le r, \qquad \sum_{i=-\infty}^{\infty} \frac{|\omega_i^{(0)} - \omega_i|}{2^i} < \varepsilon.$$

The set of (r, ε)-neighborhoods generates the topology on $\Sigma_2(E_2)$. Let $\sigma : \Sigma_2(E_2) \to \Sigma_2(E_2)$ denote the map

$$\sigma\left([\cdots ; (t_{-1}, \omega_{-1}); \underbrace{(t_0, \omega_0)}; (t_1, \omega_1); \cdots]\right) = [\cdots ; \underbrace{(t_{-1}, \omega_{-1})}; (t_0, \omega_0); (t_1, \omega_1); \cdots].$$

Then one proves in the standard way that σ is a homeomorphism.

Lemma 4 *The homeomorphism $\sigma : \Sigma_2(E_2) \to \Sigma_2(E_2)$ is transitive and its set of periodic points is dense.*

Proof The proof is just the compilation of the well known proofs of the similar statements for the classic Smale horseshoe and the Smale solenoid [13, 11, 9], therefore we omit it.□

Theorem 1 *The restriction $f|_\Omega : \Omega \to \Omega$ of the diffeomorphism f to Ω is conjugate to $\sigma : \Sigma_2(E_2) \to \Sigma_2(E_2)$.*

Proof Denote by $\vartheta : \Omega \to \Sigma_2(E_2)$ the map which assigns to a point $P \in \Omega$ its code $\widehat{P} \in \Sigma_2(E_2)$. Since the rectangles of any fixed degree are pairwise disjoint, the map ϑ is well defined and it is single-valued, i.e. if $P_1 \neq P_2$ then $\vartheta(P_1) \neq \vartheta(P_2)$. It is clear that the rectangles of the fixed degree continuously depend (in the Hausdorff topology in the space of compact sets) on the parameter t of the square D_t, therefore ϑ is continuous. Since the space $\Sigma_2(E_2)$ is compact, ϑ is a homeomorphism. It is immediate from the definition of the coding that the diagram

$$\begin{array}{ccc} \Omega & \xrightarrow{f|_\Omega} & \Omega \\ \downarrow \vartheta & & \downarrow \vartheta \\ \Sigma_2(E_2) & \xrightarrow{\sigma} & \Sigma_2(E_2) \end{array}$$

is commutative. This means that the maps $f|_\Omega$ and σ are conjugate.□

Corollary 1 *The homeomorphism $f|_\Omega : \Omega \to \Omega$ is transitive and its set of periodic points is dense.*

Consider on $S^1 \times B^2$ a magnetic field $\vec{\boldsymbol{B}}$ of unit vectors tangent to the curves $S^1 \times \{z\}$, $z \in B^2$. One assumes the curves $S^1 \times \{z\}$ to be oriented in the direction of the parameter increase. It is clear that $\vec{\boldsymbol{B}}$ can be extended to the unit vector field (therefore, divergence-free) of the entire sphere S^3. The following result shows that the energy of the magnetic field $\vec{\boldsymbol{B}}$ grows exponentially with the exponent $\mu > 0$.

Theorem 2 *The diffeomorphism $f : S^3 \to S^3$ is fast nondissipative kinematic dynamo with respect to the magnetic field $\vec{\boldsymbol{B}}$.*

Proof Since f stretches the length of the curves $S^1 \times \{z\}$ two-fold, f transforms the field $\vec{\boldsymbol{B}}$ to the field $f_*(\vec{\boldsymbol{B}})$ with the following property: there is a constant $\lambda > 1$ such that the vectors of $f_*(\vec{\boldsymbol{B}})$ are longer by at least factor λ then those of the field $\vec{\boldsymbol{B}}$. The same holds for the lengths of the vectors of the field $f_*^{n+1}(\vec{\boldsymbol{B}})$ with respect to the filed $f_*^n(\vec{\boldsymbol{B}})$. If we ignore the energy dissipation then the energy of the vector field $f_*^n(\vec{\boldsymbol{B}})$ grows exponentially with the exponent $\ln \lambda > 0$. Notice that this result also follows from Lemma 3 and the more general result of the paper [8], in which it is shown that a typical magnetic field is a nondissipative fast dynamo if and only if the diffeomorphism f has a nonzero topological entropy.□

Acknowledgments

This work is supported by the Russian Science Foundation under grant 17-11-01041, except for the proof of Lemma 2 (the extention of the diffeomorphism to a conservative diffeomorphism). The latter proof was obtained within the HSE Basic Research Program (project no. 90) in 2017.

References

[1] Arnold V and Khesin B 1999 *Topological methods in hydrodynamics* (Springer)
[2] Vainstein S and Zeldovich Y 1972 *Uspekhi Phys. Nauk (Russian)* **106** 431–457
[3] Moffatt H 1978 *Magnetic Field Generation in Electrically Conducting Fields* (Cambridge University Press)
[4] Landau L D and Lifshitz E M 1984 *Course of theoretical physics, Vol. 8. Electrodynamics of Continuous Media* (Butterworth-Heinemann)
[5] Alfvén H and Fälthammar C 1963 *Cosmical Electrodynamics: Fundamental Principles* (Oxford)
[6] Childress S and Gilbert A 1995 *Stretch, Twist, Fold: the Fast Dynamo* (Springer-Verlag, Berlin, Heidelberg, NY)
[7] Alfvén H 1943 *Ark. for Astronomi* **29** 1–17
[8] Klapper I and Young L 1995 *Comm. Math. Phys.* **173** 623–646
[9] Smale S 1967 *Bull. Amer. Math. Soc.* **73** 747–817
[10] Katok A and Hasselblatt B 1995 *Introduction to the modern theory of dynamical systems* (Cambridge Univ. Press)
[11] Smale S 1965 Diffeomorphisms with many periodic points *Differential and Combinatorial Topology* (Princeton, NJ) pp 63–80
[12] Anosov D and Solodov V 1995 Hyperbolic sets *Itogi Nauki Tekhniki; Sovremennye Problemy Matematiki, Fundamental'nye Napravleniya, Dynamical Systems 9, VINITI, Akad. Nauk SSSR, Vol. 66, Moscow, (1991), p.12-99 (Russian), English translation in Encyclopaedia of Mathematical Sciences, Dynamical Systems IX* vol 66 (Springer-Verlag, Berlin-Heidelberg)
[13] Robinson C 1999 *Dynamical systems: stability, symbolic dynamics, and chaos* vol 28 (CRC press Boca Raton, FL)
[14] Zhuzhoma E and Isaenkova N 2011 *Sb. Math.* **202** 351–372 ISSN 1064-5616; 1468-4802/e
[15] Shub M 1969 *Amer. J. Math.* **91** 175–199
[16] Jiang B, Ni Y and Wang S 2004 *Trans. Amer. Math. Soc.* **356** 4371–4382 ISSN 00029947 URL `http://www.jstor.org/stable/3844925`
[17] Bowen R 1970 Topological entropy and axiom A *Proc. Sympos. Pure Math* vol 14 pp 23–41

On existence of a smooth arc without bifurcations joining source-sink diffeomorphisms on 2-sphere

E. V. Nozdrinova, O. V. Pochinka

HSE, B. Pecherskaya 25, 603155 Nizhny Novgorod, Russia

E-mail: maati@mail.ru, olga-pochinka@yandex.ru

Abstract. In this paper we construct a smooth arc without bifurcation points joining source-sink diffeomorphisms on the two-dimensional sphere.

1. Introduction and statement of results

One of the fifty problems of dynamical systems compiled by J. Palis and C. Pugh is the problem of the existence of an arc with a finite or countable set of bifurcations joining two Morse-Smale systems on a smooth closed manifold [16]. In [14] Sh. Newhouse and M. Peixoto proved that every Morse-Smale vector fields can be joined by *a simple arc*. Simplicity means that the arc consists of Morse-Smale systems with the exception of a finite set of points where the vector field has the least deviation from a Morse-Smale system. For discrete dynamical systems the situation is different. Two orientation preserving Morse-Smale diffeomorphisms on a circle can be joined by a simple arc if and only if they have the same rotation numbers. As follows from the works by Sh. Matsumoto [11] and by P. Blanchard [2], any orientable closed surface admits isotopic Morse-Smale diffeomorphisms that can not be joined by a simple arc. It is said that two isotopic Morse-Smale diffeomorphisms belong to the same *simple isotopic class* if they can be joined by a simple arc. It follows from the papers [11, 2] that there exist infinitely many simple isotopy classes of Morse-Smale diffeomorphisms on any orientable surface inside an isotopy class that admits Morse-Smale diffeomorphisms.

In dimension 3, the problem of the existence of a simple arc is complicated by the presence of Morse-Smale diffeomorphisms with wildly embedded separatrices. The first "wild" example was constructed by D. Pixton [17]. This diffeomorphism belongs to the class (called in [7] *Pixton class*) consisting of three-dimensional Morse-Smale diffeomorphisms for which the non-wandering set consists of exactly four points: two sinks, a source and a saddle. As follows from the work by Ch. Bonatti, V. Grines, V. Medvedev, O. Pochinka [5], any Pixton diffeomorphism is connected by a simple arc with some source-sink diffeomorphism. This effect is due to the fact that for any diffeomorphism from the Pixton class at least one one-dimensional separatrix of the saddle point is tame [4]. Using the operation of a taking the connected sum of two 3-spheres on which Pixton diffeomorphisms with wildly embedded separatrices are given, it is easy to construct a diffeomorphism for which all the separatrices of all saddles are wildly imbedded. In the paper by V. Grines and O. Pochinka [8] it was proved that such a diffeomorphism is not joined by a simple arc with a source-sink diffeomorphism.

Gradient-like diffeomorphisms on surfaces, in contrast to the three-dimensional case, do not reveal the wild behavior of separatrices, which creates the prerequisites for an exhaustive solution of the Palis-Pugh problem. The first steps in this direction have been taken in this paper. In more detail.

Consider 2-sphere $\mathbb{S}^2 = \{(x_1, x_2, x_3) \in \mathbb{R}^3 : x_1^2 + x_2^2 + x_3^2 = 1\}$. We denote by $Diff(\mathbb{S}^2)$ the set of diffeomorphisms on 2-sphere, that is the smoothness of considered diffeomorphism more or equal than 1. Denote by $J(\mathbb{S}^2) \subset Diff(\mathbb{S}^2)$ set of orientation preserving *source-sink diffeomorphisms*, that is, diffeomorphisms whose non-wandering sets consist of precisely two hyperbolic points: a source, a sink. The existence of a smooth arc joining any orientation-preserving diffeomorphisms and, therefore, any source-sink diffeomorphisms of $\mathbb{S}^2$ is a classical result by S. Smale [18]. We show that this arc can be chosen in such a way that it consists of source-sink diffeomorphisms only.

Theorem 1. *For any diffeomorphisms $f, f' \in J(\mathbb{S}^2)$ there is a smooth arc $\{f_t \in J(\mathbb{S}^2)\}$ their joining.*

Thus, all orientation preserving source-sink diffeomorphisms on 2-sphere belong to the same simply connected component. According to [13] the boundary of this component consists of systems with the following properties: (1) all stable, strong stable, unstable, and strong unstable manifolds intersect transversally; (2) it has no cycles and has exactly one nonhyperbolic periodic orbit which is either a flip or a noncritical saddle-node.

2. Denotations

Let M be a smooth orientable manifold. Two diffeomorphisms $f, f' : M \to M$ are smoothly isotopic if there is a smooth map $F : M \times [0, 1] \to M$ (smooth isotopy) such that f_t given by the formula $f_t(x) = f(x, t)$ is a diffeomorphism for each $t \in [0, 1]$ and $f_0 = f, f_1 = f'$. We say that the family $\{f_t\}$ is a smooth arc joining diffeomorphisms f and f'. The support supp $\{f_t\}$ of the isotopy $\{f_t\}$ is the closure of the set $\{x \in M : f_t(x) \neq f_0(x)$ for some $t \in [0, 1]\}$. Let $Diff(M)$ be the space of the diffeomorphisms of M with C^1-topology. Notice that the existence of the smooth isotopy $F : M \times [0, 1] \to M$ implies the existence of the continuous path $\gamma : [0, 1] \in Diff(M)$ such that $\gamma(t) = f_t$, however the opposite is not true.

For the smooth arcs $\{f_t\}$ and $\{g_t\}$ such that $f_1 = g_0$ the usually product of the respective paths is not smooth in general, but we can define the smooth arc $\{h_t\}$ joining diffeomorphisms f_0, g_1 as follows:

$$h_t = \begin{cases} f_{2\tau(t)}, 0 \leq t \leq \frac{1}{2} \\ g_{2\tau(t)-1}, \frac{1}{2} \leq t \leq 1 \end{cases},$$

where $\tau : [0, 1] \to [0, 1]$ is a smooth monotone function such that $\tau(t) = 0$ for $0 \leq t \leq \frac{1}{3}$ and $\tau(t) = 1$ for $\frac{2}{3} \leq t \leq 1$ (see, for example, [12], Lemma 1.8). The smooth arc $\{h_t\}$ we will called a smooth product of the smooth arcs $\{f_t\}$ and $\{g_t\}$ and denoted $\{h_t\} = \{f_t * g_t\}$.

In the present paper we will work with the class $J(\mathbb{S}^2) \subset Diff(\mathbb{S}^2)$ of orientation preserving source-sink diffeomorphisms, that is, diffeomorphisms whose non-wandering sets consist of precisely two hyperbolic points: a source, a sink. Also we introduce a subclass $NS(\mathbb{S}^2) \subset J(\mathbb{S}^2)$ for which the source is the North pole $N(0, 0, 1)$ and the sink is the South pole $S(0, 0, -1)$.

Define the diffeomorphisms

$$\vartheta_- : \mathbb{S}^2 \setminus \{S\} \to \mathbb{R}^2,$$

$$\vartheta_+ : \mathbb{S}^2 \setminus \{N\} \to \mathbb{R}^2$$

(the stereographic projections) by the formulas:

$$\vartheta_-(x_1, x_2, x_3) = \left(\frac{x_1}{1 + x_3}, \frac{x_2}{1 + x_3}\right),$$

$$\vartheta_+(x_1,x_2,x_3)=\left(\frac{x_1}{1-x_3},\frac{x_2}{1-x_3}\right).$$

Then the inverse maps

$$\vartheta_-^{-1}:\mathbb{R}^2\to\mathbb{S}^2\setminus\{S\},$$
$$\vartheta_+^{-1}:\mathbb{R}^2\to\mathbb{S}^2\setminus\{N\}$$

given by the formulas:

$$\vartheta_-^{-1}(x_1,x_2)=\left(\frac{2x_1}{x_1^2+x_2^2+1},\frac{2x_2}{x_1^2+x_2^2+1},\frac{1-(x_1^2+x_2^2)}{x_1^2+x_2^2+1}\right),$$

$$\vartheta_+^{-1}(x_1,x_2)=\left(\frac{2x_1}{x_1^2+x_2^2+1},\frac{2x_2}{x_1^2+x_2^2+1},\frac{x_1^2+x_2^2-1}{x_1^2+x_2^2+1}\right).$$

Through the stereographic projections the diffeomorphisms from $NS(\mathbb{S}^2)$ connected with the contractions

$$S(\mathbb{R}^2)=\{\vartheta_+f\vartheta_+^{-1},f\in NS(\mathbb{S}^2)\}$$

and extension

$$N(\mathbb{R}^2)=\{\vartheta_-f\vartheta_-^{-1},f\in NS(\mathbb{S}^2)\}.$$

Let us define the model diffeomorphisms $g:\mathbb{S}^2\to\mathbb{S}^2$, $\bar g:\mathbb{R}^2\to\mathbb{R}^2$ by the formulas

$$g(x_1,x_2,x_3)=\left(\frac{4x_1}{5-3x_3},\frac{4x_2}{3x_3-5},\frac{5x_3-3}{5-3x_3}\right),$$

$$\bar g(x_1,x_2)=\left(\frac{x_1}{2},\frac{x_2}{2}\right).$$

It is easy to check that $g\in NS(\mathbb{S}^2)$, $\bar g\in S(\mathbb{R}^2)$, $\bar g^{-1}\in N(\mathbb{R}^2)$, $\bar g^{-1}=\vartheta_-g\vartheta_-^{-1}$ and $\bar g=\vartheta_+g\vartheta_+^{-1}$.

Denote by $E_g\subset NS(\mathbb{S}^2)$ the set of diffeomorphisms such that for each diffeomorphism $h\in E_g$ there are neighborhoods $V_h(N)$, $V_h(S)$ of the points N, S, where $h|_{V_h(N)\cup V_h(S)}=g|_{V_h(N)\cup V_h(S)}$. Denote by $E_{\bar g}\subset S(\mathbb{R}^2)$ ($E_{\bar g^{-1}}\subset N(\mathbb{R}^2)$) the set of diffeomorphisms such that for each diffeomorphism $\bar h\in E_{\bar g}$ ($\bar h\in E_{\bar g^{-1}}$) there is a neighborhood $V_{\bar h}(O)$ of the point O, where $\bar h|_{V_{\bar h}(O)}=\bar g|_{V_{\bar h}(O)}$ ($\bar h|_{V_{\bar h}(O)}=\bar g^{-1}|_{V_{\bar h}(O)}$).

3. Construction of a smooth arc

It is easy to prove Theorem 1 if f and f' are smoothly conjugated by a diffeomorphism h. In this case, due to [18], there is a smooth isotopy $h_t:\mathbb{S}^2\to\mathbb{S}^2$ such that $h_0=id$, $h_1=h$. Thus $f_t=h_t^{-1}fh_t$ is a required arc. In the opposite case we show how to construct a smooth arc $\{l_t\in J(\mathbb{S}^2)\}$ joining an arbitrary diffeomorphism $f\in J(\mathbb{S}^2)$ with the model diffeomorphism g. Similarly, we construct a smooth arc $\{l'_t\in J(\mathbb{S}^2)\}$ that connects f' and g. Then the smooth product of smooth arcs $f_t=l_t*l'_{1-t}$ is the required arc. The arc l_t is a smooth product of tree smooth arcs $\beta_t,\gamma_t,\delta_t$ ($l_t=\beta_t*\gamma_t*\delta_t$), whose construction is given in the corresponding sentences below:

(i) l_t is a diffeomorphism from the class $J(\mathbb{S}^2)$ for all $t\in[0,\frac{1}{3}]$, $l_0=f$ and $l_{\frac{1}{3}}$ is a diffeomorphism from the class $NS(\mathbb{S}^2)$ (see Proposition 1);

(ii) l_t is a diffeomorphism from the class $NS(\mathbb{S}^2)$ for all $t\in[\frac{1}{3},\frac{2}{3}]$, $l_{\frac{2}{3}}\in E_g$ (see Proposition 2);

(iii) l_t is a diffeomorphism from the class $J(\mathbb{S}^2)$ for all $t\in[\frac{2}{3},1]$, $l_1=g$ (see Proposition 3).

Proposition 1. *For any diffeomorphism $f \in J(\mathbb{S}^2)$ there exists a smooth arc $\{\beta_t \in (J(\mathbb{S}^2) \cap Diff(\mathbb{S}^2))\}$ joining $\beta_0 = f$ with a diffeomorphism $\beta_1 \in NS(\mathbb{S}^2)$.*

Proof. Let $f \in J(\mathbb{S}^2)$. We denote by α the source and through the ω sink of the diffeomorphism f. Let D_α, D_ω (D_S, D_N) be pairwise disjoint 2-disks containing α, ω (S, N). According to [9], there exists a smooth arc $\{H_t \in Diff(\mathbb{S}^2)\}$ with the following properties $H_0 = id$, $H_1(D_N) = D_\alpha$, $H_1(D_S) = D_\omega$, $H_1(N) = \alpha$ and $H_1(S) = \omega$. Then $\beta_t = H_t^{-1} f H_t$ is the desired isotopy joining the diffeomorphism $f = \beta_0$ with the diffeomorphism $\beta_1 = H_1^{-1} f H_1 \in NS(\mathbb{S}^2)$. □

Proposition 2. *For any diffeomorphism $\beta \in NS(\mathbb{S}^2)$ there exists a smooth arc $\{\gamma_t \in NS(\mathbb{S}^2)\}$ joining $\gamma_0 = \beta$ with a diffeomorphism $\gamma_1 \in E_g$.*

Proof. Define $\bar\beta^+ = \vartheta_+ \beta \vartheta_+^{-1}$, $\bar\beta^- = \vartheta_- \beta \vartheta_-^{-1}$. Then $\bar\beta^+ \in S(\mathbb{R}^2)$, $\bar\beta^- \in N(\mathbb{R}^2)$. If $\bar\beta^+ = \bar g$ in some neighborhood of the origin O, then we define $\{\bar\gamma_t^+ = \bar\beta^+\}$. In the opposite case we show below how to construct a smooth arc $\{\bar\gamma_t^+ \in S(\mathbb{R}^2)\}$ joining $\bar\beta^+$ with a diffeomorphism $\bar\gamma_1^+ \in E_{\bar g}$ such that $\bar\gamma_t^+$ for every $t \in [0,1]$ coincides with $\bar\beta^+$ out of a neighborhood $V^+ \subset \mathbb{B}^2 = \{(x_1, x_2) \in \mathbb{R}^2 : x_1^2 + x_2^2 \le 1\}$ of the origin O. Similarly if $\bar\beta^- = \bar g^{-1}$ in some neighborhood of the origin O, then we define $\{\bar\gamma_t^- = \bar\beta^-\}$. In the opposite case similarly to $\bar\gamma_t^+$ we construct a smooth arc $\{\bar\gamma_t^- \in N(\mathbb{R}^2)\}$ joining $\bar\beta^-$ with a diffeomorphism $\bar\gamma_1^- \in E_{\bar g^{-1}}$ such that $\bar\gamma_t^-$ for every $t \in [0,1]$ coincides with $\bar\beta^-$ out of a neighborhood $V^- \subset \mathbb{B}^2$ of the origin O. Thus the required arc γ_t given by the formula:

$$\gamma_t(w) = \begin{cases} \vartheta_+^{-1}(\gamma_t^+(\vartheta_+(w))), w \in \vartheta_+^{-1}(V^+); \\ \vartheta_-^{-1}(\gamma_t^-(\vartheta_-(w))), w \in \vartheta_-^{-1}(V^-); \\ \beta(w), w \in (\mathbb{S}^2 \setminus (\vartheta_+^{-1}(V^+) \cup \vartheta_-^{-1}(V^-)). \end{cases}$$

Let us construct a smooth arc $\{\bar\gamma_t^+ \in S(\mathbb{R}^2)\}$ joining $\bar\beta^+$ with a diffeomorphism $\bar\gamma_1^+ \in E_{\bar g}$ such that $\bar\gamma_t^+$ for every $t \in [0,1]$ coincides with $\bar\beta^+$ out of a neighborhood $V^+ \subset \mathbb{B}^2$ (the construction of the arc $\bar\gamma_t^-$ is similar). The arc $\bar\gamma_t^+$ will be a smooth product of arcs $\bar\rho_t^+$, $\bar\eta_t^+$ and $\bar\xi_t^+$, where

1) $\{\bar\rho_t^+ \in S(\mathbb{R}^2)\}$ joining $\bar\beta^+$ with a diffeomorphism $\bar\rho_1^+$ such that $\bar\rho_t^+$ for every $t \in [0,1]$ coincides with $\bar\beta^+$ out of a neighborhood $V_1^+ \subset \mathbb{B}^2$ of the origin O and $\bar\rho_1^+$ coincides with the differential $D\bar\beta_O^+$ (denote it by Q) in a neighborhood $U_1^+ \subset V_1^+$ of the origin O;

2) $\{\bar\eta_t^+ \in S(\mathbb{R}^2)\}$ joining $\bar\rho_1^+$ with a diffeomorphism $\bar\eta_1^+$ such that $\bar\varrho_t^+$ for every $t \in [0,1]$ coincides with $\bar\rho_1^+$ out of a neighborhood $V_2^+ \subset V_1^+$ of the origin O and $\bar\eta_1^+$ coincides with a diffeomorphism G given by the normal Jordan form of the diffeomorphism Q in a neighborhood $U_2^+ \subset V_2^+$ of the origin O;

3) $\{\bar\xi_t^+ \in S(\mathbb{R}^2)\}$ joining $\bar\eta_1^+$ with a diffeomorphism $\bar\xi_1^+ \in E_{\bar g}$ such that $\bar\xi_t^+$ for every $t \in [0,1]$ coincides with $\bar\eta_1^+$ out of a neighborhood $V_3^+ \subset V_2^+$.

1) As O is a hyperbolic sink for $\bar\beta_t^+$ then there is a metric $\|\cdot\|$ on $\mathbb{R}^2$ such that for some λ, $0 < \lambda < 1$, we have:

$$\|Q(v)\| \le \lambda \|v\|,$$

for each v from the tangent space $T_O\mathbb{R}^2 = \mathbb{R}^2$ (see, for example, [10]). Identify a small neighborhood $U_O \subset \bar\beta^+(\mathbb{B}^2)$ of O with a neighborhood of the zero-section of $T_O\mathbb{R}^2$ by the exponential map $e : U_O \to T_O\mathbb{R}^2$. Let $\tilde U = e(U_O)$ and define $\tilde\beta^+$ on $\tilde U$ by the formula $\tilde\beta^+ = e^{-1}\bar\beta^+ e$. For every $v \in \tilde U$ we have that $\tilde\beta^+(v) = Q(v) + \|v\| a(v)$, where $a(v)$ goes to 0 as $\|v\|$ goes to 0. As $\lambda < 1$ and $a(v)$ goes to 0 as $\|v\|$ goes to 0 then there is a number $\ell > 0$ such that in the neighborhood $\tilde V = \{(x,y) \in T_O\mathbb{R}^2 : x^2 + y^2 \le \ell^2\} \subset \tilde U$ of O for all non-zero $v \in \tilde V$ the following inequality holds $\|\tilde\beta^+(v)\| < \|v\|$. Let us define on $\tilde V$ an arc $\tilde\chi_t^+ : T_O\mathbb{R}^2 \to T_O\mathbb{R}^2$ by the formula

$$\tilde\chi_t^+ = (1-t)\tilde\beta^+ + tQ.$$

Then $\|\tilde{\chi}_t^+(v)\| < \|v\|$ for all non-zero $v \in \tilde{V}$. Let $U_+^1 = e^{-1}(\tilde{V})$ and $\bar{\chi}_t^+ = e^{-1}\tilde{\chi}_t^+ e$. Notice that the origin O is a fixed point for every diffeomorphism $\bar{\chi}_t^+$ and $\bar{\chi}_t^+(U_1^+) \subset int\ U_1^+$ for every $t \in [0,1]$. Let $V_+^1 = (\bar{\beta}^+)^{-1}(U_1^+)$. Let us consider the isotopy $h_t = (\bar{\beta}^+)^{-1}\bar{\chi}_t^+$ which joints the identity map with the diffeomorphism $(\bar{\beta}^+)^{-1}Q$. By the construction $h_t(U_1^+) \subset V_1^+$ for every $t \in [0,1]$. Then, using Thom's result [12, Theorem 5.8] about the extension of an isotopy there is an isotopy $H_t : \mathbb{R}^2 \to \mathbb{R}^2$ which is h_t on U_1^+ and is identity out of V_1^+. Then the required arc is given by the formula

$$\bar{\rho}_t^+ = \bar{\beta}^+ H_t.$$

2) If $Q = \bar{g}$ then let us define $\{\bar{\eta}_t^+ = \bar{\xi}_t^+ = \bar{\rho}_1^+\}$. In the opposite case for a construction of an arc $\{\bar{\eta}_t^+\}$ let us recall that the diffeomorphism Q is smoothly conjugated to a diffeomorphism G given by the normal Jordan form of Q (see, for example, [6] [Chapter 3]). Then there is a preserving orientation diffeomorphism $h : \mathbb{R}^2 \to \mathbb{R}^2$ such that $G = hQh^{-1}$. Due to [15][Proposition 5.4], h is isotopic to identity by means an isotopy h_t with $h_0 = id$ and $h_1 = h$. Let us choose $r_2 \in (0,1)$ such that $h_t(U_2^+) \subset U_1^+$ for the disc $U_2^+ = \{(x_1,x_2) \in \mathbb{R}^2 : x_1^2 + x_2^2 \le r_2\}$ and every $t \in [0,1]$. Using Thom's result [12] about the extension of an isotopy there is an isotopy $H_t : \mathbb{R}^2 \to \mathbb{R}^2$ which is h_t on U_2^+ and is identity out of $V_2^+ = U_1^+$. Thus, the required arc is given by the formula

$$\bar{\eta}_t^+ = H_t \bar{\rho}_1^+ H_t^{-1}.$$

3) If $G = \bar{g}$ then let us define $\{\bar{\xi}_t^+ = \bar{\eta}_1^+\}$. In the opposite case, without loss of generality we will assume that the eigenvalues of Q are different (in the opposite case we can a little bit extend the arc $\bar{\rho}_t^+$ to get $\bar{\rho}_1^+$ with the needed property). Then the diffeomorphism G is given by either a matrix A, or a matrix B of the following forms:

$A = \begin{pmatrix} \lambda_1 & 0 \\ 0 & \lambda_2 \end{pmatrix}$, where $0 < \lambda_1 < \lambda_2 < 1$;

$B = \begin{pmatrix} \alpha & \beta \\ -\beta & \alpha \end{pmatrix}$, where $0 < \alpha^2 + \beta^2 < 1$.

In the both cases we see that G contracts every disc of the form $\{(x_1,x_2) \in \mathbb{R}^2 : x_1^2 + x_2^2 \le r\}$. Let $V_3^+ = U_2^+$ and let us choose $r_3 \in (0, r_2)$ such that the disc $U_3^+ = \{(x_1,x_2) \in \mathbb{R}^2 : x_1^2 + x_2^2 \le r_3\}$ contains $G(V_3^+)$. Let us define an arc $\bar{\tau}_t^+ : \mathbb{R}^2 \to \mathbb{R}^2$ by the formula

$$\bar{\tau}_t^+ = (1-t)G + tg.$$

Let us consider the isotopy $\lambda_t = G^{-1}\bar{\tau}_t^+$ which joints the identity map with the diffeomorphism $G^{-1}g$. By the construction $\lambda_t(U_3^+) \subset V_3^+$ for every $t \in [0,1]$. Then, using Thom's result [12, Theorem 5.8] about the extension of an isotopy there is an isotopy $\Lambda_t : \mathbb{R}^2 \to \mathbb{R}^2$ which is λ_t on U_3^+ and is identity out of V_3^+. Then the required arc is given by the formula

$$\bar{\xi}_t^+ = \bar{\eta}_1^+ \Lambda_t.$$

□

Proposition 3. *For any diffeomorphism $\gamma \in E_g$ there is a smooth arc $\{\delta_t \in J(\mathbb{S}^2)\}$ joining $\delta_0 = \gamma$ with $g = \delta_1$.*

Proof. As $\gamma \in E_g$ then there are neighborhoods $V_\gamma(N)$, $V_\gamma(S)$ of the points N, S, where $\gamma|_{V_\gamma(N) \cup V_\gamma(S)} = g|_{V_\gamma(N) \cup V_\gamma(S)}$. Let us define a diffeomorphism $\psi_\gamma : \mathbb{S}^2 \setminus \{S\} \to \mathbb{S}^2 \setminus \{S\}$ by the formula $\psi_\gamma(w) = g^k(\gamma^{-k}(w))$, where $k \in \mathbb{Z}$ such that $\gamma^{-k}(w) \in V_\gamma(N)$ for $w \in \mathbb{S}^2 \setminus \{S\}$. Then $\gamma = \psi_\gamma^{-1} g \psi_\gamma$. If ψ_γ can be smoothly extended to S by $\psi_\gamma(S) = S$ then, due to [18] there is a smooth isotopy $\rho_t : \mathbb{S}^2 \to \mathbb{S}^2$ such that $\rho_0 = \psi$, $\rho_1 = id$. Thus $\delta_t = \rho_t^{-1} g \rho_t$ is a required arc.

In the opposite case we show that there is a smooth arc $\{\zeta_t \in E_g\}$ joining the diffeomorphism $\zeta_0 = \gamma$ with some diffeomorphism ζ_1 such that ψ_{ζ_1} can be smoothly extended to S by $\psi_{\zeta_1}(S) = S$.

Define $\mathbb{B}_r(O) = \{(x_1, x_2) \in \mathbb{R}^2 : x_1^2 + x_2^2 \leq 1\}$ and $K_r = cl(\mathbb{B}_r \setminus \mathbb{B}_{\frac{r}{2}})$ for $r > 0$. Let $\bar{\gamma} = \vartheta_+ \gamma \vartheta_+^{-1}$ and $\bar{\psi}_\gamma = \vartheta_+ \psi_\gamma \vartheta_+^{-1}$. Then $\bar{\gamma} \in E_{\bar{g}}$ and, hence, there is $r_0 > 0$ such that $\bar{\gamma} = \bar{g}$ on $\mathbb{B}_{r_0}$ and the annulus K_{r_0} is a fundamental domain of the diffeomorphism $\bar{g}$ (also $\bar{\gamma}$) on $\mathbb{R}^2 \setminus \{O\}$. Let us represent $\mathbb{T}^2$ as the orbit space $(\mathbb{R}^2 \setminus \{O\})/\bar{g}$. Denote by $p : \mathbb{B}_{r_0} \setminus \{O\} \to \mathbb{T}^2$ the natural projection. Then curves $a = p(Ox_1), b = p(\partial \mathbb{B}_{r_0})$ are generators of the fundamental group $\pi_1(\mathbb{T}^2)$. As $\bar{\psi}_\gamma$ sends the orbits of $\bar{g}$ to the orbits $\bar{\gamma}$ and K_{r_0} is the common fundamental domain for $\bar{g}, \bar{\gamma}$ on $\mathbb{R}^2 \setminus \{O\}$ then $\bar{\psi}_\gamma$ can be projected to $\mathbb{T}^2$ as a diffeomorphism given by the formula $\hat{\psi}_\gamma = p\bar{\psi}_\gamma p^{-1}$. Then, the induced isomorphism $\hat{\psi}_{\gamma *} : \pi_1(\mathbb{T}^2) \to \pi_1(\mathbb{T}^2)$ preserves the homotopic class of the generator a and, hence, given by the matrix

$$\begin{pmatrix} 1 & n_0 \\ 0 & 1 \end{pmatrix}$$

for some integer n_0. Let us introduce the polar coordinates ρ, φ on $\mathbb{R}^2$.

The arc ζ_t will be the smooth product of arcs ν_t and μ_t, where

1) $\{\nu_t \in J(\mathbb{S}^2)\}$ is a smooth arc joining the diffeomorphism $\nu_0 = \gamma$ with some diffeomorphism $\nu_1 \in E_g$ such that $\hat{\psi}_{\nu_1}$ induces the identity action in $\pi_1(\mathbb{T}^2)$;

2) $\{\mu_t \in E_g\}$ is a smooth arc joining the diffeomorphism $\mu_0 = \nu_1$ with a diffeomorphism μ_1 such that $\hat{\psi}_{\mu_1} = id$, it means that $\psi_{\mu_1} = g^k$ for some $k \in \mathbb{Z} \setminus \{0\}$.

1) Define a diffeomorphism $\bar{\theta}_t$ by the formula $\bar{\theta}_t(\rho e^{i\varphi}) = \begin{cases} \rho e^{i\varphi}, \rho > r_0, \\ \rho e^{i\left(\varphi + 4n_0\pi t\left(1 - \frac{\rho}{r_0}\right)\right)}, \frac{r_0}{2} \leq \rho \leq r_0; \\ \rho e^{i(\varphi + 2n_0\pi t)}, \rho < \frac{r_0}{2}. \end{cases}$

Let $\theta_t = \vartheta_+^{-1}\bar{\theta}_t\vartheta_+ : \mathbb{S}^2 \setminus \{S\} \to \mathbb{S}^2 \setminus \{S\}$, then θ_t can be smoothly extended to $\mathbb{S}^2$ by $\theta_t(S) = S$. Moreover, by the construction $\hat{\psi}_{\theta_1\gamma}$ induces the identity action in $\pi_1(\mathbb{T}^2)$. Thus $\nu_t = \theta_t\gamma : \mathbb{S}^2 \to \mathbb{S}^2$ the required arc.

2) Here we have the diffeomorphism $\nu_1 : \mathbb{S}^2 \to \mathbb{S}^2$ such that the diffeomorphism $\hat{\psi}_{\nu_1}$ induces the identity action in $\pi_1(\mathbb{T}^2)$. Thus, by [19, 3], the diffeomorphism $\hat{\psi}_{\nu_1}$ is smoothly isotopic to the identity map. Pick an open cover $U = \{U_1, \dots, U_q\}$ of the manifold $\mathbb{T}^2$ consisting of connected sets such that each connected component of the set $p^{-1}(U_i)$ is a subset of K_{r_i} for some $r_i < \frac{r_{i-1}}{2}$. By [1, Lemma de fragmentation] there are diffeomorphisms $\hat{w}_1, \dots, \hat{w}_q : \mathbb{T}^2 \to \mathbb{T}^2$ such that they are smoothly isotopic to the identity and

i) for each $i = \overline{1,q}$ there is $U_{j(i)} \in U$ such that for every $t \in [0,1]$ the map $\hat{w}_{i,t}$ is the identity out of $U_{j(i)}$ where $\{\hat{w}_{i,t}\}$ is a smooth isotopy between the identity map and $\hat{w}_i$;

ii) $\hat{\psi}_{\nu_1} = \hat{w}_1 \dots \hat{w}_q$.

Let $\bar{w}_{i,t} : \mathbb{R}^2 \to \mathbb{R}^2$ be the diffeomorphism which coincides with $(p|_{K_{r_i}})^{-1}\hat{w}_{i,t}p$ on K_{r_i} and coincides with the identity map outside K_{r_i}. Let $\bar{\mu}_t = \bar{\nu}_1\bar{w}_{1,t}\dots\bar{w}_{q,t} : \mathbb{R}^2 \setminus \{O\} \to \mathbb{R}^2 \setminus \{O\}$. By the construction $\bar{\mu}_t \in E_{\bar{g}}$ for every $t \in [0,1]$ and $\bar{\mu}_0 = \bar{\nu}_1$. Moreover, for the diffeomorphism $\bar{\mu}_1$ the following equality holds: $\hat{\psi}_{\mu_1} = \hat{w}_q^{-1}\dots\hat{w}_1^{-1}\hat{\psi}_{\nu_1} = \hat{w}_q^{-1}\dots\hat{w}_1^{-1}\hat{w}_1\dots\hat{w}_q = id$. Thus $\mu_t = \vartheta_+^{-1}\bar{\mu}_t\vartheta_+ : \mathbb{S}^2 \setminus \{S\} \to \mathbb{S}^2 \setminus \{S\}$ can be smoothly extended to $\mathbb{S}^2$ by $\mu_t(S) = S$ up to the desired arc. □

4. Acknowledgments

The study of a structure of an arc connecting source-sink diffeomorphism with one which locally linear was financial supported by RSF (Grant No. 17-11-01041), the construction of an arc connecting locally linear diffeomorphism with the model was supported by the fundamental research program of the HSE (project 95) in 2018.

IOP Conf. Series: Journal of Physics: Conf. Series **990** (2018) 012010 doi:10.1088/1742-6596/990/1/012010

5. References

[1] Banyaga A 1977 *The structure of the group of equivariant diffeomorphism.* 16 (Topology) pp279–83.

[2] Blanchard P R 1980 *Invariants of the NPT isotopy classes of Morse-Smale diffeomorphisms of surfaces* 47:1 (Duke Math. J.) pp 33-46.

[3] Boldsen 2009 *Different versions of mapping class groups of surfaces.*

[4] Bonatti C Grines V Z 2000 *Knots as topological invariant for gradient-like diffeomorphisms of the 3-sphere* 6:4 (J. Dynam. Control Systems) pp 579–602

[5] Bonatti Ch Grines V Z Medvedev V S and Pochinka O V 2007 *Bifurcations of diffeomorphisms Morse-Smale with wildly embedded separatrices.* Dynamical systems and optimization, Collection of articles. To the 70th anniversary of the birth of Academician Dmitry Viktorovich Anosov, Tr. Steklov Mathematical Institute, 256 (Nauka, Moscow) pp 54-69

[6] Gelfand I M 1971 *Lectures on linear algebra* (M .: Science) p 273

[7] Grines V Medvedev T and Pochinka O 2016 *Dynamical Systems on 2- and 3-Manifolds.* (Springer International Publishing Switzerland) p 364

[8] Grines V Z and Pochinka O V 2013 *On the simple isotopy class of a source-sink diffeomorphism on a 3-sphere* 94 6 (Mathematical Notes) pp 828–45.

[9] Hirsch M 1979 *Differential topology* p 280

[10] Katok A and Hasselblatt B 1994 *Introduction to the Modern Theory of Dynamical Systems,* (Encyclopedia of Math. and its Appl., Cambridge University Press)

[11] Matsumoto S 1979 *There are two isotopic Morse-Smale diffeomorphism which can not be joined by simple arcs* 51:1 (Invent. Math) pp 1-7

[12] Milnor J 1965 *Lectures on the h-cobordism theorem.* (Princeton university press)

[13] S. Newhouse J and Palis F 1983 *Takens. Bifurcation and stability of families of diffeomorphisms.* Vol. 57 (Publ. Math. I.H.E.S.) pp 5–71.

[14] Newhouse S and Peixoto M 1976 *There is a simple arc joining any two Morse-Smale flows, Trois etudes en dynamique qualitative, Asterisque* 31 (Soc. Math. France, Paris) pp 15-41

[15] Palis J and de Melo W 1986 *Geometricheskaya teoriya dinamicheskikh sistem* [Geometric Theory of Dynamical Systems]. (An Introduction M. World) p 301

[16] Palis J. Pugh C. C. *Fifty problems in dynamical systems, Dynamical Systems-Warwick.* 1974, Lecture Notes in Math., 468, Springer-Verlag, Berlin, 1975,pp 345–53

[17] Pixton D 1977 *Wild unstable manifolds.*16:2 (Topology) pp 167–72

[18] Smale S 1959 *Diffeomorphisms of the 2-sphere.* vol 10 (Proc. AMS.) pp 621–26

[19] Rolfsen D 1977*Knots and links* vol 83, 5 (Bull. Amer. Math. Soc.) pp 931–35

IOP Conf. Series: Journal of Physics: Conf. Series **990** (2018) 012011 doi:10.1088/1742-6596/990/1/012011

Allee's dynamics and bifurcation structures in von Bertalanffy's population size functions

J. Leonel Rocha[1], Abdel-Kaddous Taha[2] and D. Fournier-Prunaret[3]

[1] CEAUL, ISEL-Engineering Superior Institute of Lisbon, Rua Conselheiro Emídio Navarro 1, 1959-007 Lisboa, Portugal

[2] INSA Toulouse, Federal University of Toulouse Midi-Pyrénées, 135 Avenue de Rangueil, 31077 Toulouse, France

[3] LAAS-CNRS, INSA Toulouse, Federal University of Toulouse Midi-Pyrénées, 7 Avenue du Colonel Roche, 31077 Toulouse, France

E-mail: jrocha@adm.isel.pt

Abstract. The interest and the relevance of the study of the population dynamics and the extinction phenomenon are our main motivation to investigate the induction of Allee Effect in von Bertalanffy's population size functions. The adjustment or correction factor of rational type introduced allows us to analyze simultaneously strong and weak Allee's functions and functions with no Allee effect, whose classification is dependent on the stability of the fixed point $x = 0$. This classification is founded on the concepts of strong and weak Allee's effects to the *population growth rates* associated. The transition from strong Allee effect to no Allee effect, passing through the weak Allee effect, is verified with the evolution of the rarefaction critical density or Allee's limit. The existence of cusp points on a fold bifurcation curve is related to this phenomenon of transition on Allee's dynamics. Moreover, the "foliated" structure of the parameter plane considered is also explained, with respect to the evolution of the Allee limit. The bifurcation analysis is based on the configurations of fold and flip bifurcation curves. The chaotic semistability and the nonadmissibility bifurcation curves are proposed to this family of 1D maps, which allow us to define and characterize the corresponding Allee effect region.

1. Introduction and motivation

One of the most familiar growth equation used to describe the growth of marine populations, namely fishes, seabirds, marine mammals, invertebrates, reptiles and sea turtles is von Bertalanffy's equation, see for example [5], [8] and references therein. This equation remains one of the most popular flexible growth equations to model fish weight growth, since it was presented by von Bertalanffy for this aim in 1938, see [29] and [30]. An usual form of von Bertalanffy's growth equation is given by,

$$g(W_t) = \frac{dW_t}{dt} = \frac{V}{3} W_t^{\frac{2}{3}} \left(1 - \left(\frac{W_t}{W_\infty}\right)^{\frac{1}{3}}\right), \tag{1}$$

where W_t is the weight at age t, W_∞ is the asymptotic weight, $V > 0$ is von Bertalanffy's growth rate constant and t_0 is the theoretical age the chick would have at weight zero.

Due to various causes, the extinction of certain species of trees, plants and mammals is one of the most current and worrying problems. In fact, species extinction is usually a major focus

IOP Conf. Series: Journal of Physics: Conf. Series **990** (2018) 012011 doi:10.1088/1742-6596/990/1/012011

of ecological and biological research. The Allee effect is an important dynamic phenomenon first described by Allee in 1931, see [3]. In this work, we use the concept of demographic Allee effect, which is manifested by a reduction in the *per capita* growth rate at low population sizes. The Allee effect is described as *strong* when there is a density threshold (where the *per capita* growth rate becomes null) below which the population decrease and go to extinction. On the other hand, the Allee effect is described as *weak* when population growth is slowed down for small densities but not to the point of becoming negative. The distinction between strong and weak Allee effects is important, most authors neglect the former and almost exclusively consider the strong Allee effect.

Generically, in the context of dynamical systems or difference equations, a function representing Allee effect must have three fixed points: an asymptotically stable zero fixed point, an unsable fixed point, called the threshold point, and a bigger positive fixed point, which is asymptotically stable. For more information on Allee effect see, for example, [7], [11], [13], [17], [18], [19], [22], [23] and references therein.

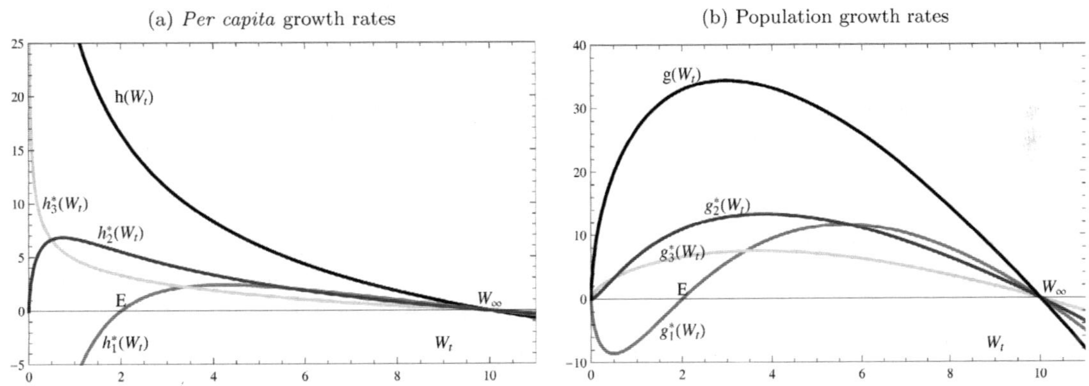

Figure 1. (a) *Per capita* growth rates defined by Eqs.(2) and (4): $h(W_t)$ at $V = 150$ and $W_\infty = 10$ (without correction); $h_1^*(W_t)$ at $V = 150$, $W_\infty = 10$, $E = 2$ and $C = 3$ (strong Allee effect); $h_2^*(W_t)$ at $V = 75$, $W_\infty = 10$, $E = 0$ and $C = 1$ (weak Allee effect); $h_3^*(W_t)$ at $V = 45$, $W_\infty = 10$, $E = -2$ and $C = 4$ (no Allee effect); (b) population growth rates defined by Eqs.(1) and (3): $g(W_t)$ at $V = 150$ and $W_\infty = 10$ (without correction); $g_1^*(W_t)$ at $V = 150$, $W_\infty = 10$, $E = 2$ and $C = 3$ (strong Allee effect); $g_2^*(W_t)$ at $V = 75$, $W_\infty = 10$, $E = 0$ and $C = 1$ (weak Allee effect); $g_3^*(W_t)$ at $V = 45$, $W_\infty = 10$, $E = -2$ and $C = 4$ (no Allee effect).

The von Bertalanffy growth model, Eq.(1), do not exhibit the Allee effect, because the *per capita* growth rate defined as

$$h(W_t) = \frac{g(W_t)}{W_t} = \frac{V}{3} W_t^{-\frac{1}{3}} \left(1 - \left(\frac{W_t}{W_\infty}\right)^{\frac{1}{3}}\right), \tag{2}$$

decreases at low densities, see Fig.1(a). In this case, the *per capita* growth rates are higher than the population growth rate at the initial time, see Fig.1(b). However, this drawback was corrected using suitable corrections in the work presented by Rocha *et al.* in [18]. One of the corrected von Bertalanffy's models proposed is defined by the differential equation,

$$g^*(W_t) = \frac{dW_t}{dt} = \frac{V}{3} W_t^{\frac{2}{3}} \left(1 - \left(\frac{W_t}{W_\infty}\right)^{\frac{1}{3}}\right) \frac{W_t - E}{W_t + C}, \tag{3}$$

with $|E| < W_\infty$ and $C > 0$, where E is the rarefaction critical density (density threshold) or Allee's limit and W_∞ is the carrying capacity. The parameter C allows us to define and study more flexible models, with variable extinction rates, see [1], [2] and [4]. Note that the purpose of introducing Allee effect in the *population growth rates*, given by Eq.(1), is satisfied under a correction factor of rational type that defines Eq.(3). The corresponding *per capita* growth rates are defined as follows:

$$h^*(W_t) = \frac{g^*(W_t)}{W_t} = \frac{V}{3}W_t^{-\frac{1}{3}}\left(1-\left(\frac{W_t}{W_\infty}\right)^{\frac{1}{3}}\right)\frac{W_t - E}{W_t + C}, \tag{4}$$

Fig.1 illustrates some examples of the parameter values for which there is strong Allee effect, weak Allee effect and no Allee effect corresponding to Eqs.(3) and (4). Similar approaches were made in [1], [2], [7], [17], [19], [25] and [26].

In Lotka's works on the logistic growth concept, presented in [14], the rate of population growth, $\frac{dW_t}{dt}$, at any moment t, is given by a function of the population size at that moment, W_t, namely, $\frac{dW_t}{dt} = g(W_t)$. This procedure leads to the Verhulst logistic equation, by expanding $g(W_t)$ as a Taylor series in a neighborhood of $W_t = 0$, see also [28]. On the other hand, many stochastic models of population growth can be approximated by a diffusion process (including a discrete time birth-death process), where the infinitesimal mean (defined also by $g(W_t)$) specifies the underlying deterministic trend, see, for example, [27]. The previous works are examples of different approaches in the use of the population size function $g(W_t)$ associated to the differential equation that defines the growth model under study. Similarly, considering population size functions, Rocha *et al.* presents probabilistic and dynamical studies in different logistic growth models, which can be seen in [18], [19], [20], [21], [22], [23], [24], [25] and [26].

1.1. Purpose of the paper and plan

In this paper, with the purpose of presenting a dynamical approach to generalized von Bertalanffy's models, given by Eq.(3), we consider the corresponding family of population size functions, defined as follows:

$$f(x; r, K, E, C) = r\, x^{\frac{2}{3}}\left(1 - x^{\frac{1}{3}}\right)\frac{Kx - E}{Kx + C},$$

which is studied in the next section. Note that this 4-parameter family of 1D maps is proportional to the right-hand side of the corrected von Bertalanffy's growth model, Eq.(3), with an adjustment or correction factor of rational type. For more details on the definition and interpretation of these parameters see, for example, [1], [2] and [4].

We remark that our approach to the Allee effect phenomenon is founded on the concepts of strong and weak Allee effects to the *population growth rates* (given by Eq.(3)) and not to the usual *per capita growth rates* (given by Eq.(2)). Our main in this work is not to study the difference equation or the discrete dynamical system associated to Eq.(3).

The plan of the work is as follows. In Section 2, we study von Bertalanffy's population size functions under a rational correction: new classes of Allee's functions and functions with no Allee effect. In particular, are also characterized subclasses of strong and weak Allee's functions and functions with no Allee effect, whose classification is dependent on the stability of the fixed point $x = 0$. The parameters E and C play a fundamental role in distinguishing these classes of functions. These families of unimodal and bimodal maps are proportional to the right hand side of the corrected von Bertalanffy's model, defined by Eq.(3). The dynamical behavior of these von Bertalanffy's population size functions is investigated, are also defined and characterized the unconditional extinction and essential extinction, Lemmas 1, 2 and 3. In Lemma 5, we discuss the dynamic behavior of the von Bertalanffy's population size functions with no Allee effect.

Section 3 is devoted to the study of the bifurcation structures of von Bertalanffy's population size functions, in the (C, r) parameters plane. This analysis is based on the configurations of fold and flip bifurcation curves. The concepts of chaotic semistability and nonadmissibility bifurcation curves to this family of maps are also considered, which allow us to define the corresponding Allee effect region. The transition from strong Allee effect to no Allee effect, passing through the weak Allee effect, is analyzed with the evolution of the Allee's limit E. To support our results, we present fold and flip bifurcations curves, numerical simulations of several bifurcation diagrams and the"foliated" structure of the (C, r) parameters plane considered. Section 4 concludes.

2. Von Bertalanffy's chaotic dynamics with strong and weak Allee effects and with no Allee effect

In the sequence of the work presented by Rocha *et al.* in [18], new one-dimensional discrete dynamical systems were established. In this section, we present a dynamical approach to von Bertalanffy's equation with the purpose of introducing Allee effect: von Bertalanffy's population size functions under a rational correction. We define new classes of von Bertalanffy's functions with different types of Allee effect, designated by strong and weak Allee's functions, which describe the growth of a population with an Allee effect and the existence of extinction. Dependent on the variation of the parameters, the von Bertalanffy's population size functions also includes another class of important functions: functions with no Allee effect. Generically, these functions are unimodal or bimodal maps for which the extinction is inevitable for too high or too low initial population densities. The transition from strong Allee effect to no Allee effect, passing through the weak Allee effect, involve the several parameters considered in the models, standing out the Allee's limit. The complex dynamical behavior of these von Bertalanffy's population size functions is investigated in detail.

2.1. Von Bertalanffy's population size functions of rational type

We consider a 4-parameter family of 1D maps, $f : [0, 1] \to]-\infty, 1]$, defined by

$$f\left(x; r, K, E, C\right) = r\ x^{\frac{2}{3}}\left(1 - x^{\frac{1}{3}}\right)\frac{Kx - E}{Kx + C}, \tag{5}$$

with $x = \frac{W_t}{W_\infty} = \frac{W_t}{K} \in [0, 1]$ the normalized weight, $r = \frac{V}{3} \times K^{\frac{2}{3}} > 0$ an intrinsic growth rate of the individual weight, the Allee's limit E and the carrying capacity K satisfying $|E| < K$ and $C > 0$. Note that in [25] and [26] the parameter C can be null or negative, that because the correction factors are of the polynomial type. See some examples of graphics of f in Figs.2, 3 and 4.

Let $A_r = f\left(A_r; r, K, E, C\right)$ be the first positive fixed point of f and $A_r^* = max\{f^{-1}\left(A_r; r, K, E, C\right)\}$. From Eq.(5), we have that $E/K < A_r$. The interval $[0, E/K]$ will be designated by extinction interval. The following properties are verified:

(A1) f is continuous on $[0, 1]$;

(A2) there exists $[A_r, A_r^*] \subset [0, 1] : f\left(x; r, K, E, C\right) \leq A_r,\ \forall x \in [0, 1] \setminus]A_r, A_r^*[$;

(A3) f has an unique critical point $c \in]A_r, A_r^*[$;

(A4) $f'(x; r, K, E, C) \neq 0,\ \forall x \in]A_r, A_r^*[\setminus \{c\}$, $f'(c; r, K, E, C) = 0$ and $f''(c; r, K, E, C) < 0$;

(A5) $f \in C^3\left(]A_r, A_r^*[\right)$;

(A6) the Schwarzian derivative of f given by

$$S\left(f(x; r, K, E, C)\right) = \frac{f'''(x; r, K, E, C)}{f'(x; r, K, E, C)} - \frac{3}{2}\left(\frac{f''(x; r, K, E, C)}{f'(x; r, K, E, C)}\right)^2$$

verifies the next conditions,

$$S\left(f(x; r, K, E, C)\right) < 0, \forall x \in]A_r, A_r^*[\setminus \{c\} \text{ and } S\left(f(c; r, K, E, C)\right) = -\infty.$$

Properties (A1)-(A6) and more particularly the negative Schwarzian derivative ensures a "good" dynamic behavior of the models: continuity and monotonicity of topological entropy, order in the succession of bifurcations, the non-existence of wandering intervals and the existence of an upper limit to the number of stable orbits, see [15]. In the case of unimodal maps there is at most one stable orbit, and in the case of bimodal maps there are at most two stable orbits. In general, the growth models studied have negative Schwarzian derivative and the use of unimodal maps is also frequent, see for example [19], [20], [21], [22] and [28]. Depending on the parameters, to von Bertalanffy's population size functions f, we define a class of functions that describes the growth of a population with Allee effect.

Definition 1 *Consider* $0 \leq E < K$, $C > 0$, $r > 0$ *and* f *defined by Eq.(5). The von Bertalanffy's population size functions* $f : [0,1] \to]-\infty, 1]$ *attaining its maximum at* c *are Allee's functions, if there are numbers* A_r *and* B_r *such that:*

(i) $0 < A_r < c < B_r < 1$;

(ii) $\forall x \in [0,1] \setminus [A_r, B_r] \Rightarrow f(x; r, K, E, C) < x$;

(iii) $\forall x \in]A_r, B_r[\Rightarrow f(x; r, K, E, C) > x$.

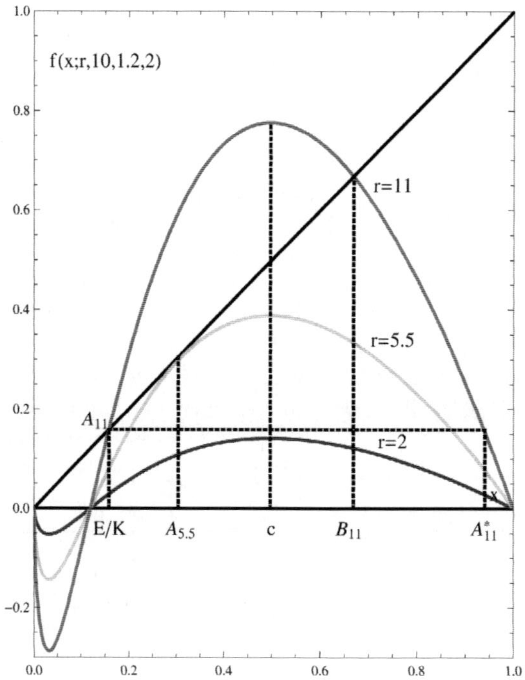

Figure 2. Von Bertalanffy's population size functions with strong Allee effect $f(x; r, K, E, C)$ at $K = 10$, $E = 1.2$, $C = 2$ and $r = 2$ (1 fixed point: 0), $r = 5.5$ (2 fixed points: 0 and $A_{5.5}$) and $r = 11$ (3 fixed points: $0, A_{11}$ and B_{11}). E/K is the Allee's threshold.

Note that,

$$\lim_{x \to 0^+} f'(x; r, K, E, C) = \begin{cases} -\infty, & \text{if } 0 < E < K \\ 0, & \text{if } E = 0 \\ +\infty, & \text{if } -K < E < 0 \end{cases} . \tag{6}$$

Consequently, this class of Allee's functions is divided into two subclasses of functions that verify the following properties. We remark that this classification is dependent on the evolution of the Allee's limit E.

Property 1 *Under the conditions of Definition 1, it is established that:*

(i) If $0 < E < K$, the von Bertalanffy population size functions f are a family of strong Allee's functions, see Fig.2;

(ii) If $E = 0$, the von Bertalanffy population size functions f are a family of weak Allee's functions, see Fig.3.

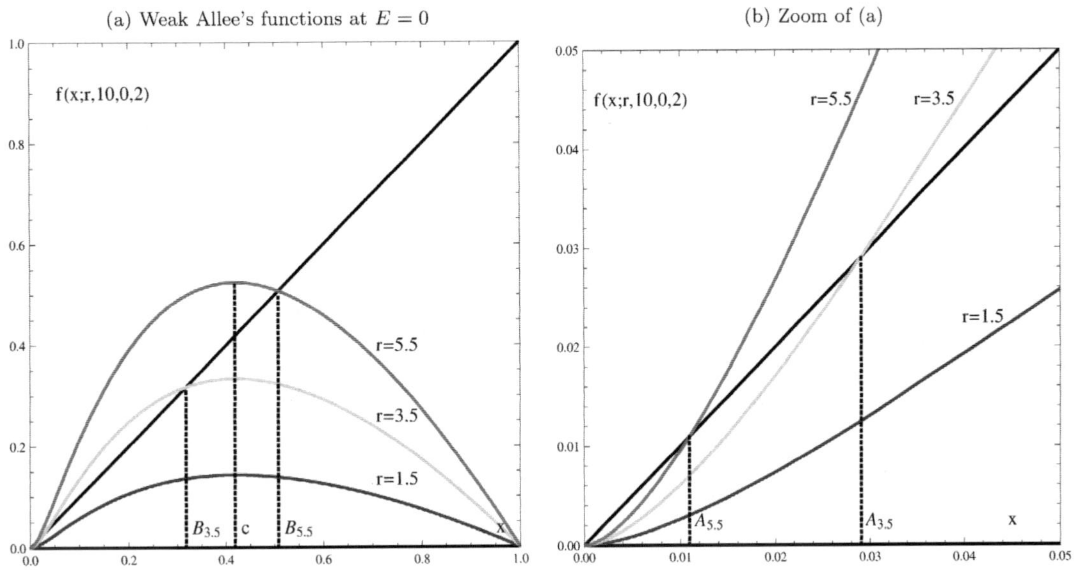

Figure 3. Von Bertalanffy's population size functions $f(x; r, K, E, C)$: (a) weak Allee's functions at $K = 10$, $E = 0$, $C = 2$ and $r = 3.5$, 5.5 (3 fixed points: $0, A_r$ and B_r), at the same parameters values and $r = 1.5$ (1 fixed point: 0); (b) Zoom of (a), where the fixed points $A_{3.5}$ and $A_{5.5}$ are well highlighted.

This classification is based on the concepts of strong and weak Allee effects to *population growth rates* associated, see Eq.(3). In this context, the Allee effect is described as *strong* when there is a density threshold, where the population growth rate becomes null and below which are negative. The Allee effect is described as *weak* when the populations do not exhibit any thresholds, the point $x = 0$ becomes stable and the population growth rate are positive. See also [19]. The von Bertalanffy's population size functions f also includes another class of important functions, for which does not exist Allee effect.

Definition 2 *Consider $-K < E < 0$, $C > 0$, $r > 0$ and f defined by Eq.(5). The von Bertalanffy's population size functions $f : [0, 1] \to [0, 1]$ attaining its maximum at c are functions with no Allee effect, if there are numbers O_r, A_r and B_r such that:*

(i) $0 < O_r < A_r < c < B_r < 1$;

(ii) $\forall x \in [0, 1] \setminus ([0, O_r] \cup [A_r, B_r]) \Rightarrow f(x; r, K, E, C) < x$;

(iii) $\forall x \in (]0, O_r[\cup]A_r, B_r[) \Rightarrow f(x; r, K, E, C) > x$.

The class of functions with no Allee effect verifies $\lim_{x \to 0^+} f'(x; r, K, E, C) > 1$, i.e., the fixed point $x = 0$ is always unstable. It is for this reason that this class of functions does not include extinction. Note that from Definition 1 it follows that, in the cases of strong and weak Allee's functions, we will have at maximum three fixed points $0, A_r$ and B_r, see Figs.2 and 3. Without loss of generality, we also consider the case where $x = 0$ is the only fixed point, see Figs.2, 3 and Lemma 1. On the other hand, from Definition 2, in the case of functions with no Allee effect, we will have at least two fixed points 0 and O_r, and at most four fixed points $0, O_r, A_r$ and B_r, depending on the evolution of the parameter r, see Figs.4 and 5. See also the definitions of Allee's functions, for example, at [13], [19], [22] and [28].

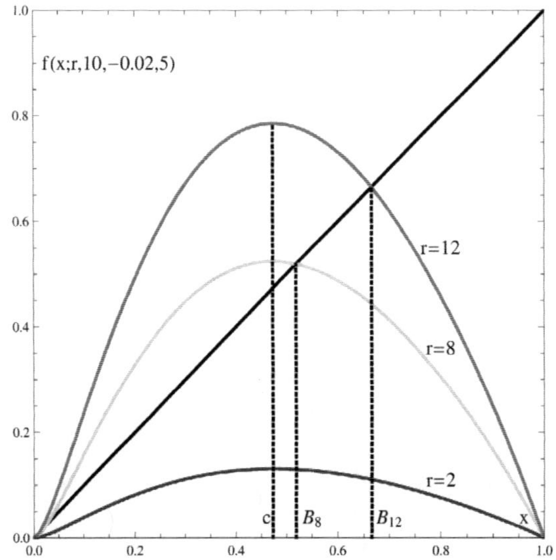

Figure 4. Von Bertalanffy's population size functions with no Allee effect $f(x; r, K, E, C)$ at $K = 10$, $E = -0.02$, $C = 5$ and $r = 2$ (2 fixed points: 0 and O_r), $r = 8$ and $r = 12$ (4 fixed points: $0, O_r, A_r$ and B_r). See Fig.5 for zoom of the fixed points O_r and A_r.

2.2. Unconditional extinction and essential extinction in von Bertalanffy's population size functions with strong and weak Allee effects

This section is devoted to the study of one of the most important phenomena in population dynamics: the extinction and the existence of Allee effect. In the next results, we study the cases where the extinction is inevitable in finite time, for all initial densities.

Lemma 1 *(Unconditional extinction) Let $f(x; r, K, E, C)$ be the von Bertalanffy population size functions, Eq.(5), with $0 \leq E < K$, $C > 0$, $r > 0$, satisfying (A1) and (A4). Consider that $f(x; r, K, E, C)$ has $x = 0$ as the unique fixed point.*

(i) If $0 < E < K$, then $\exists\ n_0 \in \mathbf{N} : n \geq n_0,\ f^n(x; r, K, E, C) < 0,\ \forall x \in]0, 1[\setminus \{E/K\}$;

(ii) If $E = 0$, then $\lim_{n \to \infty} f^n(x; r, K, E, C) = 0,\ \forall x \in [0, 1]$.

Proof 1 *Consider that $0 < E < K$, from Definition 1 and Property 1 (i) we have that strong Allee's functions are defined as $f : [0, 1] \to]-\infty, 1]$, see Fig.2 (blue graphic for $r = 2$). If $x = 0$ is the only fixed point of f and considering (A1), then $f(x; r, K, E, C) < x, \forall x \in [0, 1]$. So, we obtain $f(x; r, K, E, C) < 0,\ \forall x \in]0, E/K[$. However, when given $\forall x \in]E/K, 1[$, we have two different cases:*

(a) $\exists\ r > 0 : f(c; r, K, E, C) \leq E/K$. Considering conditions (A1) and (A4), we have that $f^2(x; r, K, E, C) < 0, \forall x \in]E/K, 1[$. In particular, $f^2(c; r, K, E, C) = 0$.

(b) $\exists\, r > 0 : f\,(c; r, K, E, C) > E/K$. *Consider*

$$Q_r = min\{f^{-1}\,(E/K; r, K, E, C)\} \quad and \quad Q_r^* = max\{f^{-1}\,(E/K; r, K, E, C)\},$$

such that $]Q_r, Q_r^*[\, \subset\,]E/K, 1[$. *Given conditions (A1) and (A4), we have that*

$$f^2\,(x; r, K, E, C) < 0, \forall x \in (]E/K, Q_r[\, \cup\,]Q_r^*, 1[)\,.$$

Otherwise, the sequence $f^n\,(x; r, K, E, C)$ *is decreasing and finite,* $\forall x \in\,]Q_r, Q_r^*[$, *where exist an order* $n_0 - 1$ *such that the sequence is bounded below by* E/K. *This implies that, for* $n \geq n_0$, *the iterates* $f^n\,(x; r, K, E, C) < 0$, $\forall x \in\,]E/K, 1[$. *So, the first claim is proved.*

Assume that $E = 0$, *whence from Definition 1 and Property 1 (ii) weak Allee's functions are defined as* $f : [0, 1] \to [0, 1]$ *and* $x = 0$ *is the only fixed point of* $f\,(x; r, K, 0, C)$, *see Fig.3 (blue graphic for* $r = 1.5$*). By (A1), continuity of* f, *we have that* $f\,(x; r, K, 0, C) < x, \forall x \in\,]0, 1]$. *Considering that* $f^n\,(x; r, K, 0, C)$ *is a decreasing sequence that is bounded below by* $x = 0$, *then* $f^n\,(x; r, K, 0, C)$ *converges to some point* $a \in [0, 1]$. *By (A1), this point* a *is a fixed point. So, the claim (ii) is proved.* ■

Lemma 2 *(Unconditional extinction) Let* $f\,(x; r, K, E, C)$ *be the von Bertalanffy population size functions, Eq.(5), with* $0 \leq E < K$, $C > 0$, $r > 0$, *satisfying (A1)-(A4). Consider that* $f\,(x; r, K, E, C)$ *has exactly three fixed points.*

(i) If $0 < E < K$, *then*

$$f\,(x; r, K, E, C) < 0, \ \forall x \in\,]0, E/K[\quad and \quad f^2\,(x; r, K, E, C) < 0, \ \forall x \in\,]E/K, A_r[\, \cup\,]A_r^*, 1[\,;$$

(ii) If $E = 0$, *then* $\lim_{n \to \infty} f^n\,(x; r, K, E, C) = 0, \ \forall x \in [0, 1] \setminus [A_r, A_r^*]\,.$

Proof 2 *Consider now that strong Allee's functions* f *has three fixed points, see Fig.2 (red graphic for* $r = 11$*). The first part of the claim (i) follows from Definition 1 and Property 1 (i). On the other hand, given (A3) and (A4), we conclude that* $E/K < A_r < c < A_r^*$. *Furthermore, by (A4) we have*

$$f'(x; r, K, E, C) > 0, \ \forall x \in\,]E/K, A_r[\quad and \quad f'(x; r, K, E, C) < 0, \ \forall x \in\,]A_r^*, 1[\,.$$

Considering $(A2)$ *it follows that,*

$$f\,(]E/K, A_{r_i}[\,; r, K, E, C) \subseteq\,]0, A_r[\quad and \quad f\,(]A_r^*, 1[\,; r, K, E, C) \subseteq\,]0, A_r[\,.$$

Therefore, we have that $f^2\,(x; r, K, E, C) < 0$, $\forall x \in\,]E/K, A_r[\, \cup\,]A_r^*, 1[$. *The second claim is proved.*

To prove (ii) assertion, consider that weak Allee's functions $f\,(x; r, K, 0, C)$ *has three fixed points, see Fig.3 (green graphic for* $r = 3.5$ *and red graphic for* $r = 5.5$*). Given (A3) and (A4), we have that* $A_r < c < A_r^*$. *Considering (A4) it follows that,*

$$f'(x; r, K, 0, C) > 0, \ \forall x \in\,]0, A_r[\quad and \quad f'(x; r, K, 0, C) < 0, \ \forall x \in\,]A_r^*, 1[\,.$$

Moreover, by (A2) we have that $f\,(x; r, K, 0, C) \leq x$, $\forall x \in [0, 1] \setminus [A_r, A_r^*]$. *Conclusively, since* $x = 0$ *is the only fixed point in the set* $[0, 1] \setminus [A_r, A_r^*]$, *we have that* $f^n\,(x; r, K, 0, C)$ *is a decreasing sequence that converges to* $x = 0$, $\forall x \in [0, 1] \setminus [A_r, A_r^*]$. *The desired result follows.* ■

See the unconditional extinction region in the (C, r) parameters plane at Fig.9. Other kind of extinction, also extremely important, is characterized by the essential extinction, which is defined when the maximum size growth of the population exceeds the critical density and the populations are almost surely doomed to extinction, see [28]. This populational occurrence is also characterized as a randomly choice of the initial densities leading to extinction with probability one. The following result characterizes this extinction in finite time, for a subset of all initial densities. We remark that, in [28] is proved a general result for similar models to Lebesgue almost every initial population densities, see also [19]. To simplify the writing of the next results, we consider c the positive critical point and A_r the first positive fixed point of f,

$$A_r^* = max\{f^{-1}(A_r; r, K, E, C)\}, P_r = min\{f^{-1}(A_r^*; r, K, E, C)\}, P_r^* = max\{f^{-1}(A_r^*; r, K, E, C)\},$$

such that $\tilde{I} =]P_r, P_r^*[\subset]A_r, A_r^*[$. We note that in [21], the point A_r is designated by Allee's point and this type of extinction is associated to Allee's effect region. In Section 3.2 we present an approach to the bifurcation structure associated to this phenomenon, in the (C, r) parameters plane, which is also designated by Allee's effect region.

Lemma 3 *(Essential extinction) Let $f(x; r, K, E, C)$ be the von Bertalanffy population size functions, Eq.(5), with $0 \leq E < K$, $C > 0$, $r > 0$, satisfying (A1)-(A4). Consider that $f(x; r, K, E, C)$ has exactly three fixed points and $f^2(c; r, K, E, C) < A_r$.*

(i) If $0 < E < K$, then $\exists\ n_0 \in \mathbf{N} : n \geq n_0,\ f^n(x; r, K, E, C) < 0,\ \forall x \in \tilde{I}$;

(ii) If $E = 0$, then $\lim_{n\to\infty} f^n(x; r, K, 0, C) = 0,\ \forall x \in \tilde{I}$.

Proof 3 *If $f(x; r, K, E, C)$ is a family of strong or weak Allee's functions, then by conditions (A1)-(A4), it follows that $A_r < P_r < c < P_r^* < A_r^*$, see Figs.4 and 5. Given (A4), we have*

$$f'(x; r, K, E, C) > 0,\ \forall x \in [A_r, c[\quad and \quad f'(x; r, K, E, C) < 0,\ \forall x \in]c, 1].$$

Considering Definition 1 (iii), we verify that there exists $[P_r, P_r^*] \subset]A_r, B_r[$, such that,

$$f(x; r, K, E, C) > x,\ \forall x \in [P_r, P_r^*].$$

By hypothesis, we have that $f^2(c; r, K, E, C) < A_r \Leftrightarrow A_r^* < f(c; r, K, E, C) \leq 1$. Consequently, we obtain $A_r^* \leq f(x; r, K, E, C) \leq 1$, $\forall x \in [P_r, P_r^*]$. By (A4), monotonicity of f, we have $0 \leq f^2(x; r, K, E, C) \leq A_r$, $\forall x \in [P_r, P_r^*]$. If $f(x; r, K, E, C)$ is a family of strong Allee's functions, then by Lemma 2 (i) follows the desired result of (i). On the other hand, if $f(x; r, K, 0, C)$ is a family of weak Allee's functions, then by (ii) of Lemma 2 follows (ii). ■

See the essential extinction region in the (C, r) parameters plane at Fig.9. However, considering (i) and (ii) of Lemma 2, Lemma 3 and (4) of Theorem 1 proved in [28], we can state that:

Lemma 4 *Let $f(x; r, K, E, C)$ be the von Bertalanffy population size functions, Eq.(5), with $0 \leq E < K$, $C > 0$, $r > 0$, satisfying (A1)-(A4). Consider that $f(x; r, K, E, C)$ has exactly three fixed points and $f^2(c; r, K, E, C) < A_r$.*

(i) If $0 < E < K$, then $\exists\ n_0 \in \mathbf{N} : n \geq n_0,\ f^n(x; r, K, E, C) < 0$, for Lebesgue almost every $x \in]A_r, P_r[\cup]P_r^*, A_r^*[$. The basin of attraction of the extinction interval $[0, E/K]$ is given by

$$\Omega_{[0,E/K]} = [E/K, A_r[\bigcup]P_r, P_r^*[\bigcup]A_r^*, 1]\bigcup\left\{\bigcup_{n\geq1} f^{-n}\left(\tilde{I}; r, K, E, C\right)\right\};$$

(ii) If $E = 0$, then $\lim_{n\to\infty} f^n(x; r, K, 0, C) = 0$, for Lebesgue almost every $x \in]A_r, P_r[\cup]P_r^*, A_r^*[$. The origin's basin of attraction is given by

$$\Omega_0 = [0, A_r[\bigcup]P_r, P_r^*[\bigcup]A_r^*, 1]\bigcup\left\{\bigcup_{n\geq 1} f^{-n}\left(\tilde{I}; r, K, 0, C\right)\right\}.$$

2.3. Chaotic dynamics of von Bertalanffy's population size functions with no Allee effect

In this section we study von Bertalanffy's population size functions with no Allee effect, which require special attention, because we no longer have three fixed points but four fixed points, at maximum, see Figs.4 and 5. Considering Definition 2 (i), it is verified that,

$$0 < O_r < A_r < c < B_r < 1, \tag{7}$$

where now O_r is the first positive fixed point, extremely near zero, see Fig.5 (b). In this case, the fixed point $x = 0$ becomes unstable. The next result characterizes the chaotic dynamics of this class of functions.

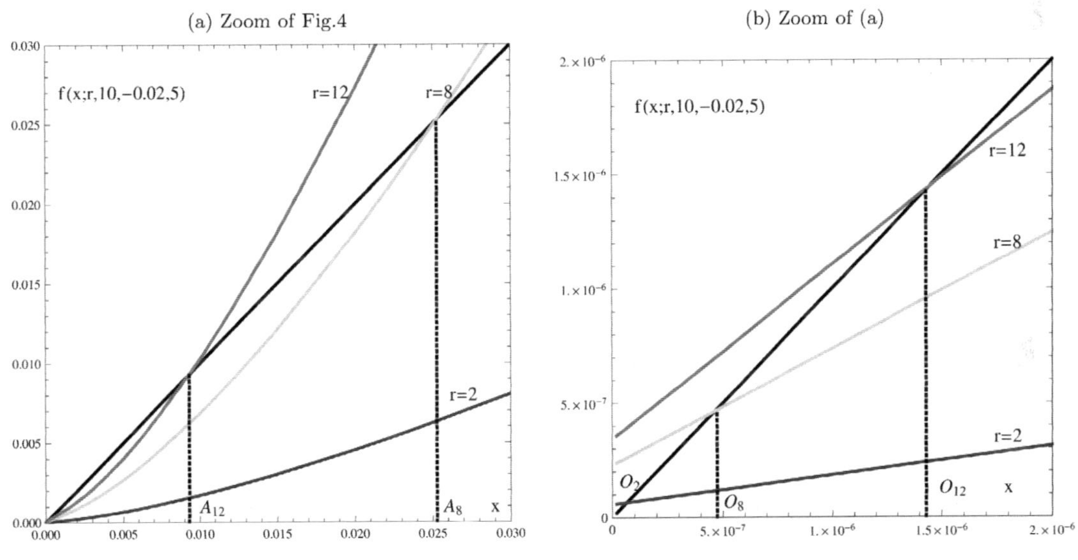

Figure 5. Von Bertalanffy's population size functions with no Allee effect $f(x; r, K, E, C)$ at $K = 10$, $E = -0.02$, $C = 5$ and $r = 2$, $r = 8$ and $r = 12$: (a) Zoom of Fig.4, where the fixed points A_8 and A_{12} are well highlighted; (b) Zoom of (a), where the fixed points O_2, O_8 and O_{12} are well highlighted.

Lemma 5 *Let $f(x; r, K, E, C)$ be the von Bertalanffy population size functions, Eq.(5), with $-K < E < 0$, $C > 0$, $r > 0$, satisfying (A1)-(A6).*

(i) If the functions $f(x; r, K, E, C)$ with no Allee effect have two fixed points 0 and O_r, then

$$\lim_{n\to\infty} f^n(x; r, K, E, C) = O_r, \ \forall x \in]0, 1[;$$

(ii) If the functions $f(x; r, K, E, C)$ with no Allee effect have four fixed points $0, O_r, A_r$ and B_r, then:

- $\lim_{n\to\infty} f^n(x; r, K, E, C) = O_r, \ \forall x \in]0, 1[\setminus [A_r, A_r^*];$

- *there is a linearly stable fixed point* $B_r \in]A_r, A_r^*[$ *whose basin of attraction is* $]A_r, A_r^*[$ *or the interval* $[f^2(c;r,K,E,C), f(c;r,K,E,C)]$ *is forward invariant with basin of attraction* $]A_r, A_r^*[$.

Proof 4 *Given that* $\lim_{x\to 0^+} f'(x;r,K,E,C) > 1$, *by (A1) f is continuous, and that* $x = 0$ *and* $x = O_r$ *are two fixed points satisfying Eq.(7), we have*

$$f(x;r,K,E,C) > x,\ \forall x \in]0, O_r[\text{ and } f(x;r,K,E,C) < x,\ \forall x \in]O_r, 1].$$

Considering that $f^n(x;r,K,E,C)$ *is an increasing sequence,* $\forall x \in]0, O_r] \cup [O_r^*, 1[$, *with*

$$O_r^* = max\{f^{-1}(O_r;r,K,E,C)\},$$

that is bounded above by O_r, *then* $f^n(x;r,K,E,C)$ *converges to some point* $a \in]0, O_r]$. *By (A1), this point is the first positive fixed point* O_r, *because the fixed point* $x = 0$ *is unstable.*

On the other hand, we have $f(x;r,K,E,C) < x$, $\forall x \in]O_r, O_r^*[$. *By Eq.(7), follows that* $0 < O_r < c < 1$, *and by (A4), we have that*

$$f'(x;r,K,E,C) > 0,\ \forall x \in]O_r, c[\text{ and } f'(x;r,K,E,C) < 0,\ \forall x \in]c, A_r^*[.$$

So, $f^n(x;r,K,E,C)$ *is a decreasing sequence, that is bounded below by* O_r, *whence* $f^n(x;r,K,E,C)$ *converges to some point which belongs to* $[O_r, O_r^*[$. *By (A1), this point is the fixed point* O_r. *Thus, the fixed point* O_r *is stable. The claim (i) is proved.*

If $x \in]0, O_r] \cup [O_r^*, 1[$, *then by (i) we obtain the desired result. Moreover, if* $x \in [O_r, A_r[\cup]A_r^*, O_r^*]$, *then by Lemma 1 (iv) and considering* $0 \equiv O_r$, *we obtain the desired outcome. So, the first equality of (ii) is proved. Now, if we are restricted to* $x \in [A_r, A_r^*]$, *then we have an unimodal map that satisfies (A1)-(A6) conditions,* $f^2(c;r,K,E,C) > A_r$ *and the Schwarzian derivative of f verifies*

$$S(f(x;r,K,E,C)) < 0, \forall x \in]A_r, A_r^*[\backslash\{c\} \text{ and } S(f(c;r,K,E,C)) = -\infty.$$

Depending on the evolution of the parameter r, if $|f'(B_r;r,K,E,C)| < 1$, *then there is a linearly stable fixed point* $B_r \in]A_r, A_r^*[$, *whose basin of attraction is* $]A_r, A_r^*[$. *Otherwise, it is verified that*

$$f\left([f^2(c;r,K,E,C), f(c;r,K,E,C)]\right) \subseteq [f^2(c;r,K,E,C), f(c;r,K,E,C)],$$

i.e., the interval $[f^2(c;r,K,E,C), f(c;r,K,E,C)]$ *is forward invariant with basin of attraction* $]A_r, A_r^*[$. *For more details of the proof of this claim see Theorem 1 (2) at [28]. Therefore, the claim (ii) is proved.* ■

Note that the von Bertalanffy's population size functions of rational type have the same dynamical behavior concerning to the fixed point B_r.

Remark 1 *The fixed point* O_r *exists analytically, although extremely near zero, consider its valid existence is debatable from a practical (or physical) point of view. Indeed, it implies the non existence of Allee effect. Only the biological or ecological datas might clarify whether or not we should consider the existence of this fixed point* O_r. *For example, see [10] to observe the dimension of microbiological data. In Fig.5 (b), the values are* $O_2 = 5.00277 \times 10^{-7}$, $O_8 = 0.0000311786...$ *and* $O_{12} = 0.000112243....$

3. Bifurcation structures of von Bertalanffy's population size functions

In this section we investigate in detail the bifurcation structures of von Bertalanffy's population size functions, in the (C, r) two-dimensional parameters space. We will make use of the classical fold and flip bifurcations. We study the behavior of such curves related with some cycles of order $n \in \mathbf{N}$. For more details about bifurcation theory see for example [9] and [16]. We introduce the notions of chaotic semistability curve and Allee's effect region. We also analyze the existence of cusp points and the "foliated" structure of the (C, r) parameters plane, dependent on the evolution of the parameter E.

3.1. *Fold and flip bifurcations of von Bertalanffy's population size functions*

Generically, to von Bertalanffy's population size functions $f(x; r, K, E, C)$, defined by Eq.(5), with $r > 0$, $|E| < K$ and $C > 0$, the fold and flip bifurcation curves relative to a cycle of order n are determined as follows. In the (C, r) parameters plane, if $x \in [0, 1[$ is a point of an order n cycle that satisfies the equations

$$f^n(x; r, K, E, C) = x \quad \text{and} \quad \frac{\partial f^n}{\partial x}(x; r, K, E, C) = 1 \tag{8}$$

then there exists a solution φ_n, such that the fold bifurcation curves relative to a cycle of order $n \in \mathbf{N}$ are given by $r(C) = \varphi_n(x; K, E, C)$, and are denoted by $\Lambda^j_{(n)_0}$, where j is the number of the curve, which differentiates cycles of the same order, see Figs.6, 7 and 8. On the other hand, if $x \in [0, 1[$ is such that,

$$f^n(x; r, K, E, C) = x \quad \text{and} \quad \frac{\partial f^n}{\partial x}(x; r, K, E, C) = -1 \tag{9}$$

then exists a solution ψ_n, such that the flip bifurcation curves relative to a cycle of order $n \in \mathbf{N}$ are given by $r(C) = \psi_n(x; K, E, C)$, and are denoted by Λ_n, see also Figs.6, 7 and 8.

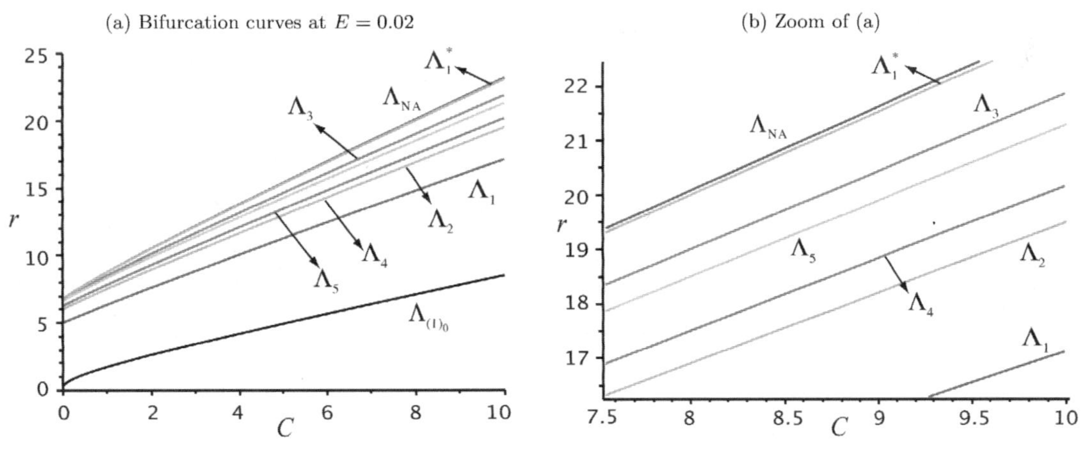

Figure 6. (a) Bifurcation curves of von Bertalanffy's population size functions with strong Allee effect $f^n(x; r, K, E, C)$, with $n = 1, 2, 3, 4, 5$, in the (C, r) parameters plane at $K = 10$ and $E = 0.02$. $\Lambda_{(1)_0}$ is the fold bifurcation curve; Λ_1, Λ_2, Λ_3, Λ_4 and Λ_5 are the flip bifurcation curves of the cycles of order $n = 1, 2, 3, 4, 5$, respectively; Λ_1^* is the chaotic semistability curve and Λ_{NA} is the bifurcation curve of nonadmissibility, in this case $\Lambda_1^* \neq \Lambda_{NA}$, according to Property 2. (b) Zoom of (a), where the region between the bifurcation curves Λ_1^* and Λ_{NA} is very well highlighted.

In particular, to von Bertalanffy's population size functions $f(x; r, K, E, C)$, defined by Eq.(5), with $r > 0$, $|E| < K$ and $C > 0$, the Eqs.(8) and (9) for $n = 1$ are given by the

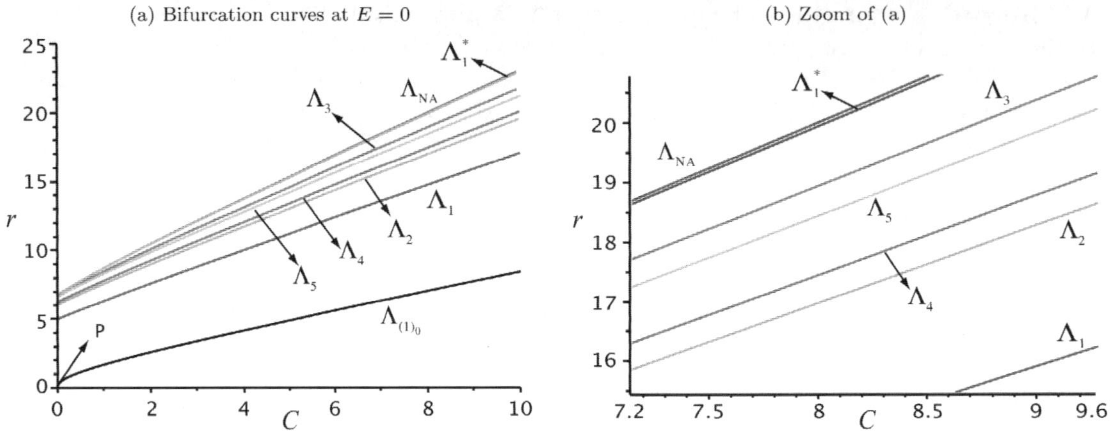

Figure 7. (a) Bifurcation curves of von Bertalanffy's population size functions with weak Allee effect $f^n(x; r, K, E, C)$, with $n = 1, 2, 3, 4, 5$, in the (C, r) parameters plane at $K = 10$ and $E = 0$. $\Lambda_{(1)_0}$ is the fold bifurcation curve; Λ_1, Λ_2, Λ_3, Λ_4 and Λ_5 are the flip bifurcation curves of the cycles of order $n = 1, 2, 3, 4, 5$, respectively; Λ_1^* is the chaotic semistability curve and Λ_{NA} is the bifurcation curve of nonadmissibility, in this case $\Lambda_1^* \neq \Lambda_{NA}$, according to Property 2. P is the cusp point, given by Eq.(15). (b) Zoom of (a), where the region between the bifurcation curves Λ_1^* and Λ_{NA} is very well highlighted.

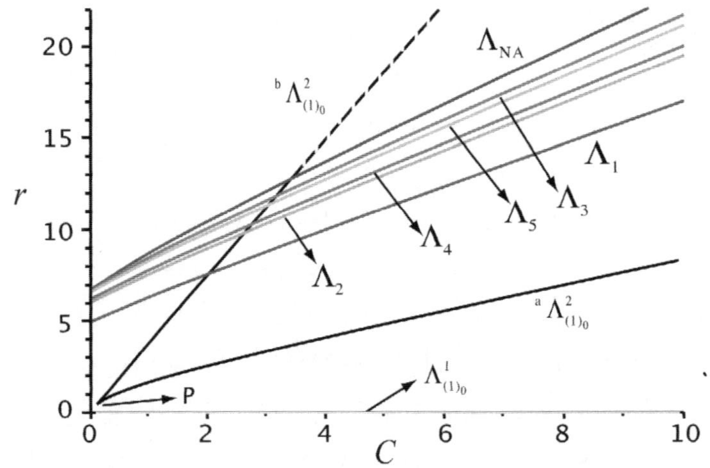

Figure 8. Bifurcation curves of von Bertalanffy's population size functions with no Allee effect $f^n(x; r, K, E, C)$, with $n = 1, 2, 3, 4, 5$, in the (C, r) parameters plane at $K = 10$ and $E = -0.02$. $\Lambda^1_{(1)_0}$, ${}^a\Lambda^2_{(1)_0}$ and ${}^b\Lambda^2_{(1)_0}$ are the fold bifurcation curves of different cycles of order $n = 1$; Λ_1, Λ_2, Λ_3, Λ_4 and Λ_5 are the flip bifurcation curves of the cycles of order $n = 1, 2, 3, 4, 5$, respectively; Λ_{NA} is the bifurcation curve of nonadmissibility, in this case not exists Λ_1^*, according to Property 2. P is the cusp point, given by Eq.(15).

following expressions,

$$\begin{cases} x = 0 \vee x = \left(r\left(1 - x^{\frac{1}{3}}\right) \frac{Kx-E}{Kx+C} \right)^3 \\ \varphi_1(x; K, E, C) = \dfrac{3\, x^{\frac{1}{3}}\, (Kx+C)^2}{3Kx\left(1-x^{\frac{1}{3}}\right)(C+E)+(Kx-E)(Kx+C)\left(2-3x^{\frac{1}{3}}\right)} \\ \psi_1(x; K, E, C) = -\varphi_1(x; K, E, C) \end{cases} . \tag{10}$$

We remark that in the case where $0 < E < K$ and $C > 0$, we consider

$$f^n(x; r, K, E, C) = \begin{cases} f(x; r, K, E, C), & \text{if } x \in [0, E/K] \\ f\left(f^{n-1}(x; r, K, E, C)\right), & \text{if } x \in]E/K, 1] \end{cases}.$$

Note that in Fig.8, ${}^a\Lambda^2_{(1)_0}$ is the fold bifurcation curve of the fixed points A_r and B_r, and ${}^b\Lambda^2_{(1)_0}$ is the fold bifurcation curve of the fixed points O_r and A_r.

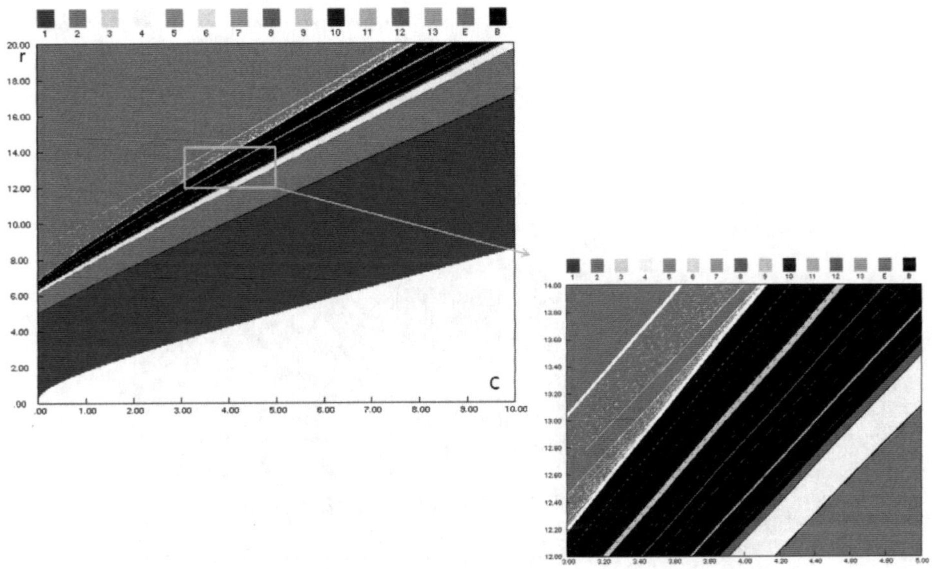

Figure 9. (on the left) Bifurcation diagram of von Bertalanffy's population size functions with strong Allee effect $f(x; r, K, E, C)$, at $K = 10$ and $E = 0.02$, in the (C, r) parameters plane. The white region is the unconditional extinction region, established in Lemmas 1 and 2. The blue region is the stability region of the fixed point A_r. The region between the blue, the gray and the white ones corresponds to period doubling region and chaotic region (existence region of cycles as shown on top of figure). The white region is the Allee effect region R_{AE} or essential extinction region, stated in Lemma 3 (i). The gray region is the no admissible region. (on the rights) Zoom of the bifurcation diagram where the Allee effect region is well highlighted.

Remark 2 *There follows a presentation of some fundamental properties of the fold and flip bifurcation curves, which allow a better understanding of the bifurcation curves and the bifurcation diagrams of von Bertalanffy's population size functions, Figs.6, 7, 8, 9 and 10, in the (C, r) parameters plane:*

(i) in the cases of strong and weak Allee's functions, the fold bifurcation curve $\Lambda_{(1)_0}$ of the fixed points A_r and B_r, given by Eq.(10), is the bifurcation curve which defines the transition between the unconditional extinction region, stated in Lemmas 1 and 2, and the stability region of the fixed point B_r. See Figs.6, 7 and 9;

(ii) in the case of functions with no Allee effect the fold bifurcation curves ${}^a\Lambda^2_{(1)_0}$ and ${}^b\Lambda^2_{(1)_0}$ characterize the stability of the fixed points O_r and B_r, and the instability of the fixed points 0 and A_r. Reason why for this class of functions does not exist an unconditional extinction region, such as established in Lemma 5. See Figs.8 and 10;

(iii) in all cases, the flip bifurcation curve Λ_1, of the stable fixed point B_r, given by Eq.(10), is the bifurcation curve which defines the transition between the stability region and the period

IOP Conf. Series: Journal of Physics: Conf. Series **990** (2018) 012011 doi:10.1088/1742-6596/990/1/012011

doubling region. The upper limit of the period doubling region is defined by the accumulation value of the flip bifurcation curves of the cycle of order 2^n, of the stable fixed points nonzero, see [16]. This bifurcation curve is denoted by Λ_∞, from Eq.(9) and considering $x \in [0, 1[$ a fixed point, we have $\Lambda_\infty = \lim_{n\to\infty} \psi_{2^n}(x; K, E, C)$;

(iv) the chaotic region is bounded below by Λ_∞ and is upper bounded by the chaotic semistability curve Λ_1^* (in the cases of strong and weak Allee's functions) or is upper bounded by the curve of nonadmissibility Λ_{NA} (in the case of functions with no Allee effect), both of these bifurcation curves are studied in Section 3.2, see also Property 2. In this region are observed all fold and flip bifurcation curves of the cycle of order different than 2^n, Λ_k, with $k \neq 2^n$, identified and ordered in the "box-within-a-box" bifurcation structure, see [9] and [16]. See Figs.9 and 10.

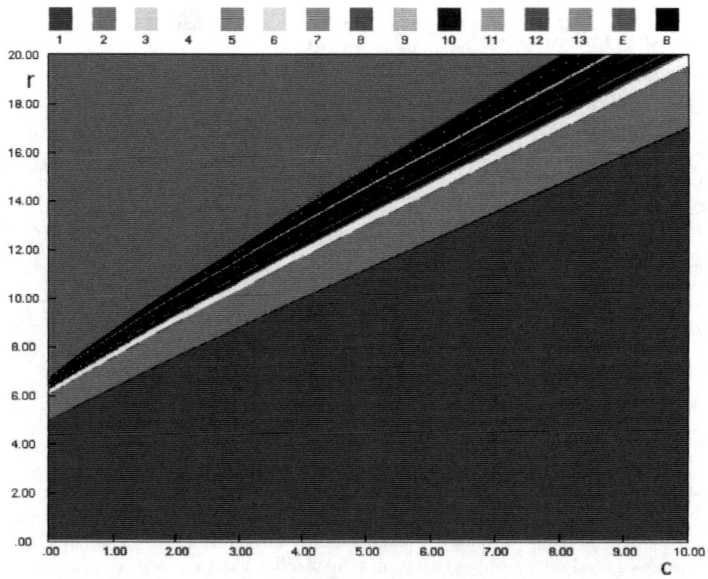

Figure 10. Bifurcation diagram of von Bertalanffy's population size functions with no Allee effect $f(x; r, K, E, C)$, at $K = 10$ and $E = -0.02$, in the (C, r) parameters plane. The blue region is the stability region of the fixed point O_r, stated in Lemma 5 (*ii*). The region between the blue and the gray ones corresponds to period doubling region and chaotic region (existence region of cycles as shown on top of figure), stated in Lemma 5 (*ii*). The gray region is the no admissible region.

3.2. Chaotic semistability and nonadmissibility bifurcation curves: Allee's effect region

However from Section 2.2, for all von Bertalanffy's population size functions with strong and weak Allee effects f, with $0 \leq E < K$, $C > 0$ and initial states in $[A_r, A_r^*]$, given (A2) and (A3), it is verified that

$$c \in]A_r, A_r^*[: \ f(c; r, K, E, C) = A_r^* \Leftrightarrow \ f^2(c; r, K, E, C) = A_r, \tag{11}$$

i.e., the maximum size growth of the population is equal to the critical density. In this case, the populations can persist in a semistable chaotic interval. In the (C, r) parameters plane, the densities to which is verified the chaotic behavior previously described by Eq.(11) characterize the *chaotic semistability curve.*

Definition 3 *Let $f(x; r, K, E, C)$ be the von Bertalanffy population size functions, Eq.(5), with $0 \leq E < K$, $C > 0$, $r > 0$, satisfying (A1)-(A4). The chaotic semistability curve, denoted by Λ_1^*, is defined by*

$$\Lambda_1^* = \left\{(C, r) \in \mathbf{R}^2 : f^2(c; r, K, E, C) = A_r\right\}, \tag{12}$$

where A_r is the first positive fixed point of f and c is the critical point defined in $(A3)$.

This bifurcation curve defines where the chaotic region finishes and begins another region with a different kind of dynamics, the *Allee effect region*, see Figs.6, 7, 8 and 9. The Allee effect region is characterized by an essential extinction, i.e., a populational occurrence where the maximum size growth of the population exceeds the critical density and the populations are almost surely doomed to extinction, verifying $f^2(c; r, K, E, C) < A_r$, see Lemma 3.

Definition 4 *Let $f(x; r, K, E, C)$ be the von Bertalanffy population size functions, Eq.(5), with $0 \leq E < K$, $C > 0$, $r > 0$, satisfying (A1)-(A4). The Allee effect region, denoted by R_{AE}, is defined by*

$$R_{AE} = \left\{(C, r) \in \mathbf{R}^2 : A_r^* < f(c; r, K, E, C) < 1\right\},$$

where $A_r^* = max\{f^{-1}(A_r; r, K, E, C)\}$, A_r is the first positive fixed point of f and c is the critical point defined in (A3).

If f is a family of strong or weak Allee's functions, then the Allee effect region is limited superiorly by the fullshift curve or curve of nonadmissibility. Otherwise, if f is a family of functions with no Allee effect, then the chaotic region is also limited superiorly by the fullshift curve or curve of nonadmissibility, which is defined as follows.

Definition 5 *Let $f(x; r, K, E, C)$ be the von Bertalanffy population size functions, Eq.(5), with $|E| < K$, $C > 0$, $r > 0$, satisfying (A1)-(A4) and under the conditions of Definitions 1 and 2. The nonadmissibility curve, denoted by Λ_{NA}, is defined by*

$$\Lambda_{NA} = \left\{(C, r) \in \mathbf{R}^2 : f(c; r, K, E, C) = 1\right\}, \tag{13}$$

where c is the critical point defined in $(A3)$.

This bifurcation curve separates the Allee effect region or the chaotic region and the no admissible region, see Figs.6(b), 7(b), 9 and 10. In the no admissible region the graphic of any von Bertalanffy's population size function is no longer totally in the invariant set $[0, 1]$. Almost all trajectories of f (besides a hyperbolic set of zero measure) leave the interval $[0, 1]$ and either escape to infinity. The maps under these conditions are not good models for populations dynamics. For more details about the bifurcation structure of similar models to this one see for example [18], [19], [20], [21] and [22].

From the above definitions, we can claim the following properties:

Property 2 *Under the conditions of Definitions 3, 4 and 5, it is established that the existence of Allee's effect region R_{AE} implies that:*

(i) if $0 \leq E < K$ and $C > 0$, then exists Λ_1^* and $\Lambda_1^* \neq \Lambda_{NA}$, see Figs.6, 7 and 9;

(ii) if $-K < E < 0$ and $C > 0$, then not exists Λ_1^*, see Figs.8 and 10.

Property 3 *Under the conditions of Definition 5, it is established that the existence of the nonadmissibility curve Λ_{NA} implies that:*

(i) if $0 \leq E < K$ and $C > 0$, then Λ_{NA} is the bifurcation curve that separates the Allee effect region R_{AE} and the no admissible region, see Figs.6, 7 and 9;

(ii) if $-K < E < 0$ and $C > 0$, then Λ_{NA} is the bifurcation curve that separates the chaotic region and the no admissible region, see Figs.8 and 10.

In Figs.9 and 10, we present the bifurcation diagrams of von Bertalanffy's population size functions $f(x;r,K,E,C)$, with strong Allee effect ($E = 0.02$) and with no Allee effect ($E = -0.02$), at $K = 10$, in the (C,r) parameters plane. In these figures is quite clear the introduction of Allee effect in the models under study, the unconditional extinction region and the Allee effect region R_{AE} or essential extinction region are well highlighted. We remark that the bifurcation diagram of von Bertalanffy's population size functions with weak Allee effect ($E = 0$) is similar to the bifurcation diagram of von Bertalanffy's population size functions with strong Allee effect, such as the bifurcations curves presented in Figs.6 and 7.

3.3. Cusp points and the "foliated" structure of the (C,r) parameters plane, with respect to Allee's limit E

In [12] is given a necessary and sufficient condition for the existence of a cusp point on a fold bifurcation curve relative to a cycle of order n, in a parameter plane of a 1D map, see also [6]. Generically, to von Bertalanffy's population size functions $f(x;r,K,E,C)$, given by Eq.(5), with $|E| < K$, $C > 0$ and $r > 0$, that condition is given by the following expressions:

$$\begin{cases} H_1(x;r,K,E,C) \equiv f^n(x;r,K,E,C) - x = 0, \\ H_2(x;r,K,E,C) \equiv \frac{\partial f^n}{\partial x}(x;r,K,E,C) - 1 = 0, \\ D_r(x;r,K,E,C) = 0,\ D_C(x;r,K,E,C) = 0, \\ \begin{cases} D_x(x;r,K,E,C) \times \frac{\partial D_r}{\partial x}(x;r,K,E,C) \neq 0, \\ \text{or} \\ D_x(x;r,K,E,C) \times \frac{\partial D_C}{\partial x}(x;r,K,E,C) \neq 0, \end{cases} \end{cases} \tag{14}$$

where D_x, D_r and D_C are the minors of the Jacobian matrix $\frac{\partial(H_1,H_2)}{\partial(x,r,C)}$.

Note that, we use the parameters r and C to establish the minors, because we obtain explicit expressions for the existence of the cusp points on the fold bifurcation curves of the fixed point 0 and of the first positive fixed point A_r. For von Bertalanffy's population size functions $f(x;r,K,E,C)$, the necessary and sufficient condition given by Eq.(14) can be written in the following form:

$$\begin{cases} x = -\frac{2E}{K} \\ r = -\frac{2E\sqrt[3]{-\frac{2E}{K}}}{2\sqrt[3]{-\frac{2E}{K}}-1}, \quad \forall -K < E \leq 0 \\ C = -\frac{2E\left(\sqrt[3]{-\frac{2E}{K}}-2\right)}{2\sqrt[3]{-\frac{2E}{K}}-1} \end{cases} . \tag{15}$$

Consequently, from Eq.(15) results the next property:

Property 4 *Let $f(x;r,K,E,C)$ be the von Bertalanffy population size functions, Eq.(5), with $|E| < K$, $C > 0$ and $r > 0$, satisfying (A1)-(A6). If $0 < E < K$, then von Bertalanffy's population size functions with strong Allee effect not admit cusp points, given by Eq.(15).*

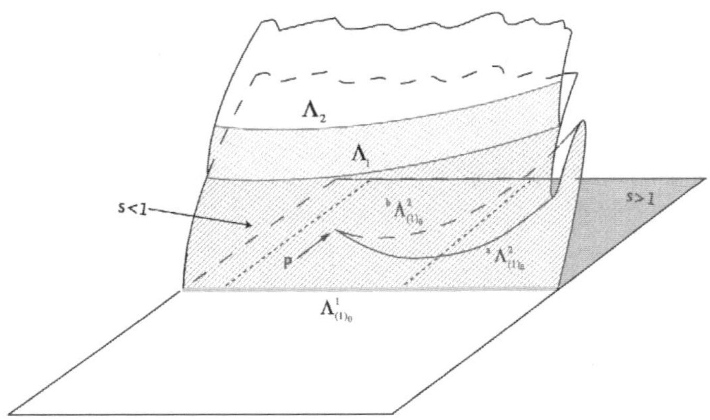

Figure 11. Foliated bifurcation structure of von Bertalanffy's population size functions $f(x;r,K,E,C)$, at $K=10$ and $E=-0.02$. $\Lambda^1_{(1)_0}$, ${}^a\Lambda^2_{(1)_0}$ and ${}^b\Lambda^2_{(1)_0}$ are the fold bifurcation curves of different cycles of order $n=1$; Λ_1 and Λ_2 are the flip bifurcation curves of the cycles of order $n=1,2$, respectively. P is the cusp point, given by Eq.(15) and $s=\frac{\partial f}{\partial x}(x;r,K,E,C)$.

Note that the existence of cusp points is dependent again of the Allee's limit E. In particular, for $E=-0.02$ and $K=10$ we obtain $x=0.004$, $r=0.4651589...$ and $C=0.1079095...$, see the cusp point P at Fig.8. In this case, for von Bertalanffy's population size functions with no Allee effect, the cusp point P corresponds to the intersection between the fold bifurcation curves ${}^a\Lambda^2_{(1)_0}$, of the fixed points A_r and B_r, and ${}^b\Lambda^2_{(1)_0}$, of the fixed points O_r and A_r, see also Fig.11. On the other hand, for $E=0$ and $K=10$ we obtain the fixed point $x=0$ and $r=C=0$, see the cusp point P at Figs.7(a) and 12. Similar phenomenon has been observed in [19] just to weak Allee effect and in [11] for a continuous time system.

In the (C,r) parameters plane, a given point is generally related to different $(n;j)$ cycles with diverse stability properties. For a better understanding of the bifurcation structures the (C,r) parameters plane must be considered as made up of sheets ("foliated" in the sense of Mira, see [16]), each one being associated with a given cycle $(n;j)$ or other singularities. A complete knowledge of the bifurcations organization implies the identification of the sheets "geometry" in a three-dimensional auxiliary qualitative space $(C,r,|z|)$. Here, the $|z|$ is a qualitative norm associated with the considered cycle. This identification amounts to seeing how to pass continuously from a sheet to another one following a continuous path in the (C,r) parameters plane, i.e., to knowing the possible communications between sheets, see for example [6].

In the simplest case, a fold bifurcation curve ($s=+1$, where $s=\frac{\partial f}{\partial x}(x;r,K,E,C)$) is the junction of two sheets, one related to an unstable (or repulsive) $(n;j)$ cycle, with $s>1$, the other to a stable (or attractive) $(n;j)$ cycle, with $-1<s<1$. A flip bifurcation curve ($s=-1$) is the junction of three sheets, one associated with a $(n;j)$ cycle with $-1<s<1$, another with an unstable (or repulsive) $(n;j)$ cycle, with $s<-1$, the third being related to a cycle $(2n;j)$ with $-1<s<1$. In Figs.11 and 12, the blue-mixed domain corresponds to the existence of the fixed point $x=0$, which is unstable in Fig.11 ($s>1$) and is stable in Fig.12 ($s<1$); the red-mixed domain corresponds to the existence of the stable fixed point A_r and the green-mixed domain corresponds to the existence of the stable order 2-cycles.

This analysis allows us to understand the transition from the weak Allee effect to no Allee effect with respect to Allee's limit E. Note that the parameter E is directly associated with the classification of von Bertalanffy's population size functions in functions with weak Allee effect and functions with no Allee effect, according to Property 1 and Definition 2, respectively. In [23]

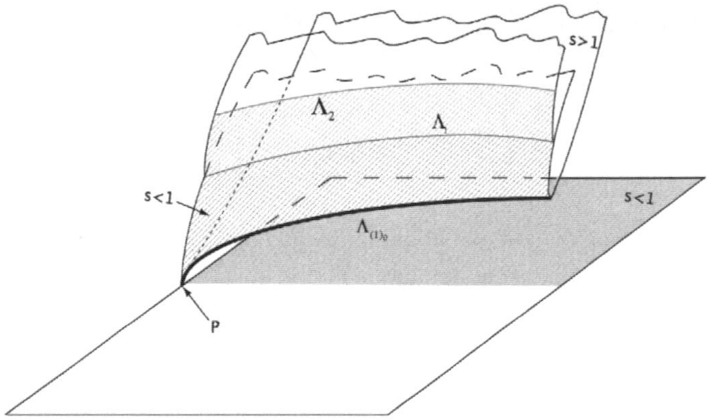

Figure 12. Foliated bifurcation structure of von Bertalanffy's population size functions $f(x; r, K, E, C)$, at $K = 10$ and $E = 0$. $\Lambda_{(1)_0}$ is the fold bifurcation curve; Λ_1 and Λ_2 are the flip bifurcation curves of the cycles of order $n = 1, 2$, respectively. P is the cusp point, given by Eq.(15) and $s = \frac{\partial f}{\partial x}(x; r, K, E, C)$.

is presented a similar analysis for Richards' growth models associated to a dovetail structure.

For the parameter values considered, we verify that when the parameter E tends toward 0, by negative values, the cusp point P tends to the cusp point $P = (0, 0)$ (evolution from Fig.11 to Fig.12). Simultaneously, the fold bifurcation curves ${}^a\Lambda^2_{(1)_0}$, of the fixed points A_r and B_r, and ${}^b\Lambda^2_{(1)_0}$, of the fixed points O_r and A_r, tend toward infinity at $E = 0$ and disappear. The bifurcation curve $\Lambda_{(1)_0}$ tends toward $(C, r) = (0, 0)$ and only remaining the cusp point $P = (0, 0)$. The cusp point $P = (0, 0)$ will make the connection between the sheet of the fixed point $x = 0$ and the sheets of the other fixed points. On the other hand, for $E > 0$ the foliated bifurcation structure is incomplete, because everything takes place in the parameters plane $C < 0$ and $r < 0$, but there is always three fixed points (0, A_r and B_r) and the foliated bifurcation structure is similar to the case $E = 0$, see Fig.12.

4. Conclusion and discussion

The distinction between strong and weak Allee effects is a very important research subject, although most authors neglect the last one and almost exclusively consider the strong Allee effect, as if it were the only one. In this sense, the main goal of this work has been the definition and investigation of new von Bertalanffy's population size functions with strong and weak Allee effects, or no Allee effect, from the point of view of stability analysis and bifurcation theory.

The introduction of strong Allee effect in von Bertalanffy's growth models was made in [18], through the ordinary differential equations defined. However, in the context of the population dynamic discrete models with the growth of the population given by a family of von Bertalanffy's population size functions, the unexpected appearance of weak Allee effect in these models is something new, obtained from the correction factor of rational type considered. In this framework, Definitions 1 and 2, and Property 1 are a significant contribution that establishes the transition from strong Allee effect to no Allee effect, involving the crossing through the weak Allee effect, for von Bertalanffy's population size functions proposed, depending on the evolution of the rarefaction critical density or Allee's limit E. Again, we highlight that we use the concepts of strong and weak Allee's effects to *population growth rates*, because we consider the population size functions associated with Eq.(3). The extinction phenomenon was analyzed in Lemmas 1, 2 and 3, providing a detailed characterization of the unconditional extinction and the essential extinction.

Another central point of our investigation was the study of bifurcation structures of von Bertalanffy's population size functions, in the (C,r) two-dimensional parameters space, dependent on the evolution of the Allee limit E. These analysis is founded on the configurations of classical fold and flip bifurcation curves. The definitions of chaotic semistability and nonadmissibility bifurcation curves were crucial in the description of Allee's effect region. In this sense, Definitions 3, 4 and 5, and Properties 2 and 3 complete the characterization and the discussion of the bifurcation structures to this family of 1D maps. Moreover, this discussion is complemented with the analysis of the existence of cusp points on a fold bifurcation curve, given by Property 4, and the corresponding "foliated" structure of the (C,r) parameters plane, with respect to Allee's limit E. A qualitative representation of the evolution of the bifurcation structure at $(C,r,|z|)$ parameters space is obtained, where the Allee limit E varies, see Figs.11 and 12. This study highlights the complex structure of von Bertalanffy's population size functions under a rational correction, as regards the number of fixed points and their dynamics.

One of the major discussions of this work is the use of the correction factor of rational type and the parameter C be positive. In [25] and [26] the parameter C can be null or negative, that because the correction factors are of the polynomial type. In the study of the bifurcation structures, in the (C,r) parameters plane, the consideration of different correction factors is evident: while in [25] and [26] are defined and characterized big bang bifurcations of the "box-within-a-box" fractal type, in the present work this special kind of bifurcation is not identified. Certainly, our work is an original contribution to the study of population dynamics and extinction phenomenon.

Acknowledgment: Research partially funded by FCT - Fundação para a Ciência e a Tecnologia, Portugal, through the project UID/MAT/00006/2013 (CEAUL), the research grant SFRH/BSAB/128144/2016 and ISEL.

[1] Aleixo S M, Rocha J L, Pestana D D 2009 *Am. Inst. of Phys.* **1124** 3–12
[2] Aleixo S M, Rocha J L 2012 *J. Comput. Inf. Technol.* **20** 201–207
[3] Allee W C 1931 *Animal Aggregations* (University of Chicago Press: Chicago)
[4] Boukal D S, Berec L 2002 *Theor. Biol.* **218** 375–394
[5] Cailliet G M, Smith W D, Mollet H F, Goldman K J 2006 *Environ. Biol. Fish* **77** 211–228
[6] Carcassès J-P 1993 *Int. J. Bifurc. Chaos* **3** (4) 869–902
[7] Duman O, Merdan H 2009 *Chaos Solit. Fract.* **41** 1218–1222
[8] Essington T E, Kitchell J F, Walters C J 2001 *Can. J. Fish Aquat. Sci.* **58** 2129–2138
[9] Fournier-Prunaret D 1991 *Int. J. Bifurc. Chaos* **1** 823–838
[10] Fujikawa H, Kai A, Morozomi S 2004 *Food Microbiol.* **21** 501–509
[11] González-Olivares E, González-Yañez B, Mena-Lorca J, Flores J D 2013 *Math. Biosci. Eng.* (10) 345–367
[12] Kawakami H 1984 *IEEE Trans. Circuits Syst.* **31** 248–260
[13] Livadiotis G, Elaydi S 2012 *J. Bio. Dyn.* **6** 959–973
[14] Lotka A J 1956 *Elements of Mathematical Biology* (Dover: New York)
[15] Melo W, van Strien S 1993 *One-Dimensional Dynamics* (Springer: New York)
[16] Mira C, Gardini L, Barugola A, Cathala J-C 1996 *Chaotic Dynamics in Two-Dimensional Noninvertible Maps* (World Scientific: Singapore)
[17] Pal P J, Saha T 2015 *Chaos Solit. Fract.* **73** 36–63
[18] Rocha J L, Aleixo S M 2012 *Discuss. Math. Probab. Stat.* **12** 35–45
[19] Rocha J L, Fournier-Prunaret D, Taha A-K 2013 *Discrete Contin. Dyn. Syst.-Ser.B* **18** (9) 2397–2425
[20] Rocha J L, Aleixo S M 2013 *Math. Biosci. Eng.* **10** (2) 379–398
[21] Rocha J L, Aleixo S M 2013 *Discrete Contin. Dyn. Syst.-Ser.B* **18** (3) 783–795
[22] Rocha J L, Fournier-Prunaret D, Taha A-K 2014 *Nonli. Dyn.* **77** (4) 1749–1771
[23] Rocha J L, Taha A-K, Fournier-Prunaret D 2015 *Nonlinear Maps and their Applications* Chp. 16, ed López-Ruiz R et al. (Springer Proc. in Maths. and Statist.) **112** 253–267
[24] Rocha J L, Taha A-K, Fournier-Prunaret D 2015 *Discrete Contin. Dyn. Syst.-Ser.B* **20** (9) 3131–3163
[25] Rocha J L, Taha A-K, Fournier-Prunaret D 2016 *J. Phys. Conf. Ser.* **692** 012007
[26] Rocha J L, Taha A-K, Fournier-Prunaret D 2016 *Nonlinear Dyn.* **82** (4) 607–626
[27] Saha B, Bhowmick A R, Chattopadhyay J, Bhattacharya S 2013 *Ecol. Model* **250** 72–80
[28] Schreiber S 2001 *J. Math. Biol.* **42** 239–260
[29] von Bertalanffy L 1938 *Hum. Biol.* **10** 181–213
[30] von Bertalanffy L 1957 *Q. Rev. of Biol.* **32** 217–231

IOP Conf. Series: Journal of Physics: Conf. Series **990** (2018) 012012 doi:10.1088/1742-6596/990/1/012012

Averaging of Random Flows of Linear and Nonlinear Maps

Vsevolod Zh. Sakbaev

Department of General Mathematics, Moscow Institute of Physics and Technology, Institutskii per. 9, Dolgoprudny, Moscow Region, 141700, Russia

E-mail: fumi2003@mail.ru

Abstract. The procedure of generalized averaging of random strongly continuous semigroups of linear operators is investigated by means of Feynman-Chernoff iterations. The main features of this procedure is the fact that the generalized mean value of random semigroup is the semigroup. The generalized mean value of the random generator of a random C_0-semigroup is defined as the generator of generalized mean value of a random semigroup. This procedure of averaging of a random semigroup of linear operators is extended onto the averaging procedure for random semiflow of nonlinear maps. For this aim the semiflows of nonlinear mappings of some linear space $\mathcal{L}$ is presented by the semigroup $\mathbf{U}_t$, $t \geq 0$, of linear operators in the space $\mathcal{H}$ of functions on the space $\mathcal{L}$ which is integrable with respect to some measure on the space $\mathcal{L}$. The conditions on the measure and on the semiflow sufficient for boundedness of the linear operators $\mathbf{U}_t$ in the space $\mathcal{H}$ are obtained. The sufficient condition of the existence of the generalized mean value of a random semiflow is investigated. The estimates for the probability of the deviation of the values of compositions of independent random semiflows from its mean value similar to the large numbers law are obtained.

1. Introduction

In this paper we investigate a random transformation of a real Hilbert space E (finite or infinite dimensional) and random maps of the real interval into a space of transformations of Hilbert space E including random semigroups of linear operators in the space E and random flows of nonlinear transformations of the space E.

A random operator-function of the real argument is defined as the measurable map ξ of some space with the probability measure $(\Omega, \mathcal{A}, \mu)$ into the topological vector space (Y, τ) of operator-valued functions of the real argument endowed with Borel σ-algebra. If the values of this map is the one-parametric group or semigroup with respect to the real parameter then this random operator function is called the random one-parameter group (semigroup) of operators in the space E or the random flow (semiflow) of maps of the space E into itself.

Let us consider some examples of arising of random semigroups.

1. A random semigroup can be realized as the regularization of singular operator of ill-posed problem by the directed set of regular operators of well-posed problems (see [1, 2, 3]). The sequence $\{\mathbf{L}_\epsilon,\ \epsilon \in (0,1),\ \epsilon \to 0\}$ of generators of semigroups is the approximation of the operator $\mathbf{L}$ of ill-posed Cauchy problem. If the set of regularized generators is endowed by the structure of the space with the measure $((0,1), \mathcal{A}, \mu)$ then the family of semigroups

$$e^{t\mathbf{L}_\epsilon},\ t \geq 0;\ \epsilon \in (0,1),$$

is the random variable with values in the set of semigroups.

2. A random group or semigroup can arise if the coefficients of a differential operator **L** or parameters of Hamilton function on the phase space are random variables (see [4, 5]).

3. The choice of the quantization procedure for defining of quantum Hamiltonian by the Hamiltonian of a classical system is not unique. If the set of quantization of classical Hamiltonian systems is endowed with the structure of probability space then, the random quantization procedure defines the random quantum Hamiltonian (see [6]) and the random unitary group.

A random variable with values in the topological vector space or in the topological space is defined by the following way.

Let (Y, τ) be a Hausdorff topological space and $\mathcal{A}_\tau$ be the minimal algebra containing any open set of the topology τ. If $\mathcal{A}_\tau$ is the σ-algebra then the symbol $\mathcal{A}_\tau$ denotes Borel σ-algebra of topological space (Y, τ).

Let $(\Omega, \mathcal{A}, \mu)$ be a probability space by the above. Y-valued random variable is the measurable map $\xi: \Omega \to Y$ of the probability space $(\Omega, \mathcal{A}, \mu)$ into the measurable space $(Y, \mathcal{A}_\tau)$, where $\mathcal{A}_\tau$ is the minimal algebra containing all open sets of the topology τ.

Definition 1. The measurable map $\xi: \Omega \to Y$ of the probability space $(\Omega, \mathcal{A}, \mu)$ into the measurable space $(Y, \mathcal{A}_\tau)$ is caled the random variable in topological space (Y, τ).

The Hausdorff topological space (Y, τ) can be realized as the space of scalar functions (random Hamiltonian), as the space of vector valued functions (random vector field) or as the space of operator valued functions (random flows in the phase space).

1a) (The random Hamiltonian). Let $P = Q = R^d$ and $E = Q \oplus P$ be the phase space of some Hamiltonian system. If the Hausdorff topological space (Y, τ) is the Banach space $C^1(E)$ of continuously differentiable real-valued functions on the phase space E then the random variable ξ with values in the space $C^1(E)$ is the random classical Hamiltonian.

1b) (The random vector field in the phase space). Let the phase space of the ordinary differential equation system be the real Hilbert space E (with finite or infinite dimension).

Let $C_b^1(E, E)$ be the Banach space of continuously differentiable (in Freshet sense) transformations of the space E endowed with the norm

$$\|\Phi\|_{C_b^1} = \sup_{\|x\|_E > 0} \frac{\|\Phi(x)\|_E}{\|x\|_E} + \sup_{x \in E} \sup_{\|h\|_E = 1} \|\Phi'(x)h\|_E.$$

Then the random vector field on the phase space E is the random variable in the topological space $Y = C_b^1(E, E)$ with the topology of Banach space.

1c) (The random flow of nonlinear transformations of the phase space). If the Hausdorff topological space (Y, τ) is the Banach space $C(R, C_b^1(E, E))$ of continuous maps of real line into the Banach space $C_b^1(E, E)$ then the random variable ξ with values in the space $C(R, C_b^1(E, E))$ is called random continuous operator-valued function. This random variable ξ is called the random flow if for any $\omega \in \Omega$ the equality $\xi_\omega(t + s, z) = \xi_\omega(t, \xi_\omega(s, z))$ holds for any $t, s \in R$ and any $z \in E$.

The mean value of random Hamiltonian is the function on the phase space. The mean value of random flow is the map $R \to E$ which can has no group property.

What relationship does exist between this mean values?

The Hausdorff topological space (Y, τ) can be realized as the space of strongly continuous functions with values in the space of bounded linear operators in Hilbert space (the random semigroup) or as the set of self-adjoint operators in Hilbert space (the random generator).

2a) *The random semigroup of linear operators.* Let H be a complex Hilbert space. Let $Y_s \equiv C_s(R, B(H))$ be the space of strongly continuous maps of the axis R into the Banach space $B(H)$ of bounded linear operators in the Hilbert space H. The topology τ_s in the space $C_s(R, B(H))$ is generated by the family of the functionals $\{\phi_{T,v},\ T \in R, v \in H\}$ where $\phi_{T,v}(f) = \sup\limits_{t\in[-T,T]} \|f(t)v\|_H\ \forall\ f \in C_s(R, B(H))$. Let (Y_s, τ_S) be the topological vector space $Y_s \equiv C_s(R_+, B(H))$ endowed with the topology τ_s. Let $\mathcal{A}_{\tau_S}$ be the minimal algebra containing the topology τ_S.

The random variable ξ with values in the space Y_s is called random strongly continuous operator valued function. The Y_s-valued random variable ξ is called the random strongly continuous unitary group (semigroup) if for any $\omega \in \Omega$ the value ξ_ω is strongly continuous unitary group (semigroup) in the space H.

A random unitary semigroup is a random variable ξ with the values in the measurable space $(Y_s, \mathcal{A}_{\tau_S})$ such that every value ξ_ω of the map ξ is a unitary semigroup.

2b) *The random Hamiltonian and the random generator of a random semigroup.* According to Stone theorem there is the one-to-one correspondence between the set of strongly continuous unitary groups in the space H and the set $SA(H)$ of self-adjoint operators in the space H. This bijection introduces the topology τ_{sa} of the set $SA(H)$ from the space $C_s(R, B(H))$ with the topology τ_s.

A random variable ξ with values in the topological space $(SA(H), \tau_{sa})$ is called a random quantum Hamiltonian. Thus there is the one-to-one correspondence between the set of random unitary groups and the set of random quantum Hamiltonians.

The plan of investigation is the following. We define the mean value and covariation of the above random variables and obtain the estimates of deviations of random variables from its mean values. Firstly, we realize this plan for the random strongly continuous unitary groups. According to bijection between the strongly continuous unitary groups and self-adjoint operators in Hilbert space the mean value of random Hamiltonian and the deviation of random Hamiltonian from its mean value are defined by means of the same characteristics of the random unitary group.

The aim of the investigation of random semigroups of linear operators and its random generators is to obtain the asymptotic properties of compositions of n independent random linear operators in a Hilbert space for $n \to \infty$. This properties had been investigated in the papers [6, 7, 8, 9]. Note that the usual mean value of the random semigroup of linear operators is the element of the space Y_s without the semigroup properties (see [6, 9]). The generalized mean value of a random semigroup and the generalized mean value of a random generator are defined and studied. This method is based on consideration of the sequence of compositions of n independent random linear operators and its mean value $(\mathbf{T}(\frac{t}{n}))^n$, $t \in R$, and passage to the limit for the sequence of iterations $\{(\mathbf{T}(\frac{t}{n}))^n,\ t \in R\}$ (so-called *Feynman-Chernoff iterations*). Thus, the method of averaging of random semigroups of linear operators is based on the Chernoff theorem ([10, 11]) and its inversion ([6, 7, 12]).

The generalized mean value of a random semigroup has the following properties: 1) it is the semigroup and 2) it coincides with the usual mean values of a random semigroup if the last operator function has the semigroup property. The generalized mean value of a random generator of a strong continuous semigroup has the following properties: 1) it is the generator of a strong continuous semigroup and 2) it coincides with the usual mean values of a random generator of a semigroup if the last mean value exists. In this paper we give the short review of the results of the papers [6, 7, 8, 9].

The main interest of this article is the extension of the methods of averaging of random linear operators and random groups of linear operators onto the random nonlinear transformations and flows of nonlinear transformations. For this aim we consider the construction of the representation of random groups of nonlinear transformations of Hilbert space E by means of random groups of linear operators in Hilbert space $\mathcal{H}(E)$. Here $\mathcal{H}(E)$ is the space of square

IOP Conf. Series: Journal of Physics: Conf. Series **990** (2018) 012012 doi:10.1088/1742-6596/990/1/012012

integrable with respect to some measure λ on the space E complex-valued functions on the space E.

Transformation $X:\ E\to E$ is called to be acceptable by measure λ on the space E if the shifted measure $\lambda\circ X$ on the space E is absolutely continuous with respect to λ, and the density of measure $\lambda\circ X$ with respect to λ is λ-measurable and λ-essentially bounded. In this case measure λ and map X are called to be coordinated with respect to each other.

We obtain the conditions on the measure λ and on the map X which are sufficient for their coordination in the above sense.

The asymptotic properties of compositions of n independent nonlinear transformations of Hilbert space E in terms of its linear representation by the compositions of n independent linear operators in Hilbert space $\mathcal{H}(E)$ are obtained.

For investigation of a random flow of nonlinear maps of the space E we endow the space E by the structure of measurable space $(E,\mathcal{A})$ with a measure λ. Then (under some additional assumptions) a flow of nonlinear maps of the space E can be represented by the group of linear operators in the space $\mathcal{H}(E)=L_2(E,\mathcal{A},\lambda,\mathbf{C})$. Therefore, a random flow of nonlinear maps of the space E can be represented as a random group of linear operators in the space $\mathcal{H}(E)$. Hence, the probabilistic characteristics of a random flow of nonlinear maps of the space E can be described by the same probabilistic characteristics of a corresponding random group of linear operators in the space $\mathcal{H}(E)$.

2. Random semigroups of linear operators and its generators

Let (Y_s,τ_s) be the topological vector space $Y_s\equiv C_s(R_+,B(H))$ of strongly continuous maps of semiaxis R_+ into the Banach space $B(H)$ of bounded linear maps of a Hilbert space H endowed with corresponding topology (see item 2a).

Definition 2 Let ξ be a random semigroup $\xi:\Omega\to Y_s$. The map $\mathrm{M}\xi:\ R_+\to B(H)$ is called the mean value of random semigroup $\xi:\ \Omega\to Y_s$ if the equality

$$\mathrm{M}\xi=\int\limits_{\Omega}\xi_\omega d\mu(\omega),$$

holds in the sense of Pettis integral, i.e. if for any $t\in R_+,\ u,v\in H$ the equality (1) holds.

$$\langle\mathrm{M}\xi(t)u,v\rangle=\int\limits_{\Omega}\langle\xi_\omega(t)u,v\rangle d\mu(\omega), \tag{1}$$

where the symbol $\langle\cdot,\cdot\rangle$ denotes the scalar product in the space H.

Theorem 1. (See [7, 8]) *If the measurable map $\xi:\ \Omega\to Y_s$ is uniformly bounded (A) and densely strongly equicontinuous (B) then $\mathrm{M}\xi\in Y_s$.*

Conditions (A) and (B) mean respectively:

(A) $\exists C>0:\ \|\xi_\omega(t)\|_{B(H)}<C\ \forall t\geq 0,\ \omega\in\Omega$

(B) $\exists D\subset H$, where D is dense subspace of H, such that $\forall u\in D;\ \forall\sigma>0\ \exists\delta>0:$ $\|\xi_\omega(t')u-\xi_\omega(t'')u\|<\sigma\ \forall\ \omega\in\Omega;\ |t'-t''|<\delta$.

Give example of the random semigroup such that its mean value has no semigroup property.

In fact, let $\mathbf{A}=\mathbf{A}^*$. Then, for the mean value of two-valued random semigroup $\xi_\pm(t)=e^{\pm it\mathbf{A}},\ t\geq 0$ such that $P(\{\xi=e^{i\mathbf{A}t}\})=\frac{1}{2}=P(\{\xi=e^{-i\mathbf{A}t}\})$ the following holds:

$$\mathrm{M}\xi(t)=cos(\mathbf{A}t)=\frac{1}{2}e^{i\mathbf{A}t}+\frac{1}{2}e^{-i\mathbf{A}t},\ t\geq 0.$$

Here $\mathrm{M}\xi(t)$ is not semigroup.

IOP Conf. Series: Journal of Physics: Conf. Series **990** (2018) 012012 doi:10.1088/1742-6596/990/1/012012

To reconstruct the semigroup property of the mean value of random semigroup and to define the generalized mean value of random semigroup which is some semigroup we use the following notion.

Definition 3. (See [13, 6]) The functions $\mathbf{F}$, $\mathbf{G} \in C_s(R_+, B(H))$ such that $\mathbf{F}(0) = \mathbf{G}(0) = \mathbf{I}$ are equivalent in Chernoff sense (denote it by $\mathbf{F} \sim \mathbf{G}$) if for any $T > 0$ and $u \in H$ the equality $\lim\limits_{n\to\infty} \sup\limits_{t\in[0,T]} \|((\mathbf{G}(\frac{t}{n}))^n - (\mathbf{F}(\frac{t}{n}))^n)u\| = 0$ holds.

The function $\mathbf{F}$ is equivalent to the semigroup $\mathbf{U}$ if $(\mathbf{F}(\frac{t}{n}))^n \to \mathbf{U}(t)$ uniformly on any segment. The representation of the semigroup $\mathbf{U}$ by the limit $\lim\limits_{n\to\infty} (\mathbf{F}(\frac{t}{n}))^n$ is called Feynman formula ([13, 6]).

Definition 4. The semigroup $\mathbf{U} \in C_s(R_+, B(H))$ is the general mean value of the random semigroup ξ if $\mathbf{U} \sim \mathrm{M}\xi$.

Theorem 2 ([12, 7]). *Let the function $\mathbf{F} \in C_s(R_+, B(H))$ satisfy the conditions $\mathbf{F}(0) = \mathbf{I}$ and $\|\mathbf{F}(t)\|_{B(H)} \leq e^{at}$, $t \in R_+$ for some $a \in R$. Let the sequence $\{\mathbf{G}_n\}$ of operator functions*

$$\mathbf{G}_n(t) = (\mathbf{F}(\frac{t}{n}))^n,\ n \in \mathbf{N},\ t \in R_+,$$

converge in the strong operator topology uniformly on any segment $[0, T]$. Then the limit function is C_0-semigroup in the space H.

Theorem 3. *([7, 6]) Let $\xi : \Omega \to C_s(R_+, B(H))$ be a random semigroup such that its set of values ξ_ω, $\omega \in \Omega$, is the strongly continuous contractive semigroup in the space H with generators $\mathbf{L}_\omega$, $\omega \in \Omega$. Let $D \subset H$ be the essential domain of linear operators $\mathbf{L}_\omega$, $\omega \in \Omega$ and $\int\limits_\Omega \|\mathbf{L}_\omega x\| d\mu(\omega) < \infty$ for any $x \in D$.*

Let the operator $\mathbf{S}u = \int\limits_\Omega \mathbf{L}_\omega u d\mu(\omega)$, $u \in D$, be essentially self-adjoint. Then

$$e^{t\mathbf{S}} \sim \mathrm{M}\xi(t),$$

where $\mathrm{M}\xi(t)$ is the usual mean value of the random semigroup.

Under the assumptions of Theorem 3 the generalized mean value of the random semigroup is the semigroup generated by the mean value of the random generator. But the result of Theorem 3 give the opportunity to define the mean value of the generator of the random semigroup

Corollary 1. *Let $\{\mu_n\}$ be a sequence of positive numbers satisfying: $\sum\limits_{n=1}^{\infty} \mu_n = 1$. Let $\{\mathbf{L}_j\}$ be uniformly bounded sequence in the space $B(X)$. Then $\sum\limits_{j=1}^{\infty} \mu_j e^{t\mathbf{L}_j} \sim e^{t\sum\limits_{j=1}^{\infty} \mu_j \mathbf{L}_j}$.*

Since there is the one-to-one correspondence Ψ between the set of C_0-semigroups in the space H as the subset of topological vector space Y_s and the set $G(H)$ of generators of C_0-semigroups as the subset of the set of closed linear operators in the space H (Hille-Yosida theorem) then, we can endow the set $G(H)$ with the structure of topological space with the topology inducing by the bijection Ψ. In particular, according to the Stone theorem there is the one-to-one correspondence between the set of unitary groups in the space H and the set $SA(H)$ of self-adjoint operators in the space H. Therefore, the topology on the set $SA(H)$ can be induced by the bijection Ψ from the topological space (Y_s, τ_s). This property, results of Theorem 3 and Corollary 1 give the opportunity to define the generalized mean value of the generator of the random semigroup in the space H.

Definition 5. The generator $\mathbf{L}$ of some semigroups of linear operators in the space H is called the generalized mean value of the random generator $\mathbf{L}_\omega$, $\omega \in \Omega$, if the function $\mathbf{G}(t) = \int\limits_\Omega e^{t\mathbf{L}_\omega} d\mu(\omega)$, $t \geq 0$ is equivalent (in Chernoff sense) to the semigroup $e^{t\mathbf{L}}$.

Corollary 1 states that Definition 5 is the extension of summation procedure from the space of bounded operators onto the set of generators $G(H)$. Now we compare the averaging in the sense of Definition 5 and averaging in the sence of quadratic forms.

Corollary 2. *Let $\mathbf{L}_1, ..., \mathbf{L}_m$ be self-adjoint operators in Hilbert space X and $\sum\limits_{j=1}^{m} p_j = 1$, $p_j > 0$. Let the quadratic form of the operator $\mathbf{L}_1$ majorize the quadratic forms of operators $\mathbf{L}_k$, $k = 2, ..., m$. Then the domain $D(\mathbf{L}_1)$ is the essential domain of operators $\mathbf{L}_k$, $k = 2, ..., m$ and the equalities $\bar{\mathbf{L}} = \sum\limits_{j=1}^{m} p_j \mathbf{L}_j$ and $\sum\limits_{j=1}^{m} p_j e^{-it\mathbf{L}_j} \sim e^{-it\sum\limits_{j=1}^{m} p_j \mathbf{L}_j}$ hold. Here $\bar{\mathbf{L}}$ is the mean operator.*

Thus, we obtain the additive analog of Trotter formula $e^{t\mathbf{A}}e^{t\mathbf{B}} \sim e^{t(\mathbf{A}+\mathbf{B})} \sim \frac{e^{2t\mathbf{A}}+e^{2t\mathbf{B}}}{2}$.

In the paper [6] we give the example of the random unitary semigroup such that the mean value of the random self-adjoint generator in the sense of Definition 5 exists, but there is no the mean value of the random generator in the sense of quadratic form. In this example the random generator takes values in the set $SA(H)$, in addition, its generalized mean value is the maximal dissipative, but not self-adjoint operator. In this example the common domain of the random generator values is smaller than the essential domain of the mean value of the random generator.

This approach to the approximation of the semigroup generated by the sum of generators of other semigroups is studied in the works [14, 15]. In the paper [3] we give the example of the random semigroup such that it has no generalized mean value in the following sense: the sequence $\{(\mathbf{G}(\frac{t}{n}))^n, t \in R_+, \}$ of iteration of the mean value $\mathbf{G}(t) = \int\limits_{\Omega} e^{t\mathbf{L}_\omega} d\mu(\omega)$, $t \in R_+$, does not converge (it has different partial limits).

3. The linear representation of the nonlinear flow

Now we describe the representation of the one-parameter family of nonlinear maps by the one-parameter family of linear operators and the representation of the flow of nonlinear maps by the group of linear operators.

Let E be a set, and $\mathcal{F}$ be some ring of subsets of the set E. Let $\mathcal{A}_{\mathcal{F}}$ be the minimal algebra containing the ring $\mathcal{F}$ (see [16]).

Definition 6. The map $X : E \to E$ of the phase space E into itself is called acceptable by the measure λ on the measurable space $(E, \mathcal{F})$ if the following conditions hold:

1) X is the measurable map of the measurable space $(E, \mathcal{F})$ into itself;

2) the measure $\lambda \circ X$ on the space E is absolutely continuous with respect to the measure λ (where $\lambda \circ X(A) = \lambda(X^{-1}(A)) \; \forall \; A \in \mathcal{F}$);

3) the density ρ of the measure $\lambda \circ X$ with respect to the measure λ satisfies the condition $\rho \in L_\infty(E, \mathcal{F}, \lambda, R)$.

In this case the map $X : E \to E$ of phase space E into itself and the measure λ on the measurable space $(E, \mathcal{F})$ are called coordinated.

The conditions of invariantness and quasi-invariantness of measures on infinitely dimensional spaces with respect to the dynamical mappings are studied in the papers [17, 16].

Let us consider the examples of the mappings $X : E \to E$ of the space E into itself with the measures λ of the space E which are accepted by this mappings.

1. Let the measure λ on a space E be invariant with respect to a transformation $X : E \to E$. Then this transformation X is accepted by this invariant measure λ.

2. Let E be a domain in Euclidean space R^d, and X be reversible continuously differentiable map $X : E \to E$ such that $|\det X'(x)| \in [\delta, \frac{1}{\delta}] \; \forall \; x \in E$ with some $\delta \in (0, 1)$. Then the transformation X is accepted by Lebesgue measure on the space R^d.

IOP Conf. Series: Journal of Physics: Conf. Series **990** (2018) 012012 doi:10.1088/1742-6596/990/1/012012

Therefore the flows of continuously differentiable maps X_t, $t \in R$, of the phase space $E = R^d$ into itself with uniformly bounded Jacobians $|\det(X_t)'(x)| \leq M \ \forall \ (x,t) \in E \times R$ with some $M > 0$ are accepted by Lebesgue measure λ on the space R^d (since $(X_t)^{-1} = X_{-t}$ and $|\det(X_t)'(x)||\det(X_{-t})'(X_t(x))| = 1 \ \forall \ (x,t) \in E \times R$).

In the case of finitely dimensional phase space E solutions representation of nonlinear ODE by means of its first integrals and solutions of linear first order PDE is well-known theory. In particular, the conditions of coordinating of the transformation of the space E with the measure on this space is well studied in the case of smooth transformations. But in the case of infinitely dimensional space E this approach is not fully investigated. The Hamiltonian approach to infinitely dimensional quantum dynamics is constructed in the work [18]. In particular, the problems of Liouville equation approach to the infinitely dimensional nonlinear ODE arises due to the absence of Lebesgue measure on the infinite dimensional space. To construct this Liouville approach to the ODE in infinitely dimensional phase space the analog of Lebesgue measure on a Hilbert space (Banach space) had been constructed in the papers [19, 20, 21]. This analog is the translation invariant, nonnegative, nontrivial, complete measure on some ring of subsets of Hilbert space. Let us consider this construction.

3. Let E be a real separable Hilbert space. Now we describe the measure on the space E which is invariant under any shift and any orthogonal mapping of the space E (see [20, 21]).

The set $\Pi \subset E$ is called rectangle if there are the orthonormal basis $\mathcal{E} = \{e_i\}$ and the elements $a, b \in l_\infty$ such that

$$\Pi = \{x \in E: \ (x, e_j) \in [a_j, b_j) \, \forall \, j \in \mathbf{N}\}. \tag{2}$$

This set can be empty. The rectangle (2) is called measurable if either $\Pi = \oslash$ or the following condition holds:

$$\sum_{j=1}^{\infty} \max\{0, \ln(b_j - a_j)\} < \infty. \tag{3}$$

Let $\mathcal{K}$ be the class of measurable rectangles in the space E and $\mathcal{R}$ be the minimal ring of subsets containing the class $\mathcal{K}$. Let λ be the function of the set on the class $\mathcal{K}$ which is defined by the condition: $\lambda(\oslash) = 0$ and

$$\lambda(\Pi) = \exp[\sum_{j=1}^{\infty} \ln(b_j - a_j)] \tag{4}$$

for any measurable nonempty rectangle (2). According to condition (3) $\lambda(\Pi) \in [0, +\infty)$ for any $\Pi \in \mathcal{K}$.

According to the papers [20, 21] the function of the set λ is additive on the class of sets $\mathcal{K}$ and has the unique additive extension onto the measure λ on the ring $\mathcal{R}$.

A set $A \subset E$ is called λ-measurable if for any $\epsilon > 0$ there are the sets $A_*, A^* \in \mathcal{R}$ such that $A_* \subset A \subset A^*$ and $\lambda(A^* \backslash A_*) < \epsilon$. The class Λ of λ-measurable sets in the space E is the ring and the measure λ has the unique extension from the ring $\mathcal{R}$ onto the ring Λ by the equality $\lambda(A) = \inf\limits_{A^* \in \mathcal{R}, A^* \supset A} \lambda(A^*) \ \forall \ A \in \Lambda$. The measure λ is invariant with respect to the shift on any vector of the space E and with respect to any orthogonal transformation of the space E (see [20, 21]).

The function $\lambda : \Lambda \to [0, +\infty)$ is the finite additive measure on the ring Λ of subsets of the space E which are invariant with respect to the shifts and orthogonal mappings. This measure is locally finite, complete, but it is not σ-additive; moreover, the algebra Λ can not contain some open balls in the space E (see [20, 21]).

Let $\mathbf{V}$ be a linear bijective operator from the space E onto itself with bounded inverse operator. Then there is the unique positive self-adjoint operator $\mathbf{S}$ and the orthogonal linear operator $\mathbf{U}$ such that $\mathbf{V} = \mathbf{US}$. This representation of a bounded linear bijective operator is called its polar decomposition.

In fact, the operator $\mathbf{V}^*\mathbf{V}$ is the bounded positive linear operator with bounded inverse operator. Then there is the unique positive linear operator $\mathbf{S} = (\mathbf{V}^*\mathbf{V})^{\frac{1}{2}}$, and the operator $\mathbf{S}$ has bounded inverse operator. Therefore, the bounded linear operator $\mathbf{U} = \mathbf{VS}^{-1}$ has the bounded linear inverse operator, and for any $x \in E$ the equality $\|\mathbf{U}x\|_E^2 = (\mathbf{VS}^{-1}x, \mathbf{VS}^{-1}x) = (x, \mathbf{S}^{-1}\mathbf{V}^*\mathbf{VS}^{-1}x) = \|x\|_E^2$ holds. Hence $\mathbf{U}$ is unitary operator.

Theorem 4. *Let $\mathbf{V}$ be a linear isomorphism of the space E onto itself, and the equality $\mathbf{V} = \mathbf{US}$ be its polar decomposition. Then the linear operator $\mathbf{V}$ is accepted by the measure λ if and only if the self-adjoint operator $\mathbf{S} - \mathbf{I}$ is the trace class operator.*

The statement of Theorem 4 is the consequence of following three lemmas. Firstly, the property of acceptability of the map $\mathbf{V} = \mathbf{US}$ for the measure λ will be investigated under the assumption that the operator $\mathbf{S}$ has the basis of eigenvectors. Further, we prove that this property of the operator $\mathbf{S}$ is the necessary condition for the acceptability.

Lemma 1. *Let $\mathbf{V}$ be a linear isomorphism of the space E onto itself and the equality $\mathbf{V} = \mathbf{US}$ be its polar decomposition. Let the operator $\mathbf{S}$ have the orthonormal basis of eigenvectors and spectrum $\sigma = \{s_j,\, j \in \mathbf{N}\}$. If $\{\alpha_j\} = \{s_j - 1\} \in l_1$. Then the linear map $\mathbf{V}$ is accepted by the measure λ.*

Let us consider the measurable rectangle $\Pi \in \mathcal{K}$ with the lengths of edges $\{d_j,\, j \in \mathbf{N}\}$. The image $\Pi' = \mathbf{V}(\Pi)$ is the rectangle in the class $\mathcal{K}$ with the lenghts of edges $\{d'_j = s_j d_j,\, j \in \mathbf{N}\}$. Hence

$$\Pi_{j=1}^{\infty} d'_j = (\Pi_{j=1}^{\infty} s_j)\lambda(\Pi) = \lambda(\Pi)\exp[\sum_{j=1}^{\infty} \ln(1+\alpha_j)] \leq \lambda(\Pi)\exp[\|\alpha\|_{l_1}]$$

and thus, $\lambda(\Pi') \leq \lambda(\Pi)\exp[\|\{s_j - 1\}\|_{l_1}]$. For the inverse mapping $\mathbf{V}^{-1}$ we obtain analogously $\lambda(\Pi) \leq \lambda(\Pi')\exp[\|\{\frac{1}{s_j} - 1\}\|_{l_1}]$.

Therefore, if $\{s_j - 1\} \in l_1$ then there is the number $\Pi_{j=1}^{\infty} s_j = \Delta \in (0, +\infty)$ such that $\lambda(\mathbf{V}(\Pi)) = \Delta\lambda(\Pi)$ for any $\Pi \in \mathcal{K}$. Since any set $A \in \Lambda$ can be approximated by a sequence of the sets from the ring $\mathcal{R}$ then $\lambda(\mathbf{V}(A)) = \Delta\lambda(A)$ for any $A \in \Lambda$.

Lemma 2. *Let $\mathbf{V}$ be a linear isomorphism of the space E onto itself, and the equality $\mathbf{V} = \mathbf{US}$ be its polar decomposition. Let the operator $\mathbf{S}$ have the orthonormal basis of eigenvectors and spectrum $\sigma = \{s_j,\, j \in \mathbf{N}\}$. If the linear operator $\mathbf{V}$ is accepted by the measure λ then $\{\alpha_j\} = \{s_j - 1\} \in l_1$.*

Let us assume the contrary: $\{\alpha_j\} = \{s_j - 1\} \notin l_1$. Then either there is the subsequence of negative numbers $\{\alpha_{j_k}\}$ such that $\sum\limits_{k=1}^{\infty} \alpha_{j_k} = -\infty$, or there is the subsequence of positive numbers $\{\alpha_{j_m}\}$ such that $\sum\limits_{m=1}^{\infty} \alpha_{j_m} = \infty$. In the second case the rectangle $\Pi' = \mathbf{V}(\Pi)$ (where Π is some rectangle with the unit edges) is not measurable since the condition (3) does not hold for Π': $\sum\limits_{m=1}^{\infty} \ln(s_{j_m}) = \sum\limits_{m=1}^{\infty} \ln(1 + \alpha_{j_m}) = +\infty$. In the first case rectangle $\Pi' = \mathbf{V}^{-1}(\Pi)$ is not measurable. In both case we obtain the contradiction since the map $\mathbf{V}$ is not admited by the measure λ.

Lemma 3. *Let $\mathbf{V}$ be a linear isomorphism of the space E onto itself, and the equality $\mathbf{V} = \mathbf{US}$ be its polar decomposition. If the linear operator $\mathbf{V}$ is accepted by the measure λ then the self-adjoint operator $\mathbf{S} - \mathbf{I}$ is the trace class operator.*

Let us assume the contrary: the self-adjoint operator $\mathbf{S} - \mathbf{I}$ is not the trace class operator. If it has the nontrivial component of continuous spectrum in the interval $(0, +\infty)$ or $(-\infty, 0)$ then there are the infinite orthonormal system $\{e_j\}$ and the number $\delta > 0$ such that either $\|\mathbf{S}e_j\|_E \geq (1+\delta)\|e_j\|$ or $\|\mathbf{S}e_j\|_E \leq \frac{1}{1+\delta}\|e_j\|$ for all $j \in \mathbf{N}$. Therefore, according to the proof of lemma 2 the map $\mathbf{V}$ is not accepted by the measure λ. Hence, the spectrum of self-adjoint operator $\mathbf{S}-\mathbf{I}$ has only discrete component and only eigenvalue 0 can have infinite multiplicity. Then $\mathbf{S}-\mathbf{I}$ is trace class operator according to the lemma 2.

Any flow of acceptable by the measure λ maps X_t, $t \in R$, in the phase space E is associated with the group $\mathbf{U}_t$, $t \in R$, of the linear maps of the Hilbert space $\mathcal{H} = L_2(E, \mathcal{F}, \lambda, \mathbf{C})$ acting by the equality

$$\mathbf{U}_t\psi(x) = \psi(X_{-t}(x)),\ x \in E. \tag{5}$$

The flows of nonlinear maps of the phase space E which is acceptable by the measure λ on the space E can be studying as the groups of linear operators in the space $L_2(E, \mathcal{F}, \lambda, \mathbf{C})$.

4. Mean values of random flows and random vector fields

Let E be a real Hilbert space and Λ be a ring of subsets of Hilbert space E.

Let λ be the nonnegative finite additive measure on the space E which is invariant with respect to shifts and rotations. Set $\mathcal{H} = L_2(E, \Lambda, \lambda, \mathbf{C})$.

Let $M(E, \Lambda)$ be the space of measurable maps of measurable space (E, Λ) into itself and $M(E, \Lambda, \lambda)$ be the subset of maps of the space $M(E, \Lambda)$ which are accepted by the measure λ.

Let $C^1(E, E)$ be the Banach space of continuously differentiable maps of the space E into itself.

Let $F(E) = C(R, C^1(E, E))$ be the Banach space of continuous maps of the segment $[0, +\infty)$ into the Banach space $C^1(E, E)$ endowed with the norm $\|f\|_{F(E)} = \sup\limits_{t \in R} \|f(t)\|_{C^1(E,E)}$. Let $F^1(E) = C^1(R, C^1(E, E)) = \{f \in F(E) :\ f' \in F(E)\}$ be the Banach space endowed with the norm $\|f\|_{F^1(E)} = \|f\|_{F(E)} + \|f'\|_{F(E)}$.

A map $X \in F(E)$ is called a flow if $X_{t+s} = X_t \circ X_s\ \forall\, t, s \in R$.

Let $\mathcal{A}_B$ be Borel sigma-algebra of subsets in the space $F(E)$.

A random flow is defined as a measurable map of the space with a measure $(\Omega, \mathcal{A}, \mu)$ into the measurable space $(F(E), \mathcal{A}_B)$ such that the values of this map are flows.

Let $F_\lambda(E)$ be the subset of the Banach space $F(E)$ which is defined by the conditions

$$F_\lambda(E) = \{X \in F(E) :\ \forall\, t \in R\ \ X_t \in M(E, \Lambda, \lambda)\}. \tag{6}$$

If $X \in F_\lambda(E)$ then the operator-function $\mathbf{U} :\ R \to B(\mathcal{H})$ acting by the equality (5) is defined. The symbol Φ_λ means the map which is defined on the set $F_\lambda(E)$, takes values in the space of maps $R \to B(\mathcal{H})$ and is defined by the equality (5).

4.1. *Chernoff equivalence and mean values of nonlinear maps*

According to O.G. Smolyanov the function $\mathbf{F} \in C_s(R_+, B(H))$ is equivalent in the space $C_s(R_+, B(H))$ to the function $\mathbf{G} \in C_s(R_+, B(H))$ iff $\mathbf{F}(0) = \mathbf{G}(0) = \mathbf{I}$, and $\lim\limits_{n\to\infty} \sup\limits_{t\in[0,T]} \|((\mathbf{F}(\frac{t}{n})^n - (\mathbf{G}(\frac{t}{n}))^n)u\|_H = 0$ for all $T > 0$, $u \in H$ (see [13, 7, 6]).

Definition 7. Operator valued functions $X, Y \in F_\lambda(E)$ are called equivalent in the space $F_\lambda(E)$ to each other in nonlinear Chernoff sense if and only if the functions $\Phi_\lambda(X), \Phi_\lambda(Y) \in C_s(R, B(\mathcal{H}))$ are equivalent to each other in Chernoff sense.

Definition 8. The flow $Y \in F_\lambda(E)$ is called the generalized mean value of the random flow $X :\ \Omega \to F_\lambda(E)$ if $\mathrm{M}\Phi_\lambda(X) \sim \Phi_\lambda(Y)$.

Does the relation $\mathrm{M}\Phi_\lambda(X) \sim \Phi_\lambda(\mathrm{M}X)$ holds if the both mean values exist?

Firstly, we present the sufficient conditions for the validity of relation $\mathrm{M}\Phi_\lambda(X) \sim \Phi_\lambda(\mathrm{M}X)$.

Theorem 5. *Let $E = R^d$, $\mathcal{H} = L_2(R^d)$, λ be Lebesgue measure on R^d. Let $X \in F^1(E)$ be the operator valued function such that for any $t \in R$ $X_t : E \to E$ be the reversible map and $X^{-1} \in F^1(E)$, where $(X^{-1})_t = (X_t)^{-1}\ \forall\, t \in R$.*

Then $X \sim S_a$ where $X'(0) = a \in C^1(R^d, R^d)$ and S_a is the flow of transformation of the space R^d generated by differential equation $\dot{x}(t) = a(x(t))$, $t \in R$.

The proof of the Theorem 5 is given in the following three lemmas.

Lemma 4. *Let the assumptions of Theorem 5 hold. Then for any $t \in R$ the map X_t is acceptable by the Lebesgue measure λ on the space E.*

Proof. In fact, $X, (X)^{-1} \in F^1(E) \equiv C^1(R, C^1(R^d, R^d))$ by the assumption of Theorem 5. According to the "replacement theorem in the integral" for any function $\psi \in L_2(R^d)$ the following equality

$$\int\limits_{R^d} |\psi(X_t^{-1}(x))|^2 dx = \int\limits_{R^d} \frac{|\psi(y)|^2}{|(X_t^{-1})'(X_t(y))|} dy = \int\limits_{R^d} |\psi(y)|^2 |(X_t)'(y)| dy$$

holds, where $|(X_t^{-1})'(x)| = \det||\frac{\partial((X_t)^{-1}(x))_j}{\partial x_k}||$ and $|(X_t)'(x)| = \det||\frac{\partial(X_t(x))_j}{\partial x_k}||$, $(t, x) \in R \times R^d$.

Let $\frac{d}{dt}(X_t(x)) = \mathbf{b}(t, x) = \mathbf{a}(t, X_t(x))$, $t \in R$, $x \in R^d$ then $\mathbf{a}(t, x) = b(t, X_t^{-1}(x)) \in F(E)$. Since the maps X_t, $t \in R$, of the flow X are continuously differentiable, the equality $\frac{d}{dt}(X_t(x)) = \mathbf{a}(t, X_t(x))$, $t \in R$, $x \in R^d$ holds. Then $\frac{d}{dt}\det||\frac{\partial(X_t(x))_j}{\partial x_k}|| = \mathrm{div}(\mathbf{a})(t, X_t(x))\det||\frac{\partial(X_t(x))_j}{\partial x_k}||$, where the function $\mathrm{div}(\mathbf{a})(t, x) = \sum\limits_{k=1}^{d} \frac{\partial \mathbf{a}_k(t,x)}{\partial x_k}$ is bounded by the condition (4).

Therefore, there is the constant $L = L(\|X\|_{F^1(E)}, \|X^{-1}\|_{F^1(E)})$ such that Jacobians of the mappings X_t, $t \in R$ admit the estimates

$$\sup_{(t,x)\in[-T,T]\times R^d} |(X_t)'(x)| \leq e^{LT}. \tag{7}$$

Thus the norms of values of the operator function $\mathbf{T}_t$, $t \geq 0$, admit the estimates $\|\mathbf{T}_t\|_{B(H)} \leq e^{Lt}$, $t \geq 0$. Therefore, the maps X_t, $t \in R$, of the flow X are accepted by the Lebesgue measure on the space R^d.

According to Lemma 4 the one-parameter family of maps $X(t)$, $t \in R$, of the space R^d generates the operator-valued function $\mathbf{T}_t = \Phi_\lambda(X(t))$, $t \in R$, with values in the Banach space of bounded linear operators $B(H)$ in the space $H = L_2(R^d)$ acting by the equality

$$\mathbf{T}_t\psi(x) = \Phi_\lambda(X(t))\psi(x) = \psi((X_t)^{-1}(x)),\ t \in R,\ x \in R^d;\ \psi \in H.$$

Lemma 5. *Let the assumptions of Theorem 5 hold. Then $\mathbf{T} = \Phi_\lambda(X) \in C_s(R, B(\mathcal{H}))$.*

Since $X \in C^1(R, C^1(R^d, R^d))$ then Cauchy problem for the system of ODE $\dot{\mathbf{x}}(t) = \mathbf{a}(t, \mathbf{x}(t))$, $t \in R$; $\mathbf{x}(0) = x_0$ for any $x_0 \in R^d$ has the unique solution $X_t(x_0) \in C^1(R, R^d)$. Thus, according to Lemma 4 the one-parameter family $\mathbf{T}_t$, $t \in R$, of linear operators of the space $H = L_2(R^d)$ is defined. This family of linear operators has the following properties:

I) the space $W_2^1(R^d)$ is invariant with respect to operators $\mathbf{T}_t$, $t \in R$;

II) for any $u \in W_2^1(R^d)$ the equality $\frac{d}{dt}\mathbf{T}_t u(x) = (\nabla\mathbf{T}_t u(x), \mathbf{a}(t, X_t(x)))$ holds.

Hence,

1) for any element $u \in W_2^1(R^d)$ and number $T > 0$ there is a number $L > 0$ such that

$$\|\mathbf{T}_{t+\tau}u - \mathbf{T}_t u\|_H \leq L|\tau|\|u\|_{W_2^1} \tag{8}$$

for all $t, t+\tau \in [-T, T]$;

2) for any $\psi \in L_2(R^d)$ and $t \in R$ the estimate holds: $\|\mathbf{T}_t\psi\|_{L_2(R^d)} \leq e^{At}\|\psi\|_{L_2(R^d)}$ where $A = \sup\limits_{x \in R^d} |\mathrm{div}\mathbf{a}(x)|$.

Since the space $W_2^1(R^d)$ is the dense subspace of the space $L_2(R^d)$ then the equality $\lim\limits_{\tau \to 0} \|\mathbf{T}_{t+\tau}u - \mathbf{T}_t u\|_H = 0$ is the consequence of the properties I), II) of the operator-function $\mathbf{T}$.

In fact, if $\psi \in L_2(R^d)$ and $\epsilon > 0$ then there is the vector $u \in W_2^1(R^d)$ such that $\|\psi - u\|_{L_2} < \frac{\epsilon}{4}$. Then $\|\mathbf{T}_t\psi - \psi\|_H \leq \|\mathbf{T}_t(\psi - u)\|_H + \|\mathbf{T}_t u - u\|_H + \|u - \psi\|_H$. According to the condition 2) the first part of the sum admits the estimate by the value $\frac{\epsilon}{4}$ for sufficiently small $|t|$, the second term of the sum admits the estimate by the value $\frac{\epsilon}{2}$ for sufficiently small $|t|$ according to (8).

Since the vector field $\mathbf{a}(0,\cdot) \in C^1(E,E)$ is continuously differentiable then Cauchy problem for the autonomic system of differential equation $\dot{x} = \mathbf{a}(0,x)$; $x(0) = x_0$, has the unique solution $\xi_t(x_0)$, $t \in R$. Therefore, the one-parameter family of maps defined by the equality $(S_t^{\mathbf{a}}(x_0) = \xi_t(x_0)$, $x_0 \in E, t \in R$, is the flow of continuously differentiable maps, moreover, $S^{\mathbf{a}} \in F^1(E)$.

According to Lemma 4 the flow of continuously differentiable maps $S^{\mathbf{a}}$ is acceptable by the Lebesgue measure λ on the space R^d. Therefore, the one-parameter family $\mathbf{F}_t$, $t \in R$, of linear operators in Hilbert space $\mathcal{H} = L_2(R^d)$ is defined by the equality $\mathbf{F} = \Phi_\lambda(S^{\mathbf{a}})$. Then the equalities $\mathbf{F}_t u(x) = u((S_t^{\mathbf{a}})^{-1}(x))$, $t \in R$, $x \in R^d$, $u \in \mathcal{H}$ define the one-parameter family $\mathbf{F}_t$, $t \in R$, of linear operators in Hilbert space $\mathcal{H} = L_2(R^d)$. The one-parameter family $\mathbf{F}_t$, $t \in R$, is one-parameter group since $\mathbf{F}_t u(x) = u(S^{\mathbf{a}}{}_{-t}(x))$, $t \in R$, $x \in R^d$, $u \in \mathcal{H}$, and family of maps $S_t^{\mathbf{a}}$, $t \in R$, is the flow. According to Lemma 5 the one-parameter group $\mathbf{F} = \Phi_\lambda(S^{\mathbf{a}})$ of linear operators is continuous in the strong operator topology.

Lemma 6. *Let the assumptions of Theorem 5 hold. Then the operator valued functions* $\Phi_\lambda(X)$, $\Phi_\lambda(S^{\mathbf{a}}) \in C_s(R, B(\mathcal{H}))$ *are equivalent to each other in Chernoff sense.*

Let us verify the condition of Chernoff theorem about the differentiability of the operator-function $\mathbf{T} = \Phi_\lambda(X)$. For any $u \in W_2^1(R^d)$ there is the derivative $(\frac{d}{dt}(\mathbf{T}_t u))|_{t=0} = (\mathbf{a}(0,\cdot), \nabla)u$. The operator $(\mathbf{a}(0,\cdot), \nabla)$ is densely defined on the space $W_2^1(R^d)$, and it is the generator of the group $\mathbf{F}_t$, $t \in R$. According to Lemma 5 $\mathbf{T} \in C_s(R, B(\mathcal{H}))$, and according to (7) the estimate $\|\mathbf{T}(t)\|_{B(\mathcal{H})} \leq e^{Lt}$, $t \geq 0$ holds. Then according to Chernoff Theorem for operator function with values in the Banach space $B(\mathcal{H})$ the relation $\mathbf{T} \sim \mathbf{F}$ holds. Therefore, $X \sim S^{\mathbf{a}}$ according to Definition 7. Theorem 5 is proved.

Corollary 5. *Let the operator valued functions* $X, Y \in F^1(E)$ *satisfy the conditions* $X^{-1}, Y^{-1} \in F^1(E)$. *If* $\frac{d}{dt}(X_t)|_{t=0} = \frac{d}{dt}(Y_t)|_{t=0}$ *then the operator valued functions* X, Y *are equivalent to each other in nonlinear Chernoff sense.*

Let $\frac{d}{dt}(X_t)|_{t=0} = \frac{d}{dt}(Y_t)|_{t=0} = \mathbf{a} \in C^1(E,E)$. Since $\mathbf{T}_1 = \Phi_\lambda(X)$, $\mathbf{T}_2 = \Phi_\lambda(Y)$ and $\mathbf{F} = \Phi_\lambda(S^{\mathbf{a}})$ then according to Theorem 5 $X \sim S^{\mathbf{a}} \sim Y$ in the sense of Definition 7.

Corollary 6. *Let* $E = R^d$, $\mathcal{H} = L_2(R^d)$ *and* X *be the random flow with a finite set of values* $X^{(1)}, ..., X^{(m)} \in C^1(R, C^1(R^d, R^d))$.

If $\mathrm{M}X_t = p_1 X_t^{(1)} + ... + p_m X_t^{(m)}$, *where* $p_1, ..., p_m \geq 0$, $p_1 + ... + p_m = 1$, *then* $\mathrm{M}X \sim S^{\mathrm{Ma}}$, *where* $\mathrm{Ma} = p_1 a_1 + ... + p_m a_m$, $a_k = \frac{d}{dt}X^{(k)}(t)|_{t=0}$, $k = 1, ..., m$.

Thus, under the assumption of Theorem 6 the mean value of random vector field generates the generalized mean value of random flow: $\mathrm{M}\Phi_\lambda(X) \sim \Phi_\lambda(\mathrm{M}X) \sim \Phi_\lambda(S^{MX'(0)})$.

4.2. The examples of violation of the property $\mathrm{M}\Phi_\lambda(X) \sim \Phi_\lambda(\mathrm{M}X)$

Example 1. Let $E = R$, $\mathcal{H} = L_2(R)$, and

$$X_t^\omega(x) = x + \sqrt{t}\omega, \; x \in E, t \geq 0,$$

where ω is E-valued random variable such that its three first moments satisfy the conditions

$$\mathrm{M}\omega = 0; \quad \mathrm{M}(\omega)^2 = \sigma^2; \quad \mathrm{M}(|\omega|^3) < \infty. \tag{9}$$

Any map X_t^ω is accepted by the Lebesgue measure on the space E and therefore,

$$\Phi_\lambda(X_t^\omega)u(x) = u(x-\sqrt{t}\omega) = e^{-\sqrt{t}\omega\mathbf{D}}u(x);$$

where $Du = u'\ \forall\ u \in W_2^1(R)$. Thus, according to the assumptions (9)

$$\mathrm{M}X_t = \mathbf{I};\quad \Phi\mathrm{M}X_t = \mathbf{I};$$

$$\mathrm{M}\Phi(X_t) = \mathrm{ch}(\sigma\sqrt{t}\mathbf{D}) \sim \exp(\frac{\sigma^2}{2}\mathbf{D}^2 t).$$

Thus, $\Phi(\mathrm{M}X)$ and $\mathrm{M}\Phi(X)$ are not equivalent in Chernoff sense.

Example 2. Let $E = H$ be a real separable Hilbert space. According to the paper [20] the space H can be endowed with the complete measure λ on the space H which is translation, rotationally invariant and defined on the minimal ring $\mathcal{R}$ of subsets of H containing the set $\mathcal{K}$ of rectangles with absolutely converging products of edges lengths. Hilbert space $\mathcal{H} = L_2(H, \mathcal{R}, \lambda, \mathbf{C})$ of square integrable with respect to the measure λ functions $f:\ H \to \mathbf{C}$ is constructed and its properties are investigated in the paper [20].

Let $\mathcal{E} = \{e_j\}$ be some orthonormal basis in the space E. Let $\mathcal{K}_\mathcal{E} \subset \mathcal{K}$ be the class of measurable rectangles in the space E with edges collinear to the vectors of the system $\mathcal{E}$. Let $\mathcal{R}_\mathcal{E}$ be the minimal ring of subsets containing the class $\mathcal{K}_\mathcal{E}$.

Let λ be the function of the set on the class $\mathcal{K}$ defined by condition (4).

The function $\lambda:\ \mathcal{K} \to [0, +\infty)$ is additive and has the unique continuation onto the measure λ on the ring $\mathcal{R}$. Then the function $\lambda_\mathcal{E} \equiv \lambda|_{\mathcal{R}_\mathcal{E}}$ is the measure on the ring $\mathcal{R}_\mathcal{E}$. A set $A \subset E$ is called $\lambda_\mathcal{E}$-measurable if for any $\epsilon > 0$ there are the sets $A_1, A_2 \in \mathcal{R}_\mathcal{E}$ such that $A_1 \subset A \subset A_2$ and $\lambda(A_2 \backslash A_1) < \epsilon$. The set $\Lambda_\mathcal{E}$ of $\lambda_\mathcal{E}$-measurable subsets of the space E is the ring and the function $\lambda_\mathcal{E}(A) = \inf\limits_{B\in\mathcal{R}:\ B\supset A} \lambda_\mathcal{E}(B)$ is the complete measure on the ring $\Lambda_\mathcal{E}$. Then $\mathcal{H} = L_2(E, \Lambda, \lambda, \mathbf{C})$ is the Hilbert space and $\mathcal{H}_\mathcal{E} = L_2(E, \Lambda_\mathcal{E}, \lambda_\mathcal{E}, \mathbf{C})$ is the closed subspace of the space $\mathcal{H}$.

Let

$$X_t^h(x) = x + \sqrt{t}h,\ x \in H, t \geq 0, \tag{10}$$

where h is H-valued random variable with distribution given by the Gaussian measure $\nu_\mathbf{D}$ with trace class covariation operator $\mathbf{D}$. Therefore, the mean value $\mathrm{M}X_t = \int\limits_H X_t^h d\nu_\mathbf{D}(h)$ of the random map X_t^h exists and $\mathrm{M}X_t = \mathbf{I}$, $t \geq 0$.

Since $\mathrm{M}X_t = \mathbf{I}$, $t \geq 0$, then $\Phi\mathrm{M}X_t = \mathbf{I}$, $t \geq 0$.

Since the measure λ is invariant with respect to shifts then any map X_t^h, $t \geq 0$, $h \in H$, is acceptable by the measure λ. Therefore, the linear operators $\Phi_\lambda(X_t^h)$, $t \geq 0$, $h \in H$, are defined on the space $\mathcal{H}$ and are given by the equality

$$\Phi(X_t^h)u = \mathbf{S}_{\sqrt{t}h}u;$$

where $\mathbf{S}_h$, $h \in H$, is the operator of the argument shift on the vector h: $\mathbf{S}_h u(x) = u(x+h)$, $x \in H$.

Since h is H-valued random variable with distribution given by Gaussian measure $\nu_\mathbf{D}$ with covariation operator $\mathbf{D}$ then the mean value $\mathrm{M}[\Phi(X_t)] = \int\limits_H \Phi(X_t^h)d\nu_\mathbf{D}(h)$ of the random linear operators $\Phi(X_t^h)$, $t \geq 0$, exists. If the measure $\nu_\mathbf{D}$ is countable additive (this condition is equivalent to the condition: $\mathbf{D} \in \sigma_1(H)$ is the trace class operator) then by the papers [20, 21] the one-parameter family of mean values $\mathrm{M}[\Phi(X_t)]$, $t \geq 0$, is the one-parametric semigroup $\mathbf{U}_\mathbf{D}(t)$, $t \geq 0$, of nonnegative contractions in the space $\mathcal{H}$. This fact is the consequence of the semigroup property with respect to the convolution for the Gaussian measures ($\nu_{t\mathbf{D}} * \nu_{s\mathbf{D}} = \nu_{(t+s)\mathbf{D}}$ for all $t, s \geq 0$).

Let $\mathcal{E}$ be the orthonormal basis of eigenvectors of the correlation operator $\mathbf{D}$ of Gaussian measure $\nu_{\mathbf{D}}$. By the paper [21] the space $\mathcal{H}_{\mathcal{E}}$ is the invariant subspace of the operators of the semigroup $\mathrm{M}[\Phi(X_t)]$, $t \geq 0$. If in addition, the operator $\mathbf{D}^{\frac{1}{4}}$ is the trace-class operator then by the papers [20, 21] the restriction of the semigroup $\mathrm{M}[\Phi(X_t)]$ on the invariant subspace $\mathcal{H}_{\mathcal{E}}$ is strongly continuous semigroup of the self-adjoint contractions of the space $\mathcal{H}_{\mathcal{E}}$. Thus, the operator-valued functions $\Phi(\mathrm{M}X) = \mathbf{I}$ and $\mathrm{M}\Phi(X) = \mathbf{U_D}$ are not equivalent in Chernoff sense.

5. The law of large numbers

Now we investigate the properties of the sequence of compositions of independent identically distributed random semigroups of linear operators. Using the concept of Chernoff equivalence and Chernoff Theorem we obtain the analog of the low of large numbers for the products of independent identically distributed random semigroups of linear operators.

The random linear operator $\mathbf{U}$ is the measurable map of the set Ω of the probability space $(\Omega, \mathcal{A}, \mu)$ into the topological vector space $(B(H), \tau_s)$ endowed with the σ-algebra $\mathcal{A}_s$ of Borel subsets. The random operators $\mathbf{U}_1, \mathbf{U}_2$ are called independent random operators if for any $A_1, A_2 \in \mathcal{A}_s$ the equality $\mu(\{\mathbf{U}_1 \in A_1,\ \mathbf{U}_2 \in A_2\}) = \mu(\{\mathbf{U}_1 \in A_1\})\mu(\{\mathbf{U}_2 \in A_2\})$ holds.

Let $\{(\Omega_n, \mathcal{A}_n, \mu_n)\}$ be the sequence of probability spaces. Let $\mathbf{U}_k : \omega_k \to Y_s = C_s(R_+, B(H))$ be the random semigroup of bounded linear operators in the space H for any $k \in \mathbf{N}$. Let $\{\mathbf{U}_n\}$ be the sequence of independent random semigroups of bounded linear operators in the space H.

This sequence $\{\mathbf{U}_n\}$ can be realized by the following way. For any $n \in \mathbf{N}$ the composition of n independent random semigroups of bounded linear operators $\mathbf{U}_1, ..., \mathbf{U}_n$ is defined as the map $\mathbf{U}_n \circ ... \circ \mathbf{U}_1$ of the space with the measure $(\Omega_1 \times ... \times \Omega_n, \mathcal{A}_1 \otimes ... \otimes \mathcal{A}_n, \mu_1 \otimes ... \otimes \mu_n)$ into the measurable space $(Y_s(H), \mathcal{A}_s)$. This map is defined by the equality $\mathbf{U}_n \circ ... \circ \mathbf{U}_1(\omega_1, ..., \omega_n) = \mathbf{U}_n(\omega_n) \circ ... \circ \mathbf{U}_1(\omega_1)$, $(\omega_1, ..., \omega_n) \in \Omega_1 \times ... \times \Omega_n$. Then for any $n \in \mathbf{N}$ the composition of n independent random semigroups of linear operators is the random operator valued function with values in the measurable space $(Y_s(H), \mathcal{A}_s)$.

Theorem 7. *Let for any $n \in \mathbf{N}$ the random operators $\mathbf{U}_1, \mathbf{U}_2, ..., \mathbf{U}_n$ be independent. Then*

$$M[\mathbf{U}_n \circ ... \circ \mathbf{U}_1] = M[\mathbf{U}_n]) \circ ... \circ M[\mathbf{U}_1] \quad \forall\ n \in \mathbf{N}.$$

This statement is the consequence of the Fubini Theorem (see [22], Theorem 3.4.4) in its application to the function

$$f_{u,v,t}(\omega_1, ..., \omega_n) = (\mathbf{U}_n(\omega_n) \circ ... \circ \mathbf{U}_1(\omega_1)u, v)_H, \ (\omega_1, ..., \omega_n) \in \Omega_1 \times ... \times \Omega_n,$$

for an arbitrary $u, v \in H$.

Corollary 6. *Let $\mathbf{U}$ be a random operator, $n \in \mathbf{N}$ and $\mathbf{U}_1, \mathbf{U}_2, ..., \mathbf{U}_n$ be independent identically distributed random operators such that for any $k \in \{1, ..., n\}$ the random operator $\mathbf{U}_k$ has the same distribution as the random operator $\mathbf{U}$. Then*

$$M[\mathbf{U}_n \circ ... \circ \mathbf{U}_1] = (M[\mathbf{U}])^n \ \forall\ n \in \mathbf{N}. \tag{11}$$

5.1. The law of large numbers for compositions of linear maps

For the sequence $\{\xi_k\}$ of independent real-valued random variables with finite mean values the law of large numbers states that

$$\lim_{n\to\infty} P(\{|\frac{\xi_n}{n} + ... + \frac{\xi_1}{n} - \mathrm{M}(\frac{\xi_n}{n} + ... + \frac{\xi_1}{n})| > \epsilon\}) = 0.$$

We extend this law onto the sequence $\{\mathbf{U}_k\}$ of independent random semigroups of bounded linear operators considering the averaging composition $(\mathbf{U}_n)^{\frac{1}{n}} \circ ... \circ (\mathbf{U}_1)^{\frac{1}{n}}$ instead of the averaged sum $\frac{\xi_n}{n} + ... + \frac{\xi_1}{n}$. The especial property of the semigroup of linear operators in a Hilbert space gives the opportunity to define the operator valued function $(\mathbf{U}(t))^{\frac{1}{n}}$, $t \geq 0$, by the equality $(\mathbf{U}(t))^{\frac{1}{n}} = \mathbf{U}(\frac{t}{n})$, $t \geq 0$. This is not the unique of root of the equation $(\mathbf{X}(t))^n = \mathbf{U}(t)$, $t \geq 0$, but the choice of the solution $\mathbf{X}(t) = \mathbf{U}(\frac{t}{n})$, $t \geq 0$, gives the dynamical definition of the unique root.

Therefore the mean value of the averaged composition $(\mathbf{U}_n(t))^{\frac{1}{n}} \circ ... \circ (\mathbf{U}_1(t))^{\frac{1}{n}}$, $t \geq 0$, of independent random semigroups is the iteration $(\mathbf{F}(\frac{t}{n}))^n$, $t \geq 0$, of operator function $\mathbf{F} = \mathrm{M}[\mathbf{U}]$. Thus, the Chernoff Theorem describes the asymptotic properties of the mean value of the composition of n independent random semigroups for $n \to \infty$.

Definition 9. 1. The sequence $\{\mathbf{U}_n\}$ of the independent random semigroups of bounded linear operators with the common mean value $\bar{\mathbf{U}}$ is said to be satisfying the large numbers law in the strong operator topology **if** the following condition

$$\lim_{n\to\infty} P(\{\|((\mathbf{U}_n(t))^{\frac{1}{n}} \circ ... \circ (\mathbf{U}_1(t))^{\frac{1}{n}} - \bar{\mathbf{U}}(t))x\|_H > \epsilon\}) = 0 \tag{12}$$

is fulfilled for every $x \in H$, $x \neq \mathbf{0}$, $t > 0$ and $\epsilon > 0$.

2. The sequence $\{\mathbf{U}_n\}$ of the independent random semigroups of bounded linear operators with the common mean value $\bar{\mathbf{U}}$ is said to be satisfying the large numbers law in the topology of operators norm **if** the following condition

$$\lim_{n\to\infty} P(\{\|(\mathbf{U}_n(t))^{\frac{1}{n}} \circ ... \circ (\mathbf{U}_1(t))^{\frac{1}{n}} - \bar{\mathbf{U}}(t)\|_{B(H)} > \epsilon\}) = 0 \tag{13}$$

is fulfilled for every $t > 0$ and $\epsilon > 0$.

To obtain estimates for compositions of independent random maps and the sufficient condition for the law of large numbers for random maps we introduce here the concept of the dispersion of random bounded linear operators and random semigroups.

The second moment of a random bounded linear operator $\mathbf{A}$ is the operator $M[\mathbf{A}^*\mathbf{A}]$.

The second moment of a random semigroups $\mathbf{U}$ is the operator valued function $M[\mathbf{U}^*\mathbf{U}]$.

Thus, the second moment of a random bounded linear operator is the nonnegative bounded linear operator. The dispersion of a random semigroup $\mathbf{U}$ is the nonnegative operator-valued function

$$D(\mathbf{U}) = M[\mathbf{U} - M[\mathbf{U}]] = M[\mathbf{U}^*\mathbf{U}] - M[\mathbf{U}]^*M[\mathbf{U}].$$

We investigate the second moments of the values of the sequence of iterations of independent random linear operators.

Let $t \geq 0$ and $n \in \mathbf{N}$. Then according to (11) the following equality for the composition of independent identically distributed random semigroups $\mathbf{U}_1, ..., \mathbf{U}_n$ holds:

$$D(\mathbf{U}_n(t) \circ ... \circ \mathbf{U}_1(t)) = M[\mathbf{U}_1^*(t) \circ ... \circ \mathbf{U}_n^*(t) \circ \mathbf{U}_n(t) \circ ... \circ \mathbf{U}_1(t)] - (M[(\mathbf{U}(t))^n])^* M[(\mathbf{U}(t))^n].$$

Let $\{\mathbf{U}_n(t), t \geq 0,\}$ be the sequence of independent identically distributed random semigroups of unitary operators. Then

$$D(\mathbf{U}_n(t) \circ ... \circ \mathbf{U}_1(t)) = \mathbf{I} - ((M[\mathbf{U}(t)])^n)^*(M[\mathbf{U}(t)])^n.$$

If, in addition, the mean value $M[\mathbf{U}]$ of the random semigroup $\mathbf{U}(t)$, $t \geq 0$, is the semigroup then

$$D(\mathbf{U}_n(t) \circ ... \circ \mathbf{U}_1(t)) = \mathbf{I} - (M[\mathbf{U}(nt)])^* M[\mathbf{U}(nt)] = D(\mathbf{U}(nt)).$$

Lemma 7. *(Chebyshev inequality). If* $\mathbf{A}$ *is a random variable with values in the Banach space* $B(H)$ *such that its dispersion is the operator* $D(\mathbf{A}) \in B(H)$ *then for any vector* $x \in H: \ \|x\|_H = 1$ *the following Chebyshev inequality is valid:*

$$P(\{\|(\mathbf{A} - M\mathbf{A})x\|_H > \epsilon\}) \leq \frac{1}{\epsilon^2}\|D(\mathbf{A})x\|_H. \tag{14}$$

Proof. Since $D(\mathbf{A}) = \int\limits_{\Omega}((\mathbf{A} - M\mathbf{A})^*(\mathbf{A} - M\mathbf{A}))d\mu$ then for any vector $x \in H$ such that $\|x\|_H = 1$ the following relations hold:

$$\|D(\mathbf{A})x\|_H = \sup_{\|u\|_H=1}(u, [\int\limits_{\Omega}(\mathbf{A} - M\mathbf{A})^*(\mathbf{A} - M\mathbf{A})d\mu]x) \geq (x, [\int\limits_{\Omega}(\mathbf{A} - M\mathbf{A})^*(\mathbf{A} - M\mathbf{A})d\mu]x) =$$

$$= \int\limits_{\Omega}(x, ((\mathbf{A} - M\mathbf{A})^*(\mathbf{A} - M\mathbf{A}))x)d\mu = \int\limits_{\Omega}\|(\mathbf{A} - M\mathbf{A})x\|_H^2 d\mu \geq$$

$$\geq \int\limits_{\{\omega\in\Omega: \ \|(\mathbf{A}-M\mathbf{A})x\|_H\geq\epsilon\}} \|(\mathbf{A} - M\mathbf{A})x\|_H^2 d\mu \geq \epsilon^2\mu(\{\omega \in \Omega: \ \|(\mathbf{A} - M\mathbf{A})x\|_H \geq \epsilon\}).$$

Therefore, Chebyshev inequality (14) holds.

Corollary 7. *Let* $\{\mathbf{U}_n\}$ *be the sequence of independent random semigroups with the common mean value* $\bar{\mathbf{U}}$ *and the bounded in the norm of the space* $B(H)$ *sequence of dispersions* $\{D(\mathbf{A}_n)\}$. *If the sequence* $\{\mathbf{U}_n\}$ *satisfies the condition*

$$\lim_{n\to\infty}\|D((\mathbf{A}_n)^{\frac{1}{n}} \circ ... \circ (\mathbf{A}_1)^{\frac{1}{n}})x\|_H = 0 \ \forall \ x \in H,$$

then the law of large numbers in the strong operator topology (12) for the sequence $\{\mathbf{U}_n\}$ *holds.*

Theorem 8. *Let* $\xi: \ \Omega \to SA(H)$ *be a random variable with the values in the set* $SA(H)$ *of self-adjoint operators in the space* H, *and*

$$\mathbf{U}_\omega(t) = \exp(i\xi(\omega)t), \ t \geq 0, \ \omega \in \Omega \tag{15}$$

be the corresponding random semigroup. Let $\mathcal{D}$ *be a dense linear manifold in the space* H *such that for any* $u \in \mathcal{D}$ *the condition* $\int\limits_{\Omega}\|\xi(\omega)u\|_H d\mu(\omega) < \infty$ *holds.*

Let $\{\mathbf{U}_k\}$ *be a sequence of independent identically distributed random semigroups such that any random semigroup* $\mathbf{U}_k$ *has the same distribution as the random semigroup (15).*

Let the linear operator $\bar{\xi}: \ \mathcal{D} \to H$ *defined by the equality* $\bar{\xi}u = \int\limits_{\Omega}\xi(\omega)ud\mu(\omega), \ u \in H$, *be essentially self-adjoint. Then the sequence* $\{\mathbf{U}_n \circ ... \circ \mathbf{U}_1\}$ *of compositions of independent identically distributed random semigroups satisfies the equality*

$$\lim_{n\to\infty}[\sup_{t\in[0,T]}\|D(\mathbf{U}_n(\frac{t}{n}) \circ ... \circ \mathbf{U}_1(\frac{t}{n}))x\|_H] = 0 \ \forall \ T > 0, \ \forall \ x \in H. \tag{16}$$

Proof. For any $n \in \mathbf{N}$ the composition $\mathbf{U}_n \circ ... \circ \mathbf{U}_1$ of n independent identically distributed semigroups is the map of the type $\{\mathbf{U}_{\omega_n} \circ ... \circ \mathbf{U}_{\omega_1}, \{\omega_1, ..., \omega_n\} \in \Omega \times ... \times \Omega\}$. Therefore, the mean value of the composition $\mathbf{U}_n \circ ... \circ \mathbf{U}_1$ satisfies (according to the theorem 7 and (11)) the equality

$$M[\mathbf{U}_n \circ ... \circ \mathbf{U}_1] = M[\mathbf{U}_n] \circ ... \circ M[\mathbf{U}_1] = (M[\mathbf{U}])^n \ \forall \ n \in \mathbf{N}.$$

Therefore, for any $t \geq 0$ and $n \in \mathbf{N}$ the following equality holds:

$$D(\mathbf{U}_n(\frac{t}{n})\circ...\circ\mathbf{U}_1(\frac{t}{n})) = M[\mathbf{U}_1^*(\frac{t}{n})\circ...\circ\mathbf{U}_n^*(\frac{t}{n})\circ\mathbf{U}_n(\frac{t}{n})\circ...\circ\mathbf{U}_1(\frac{t}{n})] - M[(\mathbf{U}^*(\frac{t}{n}))^n]M[(\mathbf{U}(\frac{t}{n}))^n].$$

Hence the equality

$$D(\mathbf{U}_n(\frac{t}{n})\circ...\circ\mathbf{U}_1(\frac{t}{n})) = \mathbf{I} - ((M[\mathbf{U}(\frac{t}{n})])^n)^*(M[\mathbf{U}(\frac{t}{n})])^n. \tag{17}$$

is fulfilled for every sequence of compositions of independent random unitary semigroups $\{\mathbf{U}_{\omega_n}\circ...\circ\mathbf{U}_{\omega_1},\ \{\omega_1,...,\omega_n\} \in \Omega\times...\times\Omega\}$

According to the Theorem 3 (see also [6]) the operator-function $M[\mathbf{U}(t)],\ t \geq 0$, is equivalent to the semigroup $\exp(i\bar{\xi}t),\ t \geq 0$ in Chernoff sense (see [13, 6]). It means that the sequence $\{(M[\mathbf{U}(\frac{t}{n})])^n,\ t \geq 0,\}$ converges to the semigroup $\exp(i\bar{\xi}t),\ t \geq 0$, in the strong operator topology uniformly on any segment of the semiaxis R_+. Therefore, the equality $\lim\limits_{n\to\infty}(\sup\limits_{t\in[0,T]}\|((M[\mathbf{U}(\frac{t}{n})])^n)^*[(M[\mathbf{U}(\frac{t}{n})])^n - \mathbf{I}]x\|_H) = 0$ holds for every $T > 0$ and $x \in H$.

Corollary 8. *Let the conditions of Theorem 8 are fulfilled. Then the law of large numbers in the strong operator topology is valid for the sequence* $\{\mathbf{U}_k\}$.

Remark. According to the estimates of Chernoff Lemma (see [11, 10]) the deviation of the mean value of compositions of the independent random semigroup $M[\mathbf{U}_n(\frac{t}{n})\circ...\circ\mathbf{U}_1(\frac{t}{n})]$ from the limit semigroup $e^{i\bar{\xi}t},\ t \geq 0)$ in any seminorm of the strong operator topology of the space $B(H)$ has the order $O(\frac{1}{\sqrt{n}})$ *as* $n \to \infty$ uniformly on any segment $[0, T]$.

We give the example of the sequence of compositions of independent identically distributed random semigroups satisfying the low of large numbers in the topology of the operator norm.

Example 3. Let us consider the example of the random semigroup of linear operators in the finitely dimensional Euclidean space H which has the random generator with uniformly bounded values. Let $\{\mathbf{U}_\omega,\ \omega \in \Omega\}$ be the random semigroup of linear operators in the finitely dimensional Euclidean space H such that $\|\mathbf{U}_\omega(t)\| \leq 1$ for all $t \geq 0$ and $\omega \in \Omega$. Let $\Omega = \{\omega_1,...,\omega_m\}$ be a finite set and $\mathbf{L}_{\omega_j},\ \omega_j \in \Omega$ be the bounded linear operator in H generating the semigroup $\mathbf{U}_{\omega_j}$ respectively. Then the mean value of the random semigroup $\mathbf{F}(t) = M[\mathbf{U}_\omega(t)],\ t \in R_+$, is the operator function which is equivalent in Chernoff sense to the semigroup $\bar{\mathbf{U}}(t) = e^{\bar{\mathbf{L}}t},\ t \geq 0$, with the mean generator $\bar{\mathbf{L}} = \sum\limits_{j=1}^{m} p_j\mathbf{L}_j$ (see [6]).

Let us consider the sequence of compositions $\{\mathbf{U}_n\circ...\circ\mathbf{U}_1\}$ of independent identically distributed semigroups such that the distribution of every component $\mathbf{U}_k,\ k \in \mathbf{N}$ of this composition is described in the example 3. Then the sequence $\{\mathbf{U}_n\circ...\circ\mathbf{U}_1\}$ of compositions of random semigroups satisfies all conditions of Theorem 8. Hence, the estimate (16) is fulfilled for this sequence. Therefore, the law of large numbers in the strong operator topology is valid for the sequence $\{\mathbf{U}_n\circ...\circ\mathbf{U}_1\}$. Since the norm of the finitely dimensional space $B(H)$ can be majorized by a sum of a finite set of seminorms of the strong operator topology of $B(H)$ then, the law of large numbers in the topology of operators norm is valid for the sequence $\{\mathbf{U}_n\}$.

5.2. The examples of violation of the law of large numbers for the sequence of random semigroups

. Firstly, we consider the example of the sequence $\{\mathbf{U}_n\circ...\circ\mathbf{U}_1\}$ of compositions of the independent identically distributed random semigroups which satisfies the law of large numbers in the strong operator topology but does not satisfy this law in the topology of the operator norm.

Example 4. (See [9]). (*Violation of the large numbers law in the topology of the norm*) Let $\Omega = \{1, 2\}$ and measure μ on the σ-algebra 2^Ω be given by the equality $\mu(\{1\}) = \mu(\{2\}) = \frac{1}{2}$. Let

$\mathbf{A}$ be the self-adjoint operator with the basis of eigenvectors $\{e_k\}$ such that $\mathbf{A}e_k = ke_k \, \forall\, k \in \mathbf{N}$. Let ξ be the random generator of the random semigroup $\mathbf{U}(t) = exp(it\xi)$, where $\xi(1) = -\mathbf{A}$, and $\xi(2) = \mathbf{A}$. Then according to Theorem 3, the law of large numbers in the strong operator topology is fulfilled. But $M(\mathbf{U}(t)) = \cos(\mathbf{A}t)$, $t \geq 0$, $M(\mathbf{U}^*(t)\mathbf{U}(t)) = \mathbf{I}$, $t \geq 0$. Then the large numbers low in the topology of operator norm is not satisfied since for any $t > 0$ the following estimates hold:

$$P(\|(\mathbf{U}(\frac{t}{n}))^n - M(\mathbf{U}(\frac{t}{n}))^n)\|_{B(H)} > \frac{1}{2}) = P(\|(\mathbf{U}(\frac{t}{n}))^n - (\cos(\mathbf{A}\frac{t}{n}))^n)\|_{B(H)} > \frac{1}{2}) \geq$$

$$\geq P(\sup_{k\in\mathbf{N}} \|(\,\mathbf{U}(\frac{t}{n}))^n - (\cos(\mathbf{A}\frac{t}{n}))^n)e_k\|_H > \frac{1}{2}) = P(\sup_{k\in\mathbf{N}} |1 - (\cos(\frac{kt}{n}))^n| > \frac{1}{2}).$$

Therefore, for any $\delta > 0$ there is a $t \in (0, \delta)$ such that $P(\|(\mathbf{U}(\frac{t}{n}))^n - M(\mathbf{U}(\frac{t}{n}))^n)\|_{B(H)} > \frac{1}{2}) = 1$ for any $n \in \mathbf{N}$. It means that the law of large numbers in the strong operator topology is violated.

5.3. The law of large numbers for the compositions of random flows of nonlinear operators.
Let E be a real separable Hilbert space. Let λ be a measure on the space E. We investigate here the operator valued functions and the flows from the set $F_\lambda(E)$ in Banach space $F(E)$ (see (6)).

Definition 10. The sequence $\{\mathbf{X}_n\}$ of independent random flows with values in the space $F_\lambda(E)$ is called satisfying the law of large numbers in the topology generated by the strong operator topology of the space $B(\mathcal{H}_\lambda)$ on the set $F_\lambda(E)$ if the equality

$$\lim_{n\to\infty} P(\{\sup_{t\in[0,T]} \|(\Phi[\mathbf{X}_n(\frac{t}{n}) \circ \ldots \circ \mathbf{X}_1(\frac{t}{n})] - \overline{\Phi(\mathbf{X}_n(\frac{t}{n}) \circ \ldots \circ \mathbf{X}_1(\frac{t}{n}))}) h\|_{\mathcal{H}} > \epsilon\}) = 0 \qquad (18)$$

holds for any $h \in \mathcal{H}$ and $\epsilon > 0$.

Example 5. Let $\{\mathbf{X}_n\}$ be the sequence of independent identically distributed random flows in the space $E = R^d$ such that the distribution of any $n \in \mathbf{N}$ random flow $\mathbf{X}_n$ satisfies the assumptions of Corollary 6. Let λ be Lebesgue measure on the space E and $F_\lambda(E)$ be the set (6) in the space $F(E)$ consisting of the operator-functions λ-acceptable values. Then, according to Corollary 6 the sequence $\{\mathbf{X}_n\}$ satisfies the law of large numbers in the topology generated by the strong operator topology of the space $B(\mathcal{H}_\lambda)$ on the set $F_\lambda(E)$.

Acknowledgments
The author thanks Prof. L.S. Efremova, O.G. Smolyanov and Yu.N. Orlov for useful discussions. The work was supported in the framework of the State Programme for the Increasing of the Competitiveness of the Leading Universities of Russian Federations among the Leading Scientific and Educational Centers of the World (Programme "5-100").

References

[1] Sakbaev V Zh 2010 *Infinite Dimensional Analysis Quantum Probability and Related Topics* **13** 65–85
[2] Sakbaev V Zh 2016 *Journal of Mathematical Sciences* **213** 287–459
[3] Sakbaev V Zh and Smolyanov O G 2016 *Doklady Mathematics* **94** 654–58
[4] Volovich I V 2015 *P-Adic Numbers Ultrametric Analysis and Applications* **7** 56–70
[5] Sakbaev V Zh and Volovich I V 2017 *P-Adic Numbers Ultrametric Anal. Appl.* **9** 39–52
[6] Orlov Yu N, Sakbaev V Zh and Smolyanov O G 2016 *Izvestiya: Mathematics* **80** 1131–58
[7] Orlov Yu N, Sakbaev V Zh and Smolyanov O G 2014 *Proceedings of the Steklov Institute of Mathematics* **285** 222–32
[8] Efremova L S and Sakbaev V Zh 2015 *Theoret. and Math. Phys.* **185** 1582–98
[9] Sakbaev V Zh 2016 *Russian Mathematics (Izvestiya VUZ. Matematika)* **60** 72–6

[10] Bogachev V I and Smolyanov O G 2009 *Real and functional analysis: a university course* (Moscow: Regular and Chaotic Dynamics)
[11] Chernoff P 1968 *J. Funct. Anal.* **84** 238–42
[12] Neklyudov A Yu 2008 *Mathematical Notes* **83** 530–8
[13] Smolyanov O G, Weizsacker H and Wittih O 2007 *Potential Anal.* **26** 1–29
[14] Butko Y A, Schilling R L and Smolyanov O G 2012 *Infin. Dimens. Anal. Quantum Probab. Relat. Top.* **15** 26–39
[15] Remizov I D 2017 *Dokl. Akad. Nauk* **476** 17-21
[16] Orlov Yu N, Sakbaev V Zh and Smolyanov O G 2017 *Doklady Mathematics* **96**(3) 574–7
[17] Kozlov V V and Smolyanov O G 2015 *Dokl. Akad. Nauk* **465** 527–31
[18] Kozlov V V and Smolyanov O G 2012 *Dokl. Akad. Nauk* **444** 607–11
[19] Baker R 1991 *Proc. Am. Math. Soc.* **113** 1023–29.
[20] Sakbaev V Zh 2017 *Theoret. and Math. Phys.* **191** 886–909.
[21] Sakbaev V Zh 2017 *Itogi Nauki i Tekhniki. Seriya " Sovr. Mat. i ee pril."* **140** 88-118 (Russian)
[22] Bogachev V I 2007 *Measure theory* vol 1 (Berlin: Springer)

IOP Conf. Series: Journal of Physics: Conf. Series **990** (2018) 012013 doi:10.1088/1742-6596/990/1/012013

Diagonalization of log-integrable operator $\mathbb{R}$-cocycles

Anatole M. Stepin
Mechanics and Mathematics Department, M. V. Lomonosov State University, Moscow, Russia
E-mail: anmist@inbox.ru

Abstract. The subject of this paper prepared as the report for NOMA conference (June 2017), is the asymptotic behaviour at large times of scalar (additive) and operator (multiplicative) cocycles in the case of $\mathbb{R}$ – action with a finite invariant measure.

Recall that a scalar additive cocycle associated with 1-parameter group $\{T^s\}_{s\in\mathbb{R}}$ of preserving measure μ transformations T^s of the probability space $(X,\,\mu)$ is a measurable mappimg $\alpha : \mathbb{R} \times X \to \mathbb{C}$ satisfying the functional equation

$$\alpha(s_1 + s_2,\, x) = \alpha(s_1, x) + \alpha(s_2, T^{s_1}x). \tag{1}$$

This term came from homology theory of G – modules. In fact, the object defined above, is a 1 – cocycle.

Here is the well known example of a scalar additive cocycle:

$$\alpha(t,x) = \int_0^t h(T^\tau x)d\tau, \tag{2}$$

where $h \in L^1(X,\mu)$. It is clear that in this case $\alpha(t,\cdot) \in L^1(X,\mu)$ for μ and a. a. t.

Our first aim is to discuss the asymptotic behaviour of the general additive scalar 1 – cocycle taking integrable values as a vector-function of t, $\alpha(t,\cdot) \in L^1(X,\mu)$. Classic source of this formulation of the problem is the Birkhoff Ergodic Theorem. For averages $t^{-1}\int_0^t h(T^\tau x)d\tau$, $h \in L^1$ this Theorem maintains the existence of their limits as $t \to \infty$.

Not every scalar additive cocycle $\alpha : \mathbb{R} \to L^1(X,\,\mu)$ is of the form (2) even if α is absolutely continuous on t. What type of asymptotic behaviour as $t \to \infty$ can have cocycle $\alpha : \mathbb{R} \to L^1$, $\alpha \to \alpha(t,\,\cdot)$, in the general case of not necessarily absolute continuous dependence on t? This question is of interest from different points of view and, primarily, in relation with the spectral theory of $\mathbb{R}$ – actionts by weighted shift operators and the metric theory of Liapoanov exponents for flows with an invariant measure. For example, V. Oseledec in his well known paper "Multiplicative ergodic theorem. Characteristic Liapounov exponents of dynamical systems" writes: It is unknown to me, whether there exists a measurable additive cocycle α with integrable values at every t such that the statement as in the Birkhoff Theorem [concerning convergence of averages $t^{-1}\alpha(t,x)$] fails.

This is why we begin, first of all, with construction of 1 – cocycle α taking values in $L^1(X,\,\mu)$ considering it as a vector-function of t such that the conclusion of Birkhoff Ergodic Theorem fails to be true. It will be done for an arbitrary measure preserving flow acting in the standart probability measure space. To this end I recall the notion of a special flow $T = \{T^s\}_{s\in\mathbb{R}}$ built by a measurable invertible ν – preserving mapping S of the base (probability) measure space (Y, ν) and a ν – measurable function $f : Y \to \mathbb{R}_+$, $\int f d\nu = 1$. The special flow T acts in the space X of pairs $x = (y, \tau)$, $\tau \in [0, f(y))$ endowed with the measure μ possessing the differential $d\mu = d\nu \cdot dt$. Here T – action is locally vertical with instant switching $(y, f(y)) \to (Sy, 0)$, i. e. the point x moves vertically up with velocity 1 till reaching graph of f (ceiling function) and so on bearing in mind identification $(y, f(y)) \sim (Sy, 0)$ made. The inverse motion is vertical down with velocity -1 till reaching base Y with instant switching $(y, 0) \mapsto (s^{-1}y, f(s^{-1}y))$. Formally $T^s(y, \tau) = (S^k y, \tau + s - \sum_0^k f(S^i y))$ if $s + \tau \in [\sum_0^{k-1} f(S^i y), \sum_0^k f(S^i y)]$. In the shightly different way but almost similarly the special action can be written for $t < 0$.

In the construction of the required cocycle we use the following fact: every ergodic flow acting on the standart probability measure space possesses the special representation, moreover, one can assume that its ceiling function f satisfies the condition $a < f < 2a$ for some $a > 0$.

To define 1 – cocycle α for the special flow $\{T^s\}_{s\in\mathbb{R}}$ it is sufficient to specify the values $\alpha(t, y)$ for points y of its base Y. Here we identify the base space Y of the special flow with the subset $\{(y, 0), y \in Y\} \subset X$, $y \leftrightarrow (y, 0)$. Moreover, the restriction of α to the base Y can be restored by its values $\alpha(t, y)$ on the intervals $[0, f(y))$, $y \in Y$. Bearing it in mind we consider associated with the special flow $\{T^s\}$ cocycle α satisfying

$$\alpha(t, y) = g(f(y) - t) - g(f(y))$$

as $t \in [0, f(y))$, where g is an integrable function with an unbounded singularity at $f(y)$, say $g(\xi) = \xi^{-1/2}$. Besides, we set $\alpha(f(y), y) = 0$. Then

$$\alpha(s, (y, \tau)) = g(\sum_0^{n-1} f(S^k y) - s - \tau) - g(f(S^n y)) - g(f(y) - \tau) - g(f(y))$$

as $\sum_0^{n-1} f(S^k y) \leq \tau + s < \sum_0^n f(S^k y)$.

As it follows from this formula, cocycle α is μ – integrable at every t. Setting $\tau = 0$ we get that $\alpha(s, y)$ as a function of $s \in (\sum_0^n f(S^k y), \sum_0^{n+1} f(S^k y))$ monotonically increases to ∞ as $s \to \sum_0^{n+1} f(S^k y)$. Hence, $\alpha(s, y)$ is (essentially) unbounded with respect to s on every interval of length $2a$. This is why there are no limit of the averages $t^{-1}\alpha(t, y)$ as $t \to \infty$ for ν – a. a. $y \in Y$. From here and cohomological relation one concludes that averages $t^{-1}\alpha(t, x)$ do not converge as $t \to \infty$ for μ – a. a. $x \in X$.

Nevertheless, for μ – a. a. $x \in X$ there is convergence of the averages $t^{-1}\alpha(t, x)$ as $t \to \infty$ along the set of density 1.

Definition. *Borel set $\tau \subset \mathbb{R}^+$ has density 1 if $\lim_{t\to\infty} \lambda(\tau \cap [0, t]) = 1$, where λ is the Lebesgue measure (for subsets of $\mathbb{R}$ the Definition is similar).*

As an acceptable formulation of ergodic theorem for 1 – cocycles $\alpha(t, x)$ having integrable values as vector-functions on t one can accept the following statement giving satisfactory description of asymptotic (t – large) behaviour of the averages $t^{-1}\alpha(t, x)$.

Theorem 1. *Let (X, μ) be the standart probability measure space, $T = \{T^s\}_{s\in\mathbb{R}}$ be μ – preserving flow on (X, μ) and α be additive scalar 1 – cocycle associated with T such that $\alpha(t, \cdot) \in L^1(X, \mu)$ for all t.*

Then there exists a measurable mapping $x \mapsto \tau(x)$ taking values in the space of Borel subsets $\mathbb{R}$ of density 1 and a measurable T – invariant function $\beta : X \to \mathbb{C}$ such that

$$\lim_{\tau(x)\ni t\to\infty} t^{-1}\alpha(t, x) = \beta(x) \ for \mu \ - \ a.e.;$$

moreover, $\int \beta d\mu = t^{-1} \int \alpha(t, \cdot) d\mu$.

Proof. Represent the flow T as the special flow built by a base automorphism S of the base space (Y, ν) and a ceiling function $f : Y \to \mathbb{R}_+$. According to the functional relation for a cocycle one has

$$t^{-1}\alpha(t, T^s y) = t^{-1}\alpha(t+s, y) - t^{-1}\alpha(s, y).$$

Hence, μ – a. e. convergence of $t^{-1}\alpha(t, x)$ as $t \to \infty$ is equivalent to ν – a. e. convergence $t^{-1}\alpha(t, y)$ and, moreover, these limits agree.
To express the average $t^{-1}\alpha(t, y)$ in terms of the ceiling function f and corresponding restriction $\alpha(f(\cdot), \cdot)$ we consider how to recover a cocycle by his restriction on the base of the special representation. Restriction on the base Y uniquely determines cocycle:

$$\alpha(t, y) + \alpha(s, T^t y) = \alpha(t+s, y).$$

The function $\alpha(t, y)$ can not be arbitrary. For correctness of the recovery procedure the following relation should be fulfilled:

$$\alpha(f(y), y) + \alpha(t, Sy) = \alpha(f(y) + \tau, y). \tag{3}$$

By values of the cocycle on the set $\{y \in Y, 0 \le t < f(y)\}$ one can restore the cocycle in the following way. Relation (3) gives the values $\alpha(t, y)$ for $t \in [f(y), f(y) + f(Sy))$, namely,

$$\alpha(t, y) = \alpha(f(y), y) + \alpha(t - f(y), Sy).$$

Inductively applying (3) one gets

$$\alpha(t, y) = \alpha(f(y), y) + \cdots + \alpha(f(S^k y), S^k y) + \alpha(t - f(y) - \cdots - f(S^k y), S^{k+1} y)$$

if $t \in [f(y)+\cdots+f(S^k y), f(y)+\cdots+f(S^{k+1}y))$. Thus, $\alpha(t, y)$ can be expressed by $\alpha(\tau, S^{k+1}y)$ for $0 \le \tau < f(S^{k+1}y)$ and S – shifts $\alpha(f(S^i y), S^i y)$ of the function $\alpha(f(\cdot), \cdot) := g(\cdot)$. Varying the base of the special representation, the time scale and applying Fubini Theorem we can assume that $g \in L^1(Y, \nu)$. It is also useful to apply here the stronger version of the special Representation Theorem due to Rudolph.

By the Ergodic Theorem (applied to S) one gets

$$k^{-1}\sum_0^k g(S^{-i}y) \to \int g d\nu, \quad k^{-1}\sum_0^k f(S^i y) \to 1$$

as $k \to \infty$. Hence,

$$2(f(y) + \cdots + f(S^k y), y)/f(y) + \cdots + f(S^k y) \to \int g d\nu$$

for ν – a. a. y's. In other words, if $t \to \infty$ taking values $\sum_0^k f(S^i y)$, then

$$t^{-1}\alpha(t, y) \to \int g d\nu.$$

We show now that for ν – a. a. $y \in Y$ there exists subset $M_y \subset \mathbb{R}_+$ of density 1 such that

$$t^{-1}\alpha(t, y) \to \int g d\nu$$

as $M_y \ni t \to \infty$.

If $f(y) + \cdots + f(S^k y) \le t < f(y) + \cdots + f(S^{k+1} y)$, where $\tau = t - \sum_0^k f(S^i y)$, and

$$\alpha(t,\, y) = \frac{g(y) + \cdots + g(S^k y) + \alpha(\tau, S^{k+1} y)}{f(y) + \cdots + f(S^k y) + \tau}.$$

It implies that the equivalence

$$t^{-1}\alpha(t,\, y) \sim \int g d\nu + k^{-1}\alpha(t - \sum_0^k f(S^i y)) \tag{4}$$

holds ν – a. a. as $t \to \infty$.

Now set

$$M_y = \bigcup_{k=0}^{\infty} \{t : |\alpha(t - \sum_0^k f(S^i y),\, S^{k+1} y)| < k^{1/2} \cdot \|\alpha(\cdot,\, S^{k+1} y)\|_{L^1(0,\, f(S^{k+1} y))}^{1/2}\}.$$

By Chebyshev Inequality

$$mes\{\tau : |\alpha(\tau,\, S^k y)| \ge k^{1/2} \cdot \|\alpha(\cdot,\, S^k y)\|_{L^1_{(0,\, f(S^k y))}}^{1/2}\} \le k^{-1/2} \cdot \|\alpha(\cdot,\, S^k y)\|_{L^1_{(0,\, f(S^k y))}}^{1/2}.$$

Since the norm $\|\alpha(\cdot,\, y)\|_{L^1_{(0,\, f(S^k y))}}$ is ν – summable (as a function of y) then by Borel – Cantelli Lemma or Birkhoff Ergodic Theorem the formula

$$k^{-1}\|\alpha(\cdot,\, S^k y)\|_{L^1_{(0,\, f(S^k y))}} \to 0 \tag{5}$$

holds ν – a. e. as $k \to \infty$. Hence, the limit equality

$$\lim_{k\to\infty} mes\{(\mathbb{R}_+ \setminus M_y) \cap [\sum_0^k f(S^i y),\, \sum_0^{k+1} f(S^i y)]\}/f(S^{k+1} y) = 0$$

is valid ν – a. e. This means that M_y has density 1 for a. a. y.

If $t \in M_y \cap [\sum_0^{k-1} f(S^i y),\, \sum_0^k f(S^i y)]$ then

$$|k^{-1} \cdot \alpha(t - \sum_0^{k-1} f(S^i y),\, S^k y)| \le k^{-1/2} \cdot \|\alpha(\cdot,\, S^k y)\|_{L^1_{(0,\, f(S^k y))}}^{1/2}.$$

Using formula (5) one has $k^{-1}\alpha(t - \sum_0^k f(S^i y),\, S^{k+1} y) \to 0$ as $M_y \ni t \to \infty$. Here k depends on t according to inclusion

$$t \in M_y \cap [\sum_0^k f(S^i y),\, \sum_0^{k+1} f(S^i y)].$$

Combining this property and equivalence (4) we get the conclusion of the Theorem 1. Moreover, if T is ergodic then

$$\lim_{M_x \ni t \to \infty} t^{-1}\alpha(t,\, x) = \int_Y \alpha(f(y),\, y) d\nu.$$

Corollary. *For any ergodic flow $T = \{T^s\}_{s\in\mathbb{R}}$ its special representation with the base space (Y, ν), a ceiling function $f : Y \to \mathbb{R}_+$, $\int f d\nu = 1$ and a scalar additive cocycle α associated with T, the quantity $\int \alpha(f(y), y)d\nu$ does not depend on the choice of its special representation.*

Remark. 1. *Used in the proof of Theorem 1 expression of cocycles associated with flows given in its special representation enables one to obtain necessary and sufficient conditions for convergence of averages $t^{-1}\alpha(t, x)$ almost everywhere as $t \to \infty$.*

In the noninvariant form (depending on the choice of a special representation of the flow) the condition is as follows

$$n^{-1} \cdot \sup_{0\le t\le f(S^n y)} |\alpha(\tau, S^n y)| \to 0 \ a.\ e.$$

In the invariant form the condition requires that

$$n^{-1} \cdot \sup_{0\le\tau<1} |\alpha(\tau, T^n x)| \to 0 \ a.\ e.$$

2. *Divergence of averages $t^{-1}\alpha(t, x)$ as $t \to \infty$ for scalar additive cocycles with values in L^1 may even be in the case of cocycles continuously depending on t. Smoothing of the construction given before Theorem 1 can be done with the help of the theorem that asserts the existence of topology making weakly mixing flow to possess the property of unique ergodicity.*

3. *There exist additive scalar cocycles that are not absolutely continuous in t such that their temporal averages converge as $t \to \infty$. M. Lipatov gave an example of a cocycle such that its averages converge as $t \to \infty$ but the Oseledec's condition*

$$\sup_{|t|\le 1} ln^t|\alpha(t, x)| \in L^1,$$

that is sufficient for such a convergence, fail to be true.

More generally (in comparison with Theorem 1) the following sub-additive ergodic proposition is valid.

Theorem 2. *Let $F = \{T^s\}_{s\in\mathbb{R}}$ be a flow acting on the standard probability space (X, μ), a measurable real-valued function α on $\mathbb{R}_+ \times X$ be such that $\alpha^t(t, \cdot) \in L^1(X, \mu)$ and $\alpha(t+s, x) \le \alpha(s, x) + \alpha(t, T^s x)$ for every $s, t \in \mathbb{R}_+$, $x \in X$.*

Then there exists a measurable function τ on X taking values in the space of Borel subsets in $\mathbb{R}_+$ having density 1 and a measurable F – invariant function $\beta : X \to \mathbb{R} \cup \{-\infty\}$ such that $\beta^+ \in L^1(X, \mu)$,

$$\lim_{\tau(x)\ni t\to\infty} t^{-1}\alpha(t, x) = \beta(x) \ \mu \ - \ a.e.,$$

$$\lim_{t\to\infty} t^{-1} \int \alpha(t, \cdot)d\mu = \inf_t t^{-1} \int \alpha(t, \cdot)d\mu = \int \beta d\mu.$$

This proposition has two precursors in the discrete time case: well known Fekete Lemma on subadditive sequences and its version for random sequences due to Kingman. Theorem 2 enables one to investigate asymptotic behaviour of operator multiplicative $\mathbb{R}$-cocycles with log-integrable values and to prove their block diagonalization.

To define generalized Liapounov exponents in Theorem 3 below we shall need the notion of F_A – invariant sub-bundle in $X \times V$, where $F = \{T^s\}_{s\in\mathbb{R}}$ is a flow on (X, μ) and A is an operator – valued cocycle associated with F and taking values in bounded operators on normed space V.

For an operator – valued cocycle A define linear extension $F_A = \{T^s_A\}_{s\in\mathbb{R}}$ of F by formulas

$$T^s_A : X \times V \circlearrowleft, \ (x, v) \mapsto (T^s x, A(s, x)v).$$

Sub-bundle L of $X \times V$, i. e., the family $\{L(x)\}$ of a subspace is invariant with respect to weighted shift operators $v(x) \mapsto A(s, x)v(T^s x)$ in the space of V – valued functions on X. For simplicity we confine ourselves to the matrix case.

Theorem 3. *Let multiplicative cocycle* $A : \mathbb{R} \times X \to GL(n, R)$, *associated with a* μ – *preserving flow* $F = \{T^s\}_{s\in\mathbb{R}}$ *on a standard probability space* (X, μ) *satisfy the condition* $log^+\|A(t, \cdot)\| \in L^1(X, \mu)$ *for every* $t \in \mathbb{R}$.

Then there exists a measurable mapping $\tau : X \to \{$*Borel subsets in* $\mathbb{R}$ *of density* $1\}$ *such that the limits*

$$\lim_{\tau(x)\ni t\to\pm\infty} (A^*(t, x)A(t, x))^{1/2|t|} =: \Lambda^{\pm}(x)$$

exist μ – *a.e., and there are* F_A – *invariant linear sub-bundles* $L_i = \{L_i(x)\}$, $i = 1, \cdots, k(x)$, $X \times \mathbb{R}^n$ *for which uniformly on* $L_i(x) \setminus \{0\}$ *the limits*

$$\lim_{\tau\ni t\to\pm\infty} |t|^{-1}log\|A(t, x)v\| = \pm\chi_i(x), \; v \in L_i(x)$$

exist μ – *a.e., where* $\{(\exp^{\pm\chi_i(x)}, dimL_i(x))\}_{i=1,\cdots,k(x)}$ *are the spectra of* $\Lambda^{\pm}(x)$ *counted with multiplicities.*

References

[1] Lyapunov A M 1892 *The general problem of the stability of motion* (Math Society of Kharkov XI) (Russian)

[2] Bylov B F, Vinograd R E, Grobman D M and Nemytsky V V 1966 *The theory of Lyapunov exponents* (M: Science) (Russian)

[3] Oseledec V I 1968 A multiplicative ergodic theorem. Lyapunov characteristic numbers for dynamical systems *Trudy Moskov Mat Obsch* **19** 179-210 (Russian)

[4] Millionshchikov V M 1969 A stability criterion for the probability spectrum of linear systems of differential equations with recurrent coefficients and a criterion for almost reducibility of systems with almost periodic coefficients *Mat Sb* **78**:2 179-201 (Russian)

[5] Raghunathan M S 1979 A proof of Oseledec's multiplicative ergodic theorem *Israel J Math* **32** 4 356- 62

[6] Stepin A M 1982 *On the homological equation of the theory of dynamical systems* (Sb: Investigations on the theory of functions of many real variables Yaroslavl Yaroslavl State University) pp 106–117 (Russian)

[7] Ruelle D 1982 Characteristic exponents and invariant manifolds in Hilbert space *Annals of Math* **115**:2 243–290

[8] Gold'sheid I Ya and Margulis G A 1989 Lyapunov indices of a product of random matrices *Uspekhi Matem Nauk* **44**:5 (269) 13–60 (Russian)

[9] Latushkin Yu D and Stepin A M 1991 Weighted translation operators and linear extensions of dynamical systems *Uspekhi Matem Nauk* **46**:2 85–143 (Russian)

[10] Lipatov M E and Stepin A M 2016 Multiplicative ergodic theorem for cocycles satisfying condition $ln\|A(t, \cdot)\| \in L^1$ *Deposited in VINITI 30 of May 2016* 75 – B2016 (Russian)

IOP Conf. Series: Journal of Physics: Conf. Series **990** (2018) 012014 doi:10.1088/1742-6596/990/1/012014

The existence of attractors of Weyl foliations modelled on pseudo-Riemannian manifolds

N.I. Zhukova

National Research University Higher School of Economics
ul. Bolshaja Pecherskaja, 25/12, Nizhny Novgorod, 603155, Russia

E-mail: nzhukova@hse.ru

Abstract. A foliation that admits a Weyl structure arising from a pseudo-Riemannian metric of any signature as its transverse structure is called a pseudo-Riemannian Weyl foliation or (for short) a Weyl foliation. We investigate codimension $q \geq 2$ Weyl foliations on (not necessarily compact) manifolds. Different interpretations of their holonomy groups are given. We prove a criterion for a Weyl foliation to be pseudo-Riemannian. We find a condition on the holonomy groups which guarantees the existence of a transitive attractor of (M, F). Moreover, if the Weyl foliation is complete, this condition implies the existence of a global transitive attractor. We describe the structure of complete Weyl foliations modelled on Riemannian manifolds.

1. Introduction

Hermann Weyl introduced a new geometry in an attempt to create a unified field theory uniting magnetism and gravitation, which is a generalization of Riemannian geometry [17]. This geometry is named now Weyl geometry.

Weyl geometry is actively used in theoretical physics, in such areas as modern cosmology, gravitation, quantum mechanics, physics of elementary particles, etc. The review of some applications of Weyl geometry in physics was made by Scholz [11].

Advanced achievements in mathematical physics show the importance of pseudo-Riemannian geometry. In this work we consider Weyl geometry arising from pseudo-Riemannian geometry. This geometry is referred to as *pseudo-Riemannian Weyl geometry*, and it includes Weyl geometry arising from Riemannian geometry.

Our purpose is to investigate the influence of the transverse geometrical structure of a foliation on its topological and dynamical properties. In the given work the role of the transverse geometrical structure plays a pseudo-Riemannian Weyl structure.

Everywhere in the given work we assume that M is an n-dimensional manifold and F is a Weyl foliation on M of codimension $q \geq 2$, unless otherwise specified. Compactness of M is not assumed.

Recall that a diffeomorphism f of a pseudo-Riemannian manifold (N, g) is called a conformal transformation if there exists a smooth positive function λ on N such that $f^*g = \lambda g$. If λ is a constant, then f is referred to as a similarity transformation or a similarity of (N, g). Two pseudo-Riemannian metrics g_1 and g_2 on M are said to be conformally equivalent if there exists a smooth positive function λ on M such that $g_2 = \lambda g_1$.

Let $\Omega^m(N)$ be the space of external forms of degree $m \geq 0$, in particular $\Omega^0(N)$ is the algebra of smooth functions on the manifold N.

Definition 1. A *pseudo-Riemannian Weyl structure* of signature (k, s) on a manifold N is called a pair $([g], f)$, where $[g]$ is a class of conformally equivalent **a conformal equivalence class of** pseudo-Riemannian metrics of signature (k, s) on N, and $f : [g] \to \Omega^1(N)$ is a map satisfying the equality $f(e^\lambda g) = f(g) - d\lambda \quad \forall \lambda \in \Omega^0(N)$.

A manifold N equipped with a pseudo-Riemannian Weyl structure of signature (k, s) is said to be a *pseudo-Riemannian Weyl geometry* or a *pseudo-Riemannian Weyl manifold of signature* (k, s) and is denoted by $(N, [g], f)$.

Weyl manifolds form a category (see Section 2.2).

A foliation (M, F) is called *pseudo-Riemannian Weyl* or (for short) Weyl if its transverse structure is modelled on a pseudo-Riemannian Weyl manifold $(N, [g], f)$. (the exact definition is given in Section 2.3).

A foliation (M, F) is referred to as *a transversally similar foliation* if all holonomy transformations are similarities of respective open subsets of the transverse pseudo-Riemannian manifold (N, g^N). If moreover, the curvature of (N, g^N) is zero, then (M, F) is called a *transversally similar pseudo-Euclidean foliation* (for the details, see Section 2.3).

First we give the following characterization of Weyl foliations.

Theorem 1. *A smooth foliation (M, F) of codimension $q \geq 2$ is a Weyl foliation modelled on a pseudo-Riemannian Weyl geometry $(N, [g], f)$ if and only if there exists $h \in [g]$ such that (M, F) is a transversally similar foliation modelled on (N, h).*

We consider Weyl foliations as a particular case of Cartan foliations in the sense of Blumenthal [2] or, that is equivalent in this case, in the sense of [19]. Denote by $\mathbb{R}^+$ the multiplicative group of positive real numbers. Let $O(k, s)$ be the pseudo-orthogonal matrix group and let $CO(k, s) := \mathbb{R}^+ \times O(k, s)$ be the conformal group. Theorem 1 implies that Weyl foliations of signature (k, s) are Cartan of the type (G, H), where $G = CO(k, s) \ltimes \mathbb{R}^q$, $q = k + s$, is the semi-direct product of the conformal group $H := CO(k, s)$ and the abelian group $\mathbb{R}^q$. The consideration of Weyl foliations as Cartan foliations allows us to apply results of the previous works of the author [19] and [22]. In particular, the foliated bundle with the lifted foliation over a Weyl foliation is the basic tool in the given work.

The following theorem contains different interpretations of the germ holonomy groups usually used in foliation theory which follow from [20, Theorem 4] and [22, Proposition 5].

Theorem 2. *Let (M, F) be a pseudo-Riemannian Weyl foliation, $\pi : \mathcal{R} \to M$ be its foliated bundle, $x \in M$, $u \in \pi^{-1}(x)$. Let $\mathcal{L} = \mathcal{L}(u)$ be the leaf of the lifted foliation $(\mathcal{R}, \mathcal{F})$ through u. Then:*

I. The holonomy group $\Gamma(L, x)$ of the leaf $L = L(x)$ through x is isomorphic to each of the following three groups:

1) the subgroup $H(\mathcal{L}, u) := \{a \in H | R_a(\mathcal{L}) = \mathcal{L}\}$ of H;

2) the group of deck transformations of the regular covering map $\pi|_{\mathcal{L}} : \mathcal{L} \to L$;

3) the linear holonomy group $D\Gamma(L, x)$ formed by the differentials of the local holonomy diffeomorphisms along leaf loops of a transversal q-dimensional disc at x.

II. If moreover, (M, F) admits an Ehresmann connection $\mathfrak{M}$, then the holonomy group $\Gamma(L, x)$ is isomorphic to the $\mathfrak{M}$-holonomy group $H_{\mathfrak{M}}(L, x)$.

Remark 1. Let u' be another point of the set $\pi^{-1}(x)$ and $\mathcal{L}' = \mathcal{L}'(u')$. Then there exists a unique $a \in H$ such that $u' = ua$, and $H(\mathcal{L}', u') = a^{-1} H(\mathcal{L}, u) a$, i.e. $H(\mathcal{L}, u)$ is replaced by a conjugated subgroup in H.

Let E_q be the unit $q \times q$-matrix. We identify the group $\mathbb{R}^+$ with the subgroup $\{\lambda E_q | \lambda > 0\}$ of the Lie group $H = CO(k, s)$, where $q = k + s$.

According to Remark 1 the following definition is correct.

Definition 2. The holonomy group $\Gamma(L, x)$ of a leaf $L = L(x)$ of a Weyl foliation (M, F) is called *inessential*, if the group $H(\mathcal{L}, u)$ satisfying Theorem 2 belongs to the subgroup $O(k, s)$ of the group $H = CO(k, s)$. Otherwise the holonomy group of L is called *essential*. The holonomy group of $L = L(x)$ is referred to as *α-essential*, if the group $H(\mathcal{L}, u)$ satisfying Theorem 2 has an element $\lambda A \in CO(k, s) = \mathbb{R}^+ \times O(k, s)$, where $\lambda \in \mathbb{R}^+$, $\lambda \neq 1$ and A belongs to a compact subgroup of $O(k, s)$. Otherwise the holonomy group of L is called *α-inessential*.

Note that any α-essential holonomy group is essential, the opposite is in general not true (see Section 5.5). We want to emphasize that for Weyl foliations of signature $(0, q)$ the concepts of essential and α-essential holonomy group coincide.

We prove the following criterion for a Weyl foliation to be pseudo-Riemannian.

Theorem 3. *Let (M, F) be a Weyl foliation modelled on a transverse pseudo-Riemannian Weyl geometry $(N, [g], f)$. Then there exists a pseudo-Riemannian metric $h \in [g]$ such that (M, F) is a pseudo-Riemannian foliation modelled on (N, h) if and only if every holonomy group of this foliation is inessential.*

Corollary 1. *Any pseudo-Riemannian Weyl foliation is a foliation with transverse linear connection.*

In accordance with [7] the automorphism group of every pseudo-Riemannian Weyl foliation in the foliation category admits a structure of infinite-dimensional Lie group modelled on LF-spaces.

Recall that a subset of a foliated manifold (M, F) is called saturated, if it is a union of some leaves of the foliation. A minimal set of (M, F) is a nonempty closed saturated subset of M that does not have proper subsets possessing these properties.

The study of minimal sets is one of fundamental problems in qualitative theory of foliations, as well as in theory of dynamical systems [1].

Let (M, F) be a foliated manifold. A nonempty closed saturated subset $\mathcal{M}$ of M for which there exists an open saturated neighbourhood $\mathcal{U}$ such that the closure of any leaf from $\mathcal{U}$ contains $\mathcal{M}$ is called an *attractor* of the foliation. The neighborhood $\mathcal{U}$ is uniquely determined and called the basin of $\mathcal{M}$. We use notation $\mathcal{U} = Attr(\mathcal{M})$. An attractor $\mathcal{M}$ is called transitive if $\mathcal{M}$ is a minimal set of (M, F), i.e., if each leaf from $\mathcal{M}$ is dense in $\mathcal{M}$. If $Attr(\mathcal{M}) = M$ the attractor $\mathcal{M}$ is said to be *global* [19].

Attractors of foliations formed by trajectories of dynamic systems are attractors in sense of [13].

A leaf of a foliation (M, F) is proper [15] if it is an embedded submanifold of M. A leaf L is referred to as closed, if L is a closed subset of M. It is known that any closed leaf is proper. A foliation (M, F) is called proper if all its leaves are proper.

Theorem 4. *Let a pseudo-Riemannian Weyl foliation (M, F) of signature (k, s) have a leaf L with an α-essential holonomy group. Then the closure $\mathcal{M} := \overline{L}$ of the leaf L is a transitive attractor.*

For a proper Weyl foliation (M, F) the attractor $\mathcal{M}$ is a closed leaf.

A Weyl foliation is called complete if it is complete as a Cartan foliation or, equivalently, as a foliation with transverse linear connection (for the exact definition, see Section 5.2).

Theorem 5. *If a complete pseudo-Riemannian Weyl foliation (M, F) has a leaf L with α-essential holonomy group, then the closure $\mathcal{M} := \overline{L}$ of L is a global transitive attractor of the foliation.*

In Section 5.5, we construct an example of a complete Weyl foliation (M, F) that admits leaves with essential holonomy groups, but has no leaves with α-essential holonomy group and no attractors.

A regular covering map $f : L_0 \to L$ onto a leaf L of a foliation is called holonomic if the group of deck transformations of f is isomorphic to the holonomy group of this leaf.

In Section 6.1 we recall the definitions of (G, X)-manifolds and (G, X)-foliations. As the application of Theorem 5 we describe the structure of complete Riemannian Weyl foliations as follows.

Theorem 6. *Let (M, F) be a complete Riemannian Weyl foliation which is not a Riemannian foliation. Then:*

(i) there exists a leaf L with an essential holonomy group, and $\mathcal{M} = \overline{L}$ is global a transitive attractor;

(ii) (M, F) is a complete transversally similar Euclidean foliation, i.e. $(Sim(\mathbb{E}^q), \mathbb{E}^q)$-foliation;

(iii) there exists a regular covering map $\kappa : \widetilde{M} \to M$ such that $\widetilde{M}$ coincides with the product of manifolds $L_0 \times \mathbb{E}^q$, where $q = k + s$, and the induced foliation $\widetilde{F} = \kappa^* F$ is formed by the fibers of the canonical projection $pr : L_0 \times \mathbb{E}^q \to \mathbb{E}^q$ and the restriction $\kappa|_{L_0 \times \{b\}}$, $b \in \mathbb{E}^q$, is a holonomic covering map onto the corresponding leaf of (M, F);

(iv) there exists an epimorphism

$$\chi : \pi_1(M, x) \to Sim(\mathbb{E}^q)$$

of the fundamental group $\pi_1(M, x)$ of M onto some subgroup $\Psi = \chi(\pi_1(M, x))$ of $Sim(\mathbb{E}^q)$;

(v) the holonomy group of any leaf $L = L(x)$, $x \in M$, is isomorphic to the isotropy subgroup Ψ_z of Ψ at $z \in pr(f^{-1}(L)) \subset \mathbb{E}^q_k$.

In the case when (M, F) is a proper Weyl foliation, the global attractor $\mathcal{M}$ is a unique closed leaf of (M, F).

Definition 3. The group Ψ satisfying Theorem 6 is referred to as the global holonomy group of (M, F).

Other properties and examples of transversally similar Euclidean foliations can be found in [19, Sections 9 and 10].

The structure of complete Riemannian foliations is well known due to works of Molino, Haefliger, Carrier and others.

Weyl manifolds and W-flows on them are studied in [18]. Pseudo-Riemannian Weyl geometry on distributions and foliations was investigated in [6]. The structure of pseudo-Riemannian foliations of signature $(1, 1)$ on closed 3-dimensional manifolds is described in [4]. The work [21] is devoted to the investigation of foliations with transverse Weyl structures arising from Riemannian geometry, i.e. Weyl structures how they were defined by Weyl ([17], see also [8] and [12]).

Notations. Following [10] we denote a principal H-bundle $p : P \to N$ by $P(N, H)$. The module of vector fields on a manifold M is denoted by $\mathfrak{X}(M)$. Let $\mathfrak{X}_{\mathfrak{M}}(M)$ be the set of vector fields tangents to some distribution $\mathfrak{M}$ on M.

2. Weyl foliations as transversally similar pseudo-Riemannian foliations

2.1. The linear connection compatible with a Weyl geometry

Definition 4. Let $(N, [g], f)$ be a Weyl manifold. A torsion free linear connection ∇ on N is called *compatible with the Weyl structure,* if

$$\nabla g + f(g) \otimes g = 0 \quad \forall g \in [g]. \tag{1}$$

We want to emphasize that a Weyl structure $([g], f)$ on a manifold N is determined by a pseudo-Riemannian metric g and 1-form φ. In fact, let $[g]$ be the conformal class of pseudo-Riemannian metrics on N. Define $f : [g] \to \Omega^1(M)$ by the following two equations: $f(g) := \varphi$ and $f(e^\lambda g) = f(g) - d\lambda$ for all $\lambda \in \Omega^0(\mathcal{N})$. In accordance to Definition 1 such pair $([g], f)$ is a Weyl geometry on the manifold N.

Proposition 1. *For the given Weyl manifold $(N, [g], f)$ there exists a unique torsion free linear connection ∇ compatible with Weyl geometry.*

A torsion free linear connection ∇ on a manifold N on which there exist a pseudo-Riemannian metric g and 1-form φ such that $\nabla g + \varphi \otimes g = 0$ is the linear connection compatible with Weyl structure determined by g and φ.

Proof. As it is known, every torsion free linear connection is uniquely determined by the covariant derivative of a symmetric non-degenerate bilinear form. Therefore, on the given Weyl manifold $(N, [g], f)$ there exists a unique torsion free linear connection ∇ compatible with Weyl structure.

The converse, let a pseudo-Riemannian metric g and 1-form φ satisfy the equality $\nabla g + \varphi \otimes g = 0$. It is easy to check that $e^\lambda g$ and $\varphi - d\lambda$ so satisfy this equality for all smooth function λ. This means that ∇ is the unique torsion free linear connection ∇ compatible with this Weyl geometry $(N, [g], f)$ determined by g and φ. □

The statement analogous to Proposition 1 for Weyl geometries modelled on Riemannian manifolds was proved in [8, Theorem 2 and Corollary].

2.2. The category of Weyl manifolds

Let $(N, [g], f)$ and $(\widetilde{N}, [\widetilde{g}], \widetilde{f})$ be two Weyl manifolds. A smooth map $h : N \to \widetilde{N}$ is called *a morphism of Weyl manifolds*, if its codifferential h^* satisfies the following equality

$$h^* \circ \widetilde{f} = f \circ h^*$$

for every pseudo-Riemannian metric from $[\widetilde{g}]$.

Proposition 1 implies that $h : N \to \widetilde{N}$ is a morphism of the Weyl manifolds $(N, [g], f)$ and $(\widetilde{N}, [\widetilde{g}], \widetilde{f})$ if and only if h is a morphism of the manifolds of linear connections (N, ∇) and $(\widetilde{N}, \widetilde{\nabla})$, where ∇ and $\widetilde{\nabla}$ are torsion free linear connections compatible with the Weyl structures $([g], f)$ and $([\widetilde{g}], \widetilde{f})$ respectively, i.e. if $h_*(\nabla_X Y) = \widetilde{\nabla}_{h_* X} h_* Y$ for any $X, Y \in \mathfrak{X}(N)$, where h_* is the differential of h.

The category whose objects are Weyl manifolds, morphisms are morphisms of Weyl manifolds, and the composition of morphisms coincides with the composition of maps is called *a category of Weyl manifolds.*

2.3. Pseudo-Riemannian Weyl foliations

Let N be a q-dimensional manifold, and the topological space of N can be disconnected. Let M be an n-dimensional manifold, and $n > q$, An N-cocycle $\eta = \{U_i, f_i, \{\gamma_{ij}\}\}_{i,j \in J}$ is given if there are the following:

(i) an open covering $\{U_i \,|\, i \in J\}$ of the manifold M and submersions $f_i : U_i \to N$ to N with connected fibers;

(ii) if $U_i \cap U_j \neq \emptyset$, there exists a diffeomorphism $\gamma_{ij} : f_j(U_i \cap U_j) \to f_i(U_i \cap U_j)$ satisfying to the equality $f_i = \gamma_{ij} \circ f_j$ on $U_i \cap U_j$;

(iii) $\gamma_{ik} = \gamma_{ij} \circ \gamma_{jk}$ for all $x \in f_k(U_i \cap U_j \cap U_k)$, where $i, j, k \in J$.

The family η is supposed to be maximal, i.e. it contains all U_i, f_i, γ_{ij} having indicated above properties and $N = \bigcup_{i \in J} f_i(U_i)$. The set of fibers of submersions $\{f_i^{-1}(x) \,|\, x \in N, i \in J\}$ forms a base of a new topology Υ in M which called the leaf topology. Connected components of the topological space (M, Υ) form a partition $F = \{L_\alpha \,|\, \alpha \in A\}$ of the manifold M which is referred to as *a foliation of codimension q determined by the N-cocycle* η. The foliation is denoted by (M, F).

Let $(N, [g], f)$ be a Weyl manifold of a signature (k, s) determined by the N-cocycle η. If each element γ_{ij} from the cocycle η is an isomorphism of the Weyl manifolds induced on the open subsets $f_j(U_i \cap U_j)$ and $f_i(U_i \cap U_j)$, then (M, F) is called *a foliation with transverse Weyl structure of the signature* (k, s) or (for short) *Weyl foliation of the signature* (k, s). It is said also that the Weyl foliation (M, F) is modelled on the transverse Weyl manifold $(N, [g], f)$.

If an N-cocycle $\eta = \{U_i, f_i, \{\gamma_{ij}\}\}_{i,j\in J}$ determines a foliation (M, F), and on N there exists a pseudo-Riemannian metric h such that any element γ_{ij} is a similarity of the pseudo-Riemannian manifolds induced on open subsets $f_j(U_i \cap U_j)$ and $f_i(U_i \cap U_j)$, then (M, F) is called *a transversally similar foliation* modelled on (N, h). If moreover, (N, h) is the pseudo-Euclidean space E^q_k, then (M, F) is called a *transversally similar pseudo-Euclidean* foliation of the signature $(k, n-k)$. In particular, when $k = 0$ and $E^n_k = E^n$ the foliation (M, F) is called a *transversally similar Euclidean* foliation.

A transversally similar foliation modelled on (N, h) is named *a pseudo-Riemannian foliation* if any element γ_{ij} is a local isometry of (N, h).

As similarity transformations of a pseudo-Riemannian manifold (N, h) preserve the Levi Civita connection ∇^h on N, transversally similar foliations may be considered as Weyl foliations modelled on $(N, [h], f)$ where $f(h)$ is an exact 1-form. In this case the Levi-Civita connection ∇^h is the unique torsion free connection compatible with this Weyl geometry which will denoted by $(N, [h], \nabla^h)$.

2.4. Proof of Theorem 1

Let (M, F) be a Weyl foliation modelled on a Weyl geometry $(N, [g], f)$. If there exists a non simply connected component N_1 of the manifold N, then we change N_1 by its universal covering manifold N_0 with the induced Weyl structure. Hence without loss a generality we assume that each component of N is simply connected. Remark that for any manifold N having only simply connected components the cohomology group $H^1(N)$ is trivial.

Since 1-forms $\varphi := f(g)$ and $f(\widetilde{g})$ for every $\widetilde{g} \in [g]$ differ by an exact form, they have a common exterior derivative. Hence they define an element of the group $H^1(N)$. As $H^1(N)$ equals to zero, there exists $h \in [g]$ such that $f(h) = \varphi_0 = 0$. This implies $\nabla g = 0$ by the definition of the compatible connection. Therefore the Weyl connection $\nabla = \nabla^h$ is the Levi-Civita connection of the pseudo-Riemannian manifold (N, h).

As every γ_{ij} from the N-cocycle $\eta = \{U_i, f_i, \{\gamma_{ij}\}\}_{i,j\in J}$ determining the foliation (M, F) is both the conformal diffeomorphism and the isomorphism of the induced Levi-Civita connections on the corresponding open subsets of the pseudo-Riemannian manifold (N, h), then [14, Chap 1, Sec. 1], γ_{ij} is a local similarity transformation of (N, h).

Thus (M, F) is a transversally similar foliation modelled on (N, h).

The converse is proved by the obvious way. □

Remark 2. On accordance with Theorem 1, without loss a generality we assume further that any Weyl foliation is modelled on a Weyl geometry $(N, [g], \nabla^g)$, where ∇^g is the Levi-Civita connection defined by some pseudo-Riemannian metric g belonging to the class of conformally equivalent metrics $[g]$.

3. A criterion for a Weyl foliation to be pseudo-Riemanniann

3.1. Weyl foliations as Cartan foliations

Recall the definition of a Cartan geometry. More information about Cartan geometries can be found in [9], [12], [5].

IOP Conf. Series: Journal of Physics: Conf. Series **990** (2018) 012014 doi:10.1088/1742-6596/990/1/012014

Let G be a Lie group with the Lie algebra $\mathfrak{g}$. Let H be a closed subgroup of G with the Lie algebra $\mathfrak{h} \subset \mathfrak{g}$. Assume that $p : P \to N$ is a principal H-bundle formed by a free right action of the group H on the manifold P. The action of an element $a \in H$ on P is denoted by R_a. A non-degenerated $\mathfrak{g}$-valued 1-form ω_0 on P is called a *Cartan connection*, if the following two conditions are satisfied:

1) $\omega_0(A^*) = A$ for $A \in \mathfrak{h}$, where A^* is the fundamental vector field on P defined by A;

2) the 1-form ω_0 is H-equivariant, i.e. $(R_a)^*\omega_0 = Ad_G(a^{-1})\omega_0 \ \forall a \in H$, where Ad_G is the adjoint representation of G on $\mathfrak{g}$.

The principal H-bundle $P(N, H)$ with the Cartan connection ω_0 is referred to as a *Cartan geometry* of the type (G, H). It is denoted by $\xi = (P(N, H), \omega_0)$.

A Cartan geometry of the type (G, H) is reductive, if the homogeneous space G/H is reductive [12]. If the group G acts effectively at the left on G/H, then the Cartan geometry of the type (G, H) is called effective.

Everywhere further in this work we denote by $G = CO(k, s) \ltimes \mathbb{R}^q$ the semi-direct product of the conformal group $H = CO(k, s)$ and the q-dimensional abelian additive group $\mathbb{R}^q$, and $\mathbb{R}^q$ is a normal subgroup of G. We write any element of G as $\langle \lambda A, a\rangle$, where $\lambda \in \mathbb{R}^+, A \in O(q), a \in \mathbb{R}^q$. The multiplication in the Lie group G is defined by the following equality

$$\langle \lambda A, a\rangle\langle \mu B, b\rangle := \langle \lambda\mu AB, \lambda Ab + a\rangle, \ \langle \lambda A, a\rangle, \langle \mu B, b\rangle \in G.$$

Note that the group G is realized as group of all similarity transformations of the pseudo-Euclidean space $\mathbb{E}^q_k$, and H is realized as its stationary group at zero $0 \in \mathbb{E}^q_k$.

Let $\mathfrak{so}(k, s)$ be the Lie algebra of the pseudo-orthogonal group $O(k, s)$ and $\mathbb{R}^1$ be the Lie algebra of $\mathbb{R}^+$. The Lie algebra $\mathfrak{g}$ of G admitted the decomposition $\mathfrak{g} = \mathfrak{h} \oplus \mathfrak{p}$, where $\mathfrak{h} = \mathfrak{co}(k, s) = \mathbb{R}^1 \oplus \mathfrak{so}(k, s)$ is the Lie algebra of the conformal Lie group $CO(k, s)$ and $\mathfrak{p} = \mathbb{R}^q$ is the abelian ideal in $\mathfrak{g}$. The vector space $\mathfrak{p}$ is $Ad_G(H)$-invariant with respect to the adjoint representation $Ad_G(H) : H \to GL(\mathfrak{g})$ of the subgroup H of G on $\mathfrak{g}$.

According to Remark 2 in the investigation of Weyl foliations we consider a Weyl geometry $(N, [g], \nabla^g)$ of signature (k, s), $k + s = q$, on q-dimensional manifold N. Such Weyl geometry defines the Cartan connection $\omega_0 = \widetilde{\omega} + \theta$ in H-bundle of conformal frames $P(N, H)$, where $\widetilde{\omega}$ is the $\mathfrak{h}$-valued 1-form on $\mathcal{R}$ and θ is the $\mathbb{R}^q$-valued canonical 1-form on P which are defined by the Levi-Civita connection ∇^g on the pseudo-Riemannian manifold (N, g). We want to emphasize that the Weyl geometry $(N, [g], \nabla^g)$ will be considered as the effective reductive Cartan geometry $\xi = (P(N, H), \omega_0)$ of the type (G, H).

Applying [19, Proposition 2] we get the following statement.

Proposition 2. *A Weyl foliation (M, F) codimension q modelled on a Weyl geometry $\xi = (P(N, H), \omega_0)$ of signature (k, s), $q = k + s \geq 2$, is a Cartan foliation of the type (G, H), where $G = CO(k, s) \ltimes \mathbb{R}^q$ and $H = CO(k, s)$, and there exist the principal H-bundle $\pi : \mathcal{R} \to M$, the H-invariant foliation $(\mathcal{R}, \mathcal{F})$ and the $\mathfrak{g}$-valued H-equivariant 1-form ω on $\mathcal{R}$ satisfying the following conditions:*

1) $\omega(A^*) = A$ for a fundamental vector field A^* corresponding to $A \in \mathfrak{h}$ for every $A \in \mathfrak{h}$;

2) the map $\omega_u : T_u\mathcal{R} \to \mathfrak{g}$, for any $u \in \mathcal{R}$ is surjective, and $ker(\omega) = TF$, where TF is the distribution tangent to the foliation $(\mathcal{R}, F)$;

3) the Lie derivative $L_X\omega$ is equal to zero for each vector field X tangent to the foliation $(\mathcal{R}, \mathcal{F})$.

The principal H-bundle $\pi : \mathcal{R} \to M$ is said to be *the foliated bundle* and $(\mathcal{R}, \mathcal{F})$ is said to be *the lifted foliation* for the Weyl foliation (M, F).

3.2. Proof of Theorem 3

Assume that all holonomy groups of a Weyl foliation (M, F) are inessential. Let $\pi : \mathcal{R} \to M$ be the projection of the foliated H-bundle over (M, F) satisfying Proposition 2. The smooth right free action of the Lie group $H = CO(k, s)$ is determined on the manifold $\mathcal{R}$, and the orbit space $\mathcal{R}/H$ coincides with the manifold M.

The smooth right free action of the normal subgroup R^+ of H on $\mathcal{R}$ is defined, and the orbit space is a smooth manifold $\widehat{\mathcal{R}} = \mathcal{R}/\mathbb{R}^+$. A smooth right free action of the factor group $H/\mathbb{R}^+ = O(k, s)$ is induced on $\widehat{\mathcal{R}}$, and the orbit space $\widehat{\mathcal{R}}/O(k, s)$ coincides with M. Thus, the canonical projections onto the orbit spaces $\alpha : \mathcal{R} \to \widehat{\mathcal{R}}$ and $\widehat{\pi} : \widehat{\mathcal{R}} \to M$ satisfy the equality $\pi = \widehat{\pi} \circ \alpha$. Since the lifted foliation $(\mathcal{R}, \mathcal{F})$ is H-invariant, it is invariant with respect to the subgroup R^+. Hence $O(k, s)$-invariant foliation $(\widehat{\mathcal{R}}, \widehat{F})$ is induced, and its leaves are images of leaves of the foliation $(\mathcal{R}, \mathcal{F})$ with respect to the map α.

By the condition of Theorem 3, any leaf of (M, F) has an inessential holonomy group. Hence the restriction α on each leaf $\mathcal{L}$ of the lifted foliation $(\mathcal{R}, \mathcal{F})$ is a diffeimorphism onto the corresponding leaf $\widehat{L}$ of the foliation $(\widehat{\mathcal{R}}, \widehat{\mathcal{F}})$.

$Ad_G(H)$-invariance of the subspace $\mathfrak{p}$ of the vector space of the Lie algebra $\mathfrak{g} = \mathfrak{co}(q) \oplus \mathfrak{p}$ implies the existence of a smooth H-invariant distribution $\mathfrak{N}$ on $\mathcal{R}$, where $\mathfrak{N}_u = \{X \in T_u(\mathcal{R}) | \omega(X) \in \mathfrak{p}\}$ for any $u \in \mathcal{R}$. Hence, $\mathfrak{N}$ is a projectable connection on the H-bundle $\mathcal{R}(M, H)$ with respect to the foliation $(\mathcal{R}, \mathcal{F})$. Therefore the distribution $\widehat{\mathfrak{N}} := \alpha_* \mathfrak{N}$ is a projectable connection on the $O(k, s)$-bundle $\widehat{\mathcal{R}}(M, O(k, s))$ with respect to $(\widehat{\mathcal{R}}, \widehat{\mathcal{F}})$.

The bundle $\mathcal{R}(\widehat{\mathcal{R}}, \mathbb{R}^1)$ has the contractible fiber, and for each leaf $\mathcal{L}$ of the foliation $(\mathcal{R}, \mathcal{F})$ the restriction $\alpha_{|\mathcal{L}}$ is a diffeomorphism of $\mathcal{L}$ onto the corresponding leaf of $(\widehat{\mathcal{R}}, \widehat{\mathcal{F}})$. According to [22, Proposition 4] in this case there exists a foliated section $\sigma : \widehat{\mathcal{R}} \to \mathcal{R}$. Thus, $\sigma(\widehat{\mathcal{R}})$ is a reduction of the bundle $\mathcal{R}(M, H)$ to the closed subgroup $O(k, s)$, and $\sigma(\widehat{\mathcal{R}})$ is a foliated $O(k, s)$-bundle for (M, F) such that $\sigma(\widehat{F}) = \mathcal{F}|_{\sigma(\widehat{\mathcal{R}})}$. It is not difficult to show that σ satisfies the equation $R_a \circ \sigma = \sigma \circ R_a$ for every $a \in O(k, s)$ and $\sigma_* \widetilde{\mathfrak{N}} = \mathfrak{N}|_{\sigma(\widehat{\mathcal{R}})}$. This means that the original Weyl foliation (M, F) with the transverse Weyl geometry $(N, [g], \nabla^g)$ is a pseudo-Riemannian foliation modelled on a pseudo Riemannian manifold (N, h) where $h \in [g]$.

The converse is fulfilled by obvious a way. □

4. A Riemannian metric and a connection adapted to the lifted foliation

Let (M, F) be a Weyl foliation of codimension q. Let $\mathcal{R}(M, H)$ be its foliated bundle with the lifted foliation $(\mathcal{R}, \mathcal{F})$ satisfying Proposition 2.

Consider a smooth q-dimensional distribution $\mathfrak{M}$ transverse to leaves of (M, F), i.e. $T_x M = \mathfrak{M}_x \oplus T_x F$ for any point $x \in M$. Let $\widetilde{\mathfrak{M}} := \pi^* \mathfrak{M}$, i.e. $\widetilde{\mathfrak{M}}_u := \{X \in T_u \mathcal{R} | \pi_{*u} X \in \mathfrak{M}_x, x = \pi(u)\}$ for all $u \in \mathcal{R}$. Denote by $\mathfrak{P}$ the smooth q-dimensional distribution on $\mathcal{R}$ which is equal to the intersection $\mathfrak{N}$ and $\widetilde{\mathfrak{M}}$, i.e. $\mathfrak{P}_u = \{X \in \widetilde{\mathfrak{M}}_u \mid \omega(X) \in \mathfrak{p}\}$ for all $u \in \mathcal{R}$. H-invariance of the distributions $\mathfrak{N}$ and $\widetilde{\mathfrak{M}}$ implies H-invariance of the distribution $\mathfrak{P}$.

Definition 5. A smooth vector field $X \in \mathfrak{X}_{\widetilde{\mathfrak{M}}}(\mathcal{R})$, for which $\omega(X) = c = const$ is said to be a $\mathfrak{g}$-*field.* If moreover $c \in \mathfrak{p}$, then X is said to be a $\mathfrak{p}$-*field,*

A piecewise smooth curve in $\mathcal{R}$ is called a $\mathfrak{g}$-*curve* (accordingly a $\mathfrak{p}$-*curve*), if each its smooth piece is an integral curve of some vector $\mathfrak{g}$-field (accordingly a $\mathfrak{p}$-*field*).

Remark, that locally anyone smooth $\mathfrak{g}$-curve σ can be represented as $\sigma(t) = \varphi_t^X(v), t \in (-\varepsilon, \varepsilon)$, where $\varepsilon > 0$, φ_t^X is the 1-parametric group of local diffeomorphisms of the manifold $\mathcal{R}$ generated by the $\mathfrak{g}$-field X for which $\sigma(t)$ is an integral curve, and $v = \sigma(0) = \varphi_0^X(v)$.

Lemma 1. *Let $g_{\mathcal{R}}$ be an arbitrary Riemannian metric on the space of the foliated bundle $\mathcal{R}$. Let d_0 be the Euclidean metric on the vector space of $\mathfrak{g}$ invariant with respect to the action of* *the compact group $Ad_G(O(k)\times O(s))$. Let $Z = Z_{\mathcal{F}} \oplus Z_{\widetilde{\mathfrak{M}}}$ be the decomposition of any vector field $Z \in \mathfrak{X}(\mathcal{R})$ corresponding to the decomposition of the tangent vector space to $\mathcal{R}$ in the direct sum of the vector subspaces $T_u\mathcal{R} = T_u\mathcal{F} \oplus \widetilde{\mathfrak{M}}_u$, $u \in \mathcal{R}$.*

Then the equality

$$d(X,Y) := g_{\mathcal{R}}(X_{\mathcal{F}}, Y_{\mathcal{F}}) + d_0(\omega(X), \omega(Y)) \ \forall X, Y \in \mathfrak{X}(\mathcal{R}),$$

defines a Riemannian metric d on $\mathcal{R}$ transversally projectable with respect to the foliation $(\mathcal{R},\mathcal{F})$ satisfying the following properties:

1) the length $l(\sigma)$ of any smooth $\mathfrak{g}$-curve σ, $\sigma(t) = \varphi_t^X(v)$, where $t \in [0,t_1], v = \sigma(0)$, is equal to $\| \omega(X) \|_{d_0} \cdot t_1$, where $\| \omega(X) \|_{d_0}^2 = d_0(\omega(X), \omega(X))$;

2) $l(\varphi_t^X(v)) = l(\varphi_t^X(v'))$, where X is a $\mathfrak{g}$-field, $t \in [0,t_1]$, for all $v, v' \in \mathcal{R}$, if $\varphi_t^X(v)$ and $\varphi_t^X(v')$ are defined at $t \in [0,t_1]$;

3) for any element $a = \lambda^{-1} \cdot A \in R^+ \times O(k) \times O(s)$, where $\lambda > 0$, and for any $\mathfrak{p}$-curve σ the curve $\widetilde{\sigma} := R_a \circ \sigma$ is a $\mathfrak{p}$-curve, with $l(\widetilde{\sigma}) = \lambda \cdot l(\sigma)$.

Proof. As $\mathfrak{g}$-valued 1-form ω is projectable, so the Riemannian metric d is also projectable with respect to the foliation $(\mathcal{R},\mathcal{F})$. As $\sigma(t) = \varphi_t^X(v)$, where $t \in [0,t_1]$, is an integral curve of some $\mathfrak{p}$-field X, that $d(X,X) = \| \omega(X) \|_{d_0}^2$ and $d(\sigma)/dt = X_{\sigma(t)}$. Therefore the length $l(\sigma)$ is calculated by the formula specified in 1).

The relation 2) is followed from 1).

Let us verify 3). First we assume that $\sigma(t) = \varphi_t^X(v), t \in [0,t_1]$, is a $\mathfrak{p}$-curve, where $\sigma(0) = v$, and $\widetilde{\sigma} := R_a \circ \sigma$. As $\widetilde{\sigma}(t) = \varphi_t^Y(v \cdot a)$, where $Y = R_{a*}(X)$, the $Ad_G(O(k) \times O(s))$-invariance of $\mathfrak{p}$ implies that Y is a $\mathfrak{p}$-field. Hence $\widetilde{\sigma}$ is a $\mathfrak{p}$-curve.

According to 1), its length is calculated by the formula $l(\widetilde{\sigma}) = \| \omega(X) \|_{d_0} \cdot t_1$. H-equivariance of the form ω implies the equality $\omega(Y) = Ad_G(a^{-1})\omega(X)$. Therefore $l(\widetilde{\sigma}) = \| Ad_G(a^{-1})\omega(X) \|_{d_0} \cdot t_1$. Direct calculations show that for any element $a = \lambda^{-1} \cdot A \in R^+ \times O(k) \times O(s)$ the following equality $\| Ad_G(a^{-1})\omega(X) \|_{d_0} = \lambda \cdot \| \omega(X) \|_{d_0}$ is valid, hence $l(\widetilde{\sigma}) = \lambda \cdot l(\sigma)$. Let now σ be a piecewise smooth $\mathfrak{p}$-curve. Then it is devised into finite number of smooth pieces $\sigma |_{I_i}, i = 1,...,m$, for each of which, as it was proved above, there is the equality $l(\widetilde{\sigma} |_{I_i}) = \lambda \cdot l(\sigma|_{I_i})$, hence $l(\widetilde{\sigma}) = \lambda \cdot l(\sigma)$. □

The following easily proved lemma takes place.

Lemma 2. *Let $E_i, i = \overline{1, dim\mathfrak{g}}$ be a basis of the Lie algebra $\mathfrak{g}$. Let X_i be such $\mathfrak{g}$-field, that $\omega(X_i) = E_i$. We shall denote by $\widetilde{\nabla}$ the Levi-Civita connection of the Riemannian manifold $(\mathcal{R}, d)$. Then the equality*

$$\nabla_Y^0 Z := Y(Z^i)X_i + \widetilde{\nabla}_Y Z_F, \qquad (*)$$

where $Z = Z_F \oplus Z_{\widetilde{\mathfrak{M}}}$, $Z_{\widetilde{\mathfrak{M}}} = Z^i X_i \in \mathfrak{X}_{\widetilde{\mathfrak{M}}}(\mathcal{R})$, $Z_F \in \mathfrak{X}_{T\mathcal{F}}(\mathcal{R})$, $Y \in \mathfrak{X}(\mathcal{R})$, defines a linear connection ∇^0 in $\mathcal{R}$, generally speaking with torsion, with respect to which all $\mathfrak{g}$-fields are parallel. Besides the parallel transfer keeps the scalar product of $\mathfrak{g}$-fields inducted by d, and integrated curves $\mathfrak{g}$-fields are geodesic lines of the connection ∇^0.

5. The existence of attractors

5.1. Proof the of Theorem 4

We use notations from Section 4. Let (M, F) be a Weyl foliation modelled on a Weyl geometry $(N, [g], \nabla^g)$.

Let $(\mathcal{R}, \mathbb{F})$ be the foliation formed by connected components of submanifolds $\pi^{-1}(L_\alpha)$ of $\mathcal{R}$, where L_α are leaves of (M, F).

Suppose that the foliation (M, F) has a leaf $L = L(x)$, $x \in M$ with an α-essential holonomy group. Pick $v \in \pi^{-1}(x)$. Let $\mathcal{L} = \mathcal{L}(v)$ be the leaf of the lifted foliation $(\mathcal{R}, \mathcal{F})$.

IOP Conf. Series: Journal of Physics: Conf. Series **990** (2018) 012014 doi:10.1088/1742-6596/990/1/012014

We emphasize that geodesics of the linear connection ∇^0 with a torsion having the Christoffel symbols Γ^k_{ij} in some coordinate neighborhood coincide with the geodesics of the torsion free linear connection $\widehat{\nabla}$ having the Christoffel symbols $\widehat{\Gamma}^k_{ij} = \frac{1}{2}(\Gamma^k_{ij} + \Gamma^k_{ji})$. Hence geodesics of a linear connection ∇^0 with a torsion has the same properties as geodesics of torsion free linear connections.

Therefore for the point v there exist $\varepsilon > 0$ and an embedded submanifold $V := D_{\widetilde{\mathfrak{M}}}(v, \varepsilon)$ formed by $\mathfrak{g}$-curves of the length $< \varepsilon$ with the origin at v. Similarly, there exists $V_1 := D_{\mathfrak{P}}(v, \varepsilon) \subset V$ formed by $\mathfrak{p}$-curves of length less than ε with the origin at v.

Since the leaf L has α-essential holonomy group, the group $H(\mathcal{L}, v)$ contains the element $a = \lambda^{-1} \cdot A$, where $\lambda \in (0, 1)$ and A belongs to a compact subgroup of $O(k, s)$. Since $O(k) \times O(s)$ is the maximal compact subgroup of $O(k, s)$, another maximal subgroup of $O(k, s)$ is conjugated with $O(k) \times O(s)$. According to Remark 1 we can take such point $v \in \pi^{-1}(x)$ that $A \in O(k) \times O(s)$.

As points v and $u := v \cdot a$ belong to the same leaf $\mathcal{L}$, we can connect them by a smooth curve $h : [0, 1] \to \mathcal{L}$, where $v = h(0), u = h(1)$. Without loss a generality we believe that the holonomy diffeomorphism Φ_h along the path h ([15]) with respect to $(\mathcal{R}, \mathcal{F})$ is defined on V.

We emphasize that the normal neighborhoods respectively to ∇^0 form a base of neighborhoods at each point of $\mathcal{R}$. Therefore for some $\varepsilon > 0$ there is a neighborhood $\mathcal{W} = \mathcal{W}(\varepsilon)$ of the point v satisfying the following conditions: 1) $\mathcal{W}$ is adapted to the foliation $(\mathcal{R}, \mathbb{F})$ and 2) the submanifold $V_1 := D_{\mathfrak{P}}(v, \varepsilon)$ is transversal to leaves of $(\mathcal{W}, \mathbb{F}_{\mathcal{W}})$, and V_1 intersects each its leaf exactly once.

Therefore any two leaves of $(\mathcal{W}, \mathbb{F}_{\mathcal{W}})$ can be connected by some smooth $\mathfrak{p}$-curve of length $< \varepsilon$. Note that $U := \pi(\mathcal{W})$ is an adapted neighborhood with respect to (M, F) and $D := \pi(V_1)$ is a transversal q-dimensional manifold in U.

Show that for the closure $\overline{L}_\alpha$ of any leaf L_α of a foliation (M, F) intersecting U we have the inclusion $\overline{L}_\alpha \supset L$.

Let $L_\alpha \cap U \neq \emptyset$, then there exist points $x_0 \in L_\alpha \cap D$ and $v_0 \in \pi^{-1}(x_0) \cap V_1$. Therefore there is a $\mathfrak{p}$-curve $\varphi^X_t(v), t \in [0, t_1]$ in V_1 connecting $v = \varphi^X_0(v)$ with $v_0 = \varphi^X_{t_1}(v)$. This means that $v_0 \in \mathbb{L} \cap W$ for the leaf $\mathbb{L} \subset \pi^{-1}(L_\alpha)$ of the foliation $(\mathcal{R}, \mathbb{F})$.

Introduce the notations $\sigma(t) := \varphi^X_t(v), t \in [0, t_1], \widetilde{\sigma} := R_a \circ \sigma$, where $a = \lambda^{-1} \cdot A$, $\lambda \in (0, 1)$, $A \in O(k) \times O(s)$. Then $\widetilde{\sigma}(t) = \varphi^Y_t(v \cdot a)$ where $Y = R_{a*}(X)$. In accordance with the statement 3) of Lemma 1 $\widetilde{\sigma}(t)$ is a $\mathfrak{p}$-curve, and $l(\widetilde{\sigma}) = \lambda \cdot l(\sigma) < \lambda \cdot \varepsilon$.

According to the statement 2) of Lemma 1, if there exists a $\mathfrak{p}$-curve $\sigma_1(t) := \varphi^Y_t(v), t \in [0, t_1]$, then it has the length equal to the length of the curve $\widetilde{\sigma}$, and $l(\sigma_1) = \lambda \cdot l(\sigma) < \lambda \cdot \varepsilon < \varepsilon$. Hence, such $\mathfrak{p}$-curve exists in the neighborhood of V_1, i.e. $\sigma_1(t) \in V_1$ for all $t \in [0, t_1]$. Therefore the holonomy diffeomorphism Φ_h is defined at all points of this curve.

Since $\mathfrak{p}$-curves are transversally projectable, by the definition of the holonomy diffeomorphism Φ_h, the curve $\varphi^Y_{t_1}(h(\tau)), \tau \in [0, 1]$, lies in the leaf $\mathcal{L}' = \mathcal{L}'(v_0)$ of the foliation $(\mathcal{R}, \mathcal{F})$. This implies that points $v_1 := \sigma_1(t_1)$ and $\widetilde{v}_1 := \widetilde{\sigma}(t_1)$ of this curve are projected to the same leaf $L_\alpha = \pi(\mathcal{L}')$ of (M, F). The point v is connected to the point of v_1 by the curve σ_1 of length $l(\sigma_1) < \lambda \cdot \varepsilon$, where $\lambda \in (0, 1)$.

Repeating the above argument with $(m-1)$ times we get a point v_m which is connected with v by the $\mathfrak{p}$-curve σ_m of length $l(\sigma_m) = \lambda^m \cdot l(\sigma) < \lambda^m \cdot \varepsilon$, and points v_m and v_0 are projected to the same leaf L_α. Since $v_m \to v$ as $m \to \infty$, $x_m := \pi(v_m) \to x = \pi(v)$ as $m \to \infty$. Since $x_m \in L_\alpha \ \forall m$ and $L = L(x)$, this means what $L \subset \overline{L}_\alpha$. This implies that $\mathcal{M} := \overline{L} \subset \overline{L}_\alpha$ for each L_α intersecting U.

The union $\mathcal{U} := \cup L_\alpha$ of leaves $L_\alpha \in F$ such that $L_\alpha \cap U \neq \emptyset$ is an open saturated neighborhood of $\mathcal{M}$ in M. Thus, $\mathcal{M}$ is an attractor of (M, F), and $\mathcal{U} = Attr\mathcal{M}$.

Consider an arbitrary leaf $L' \subset \mathcal{M}$, then $\overline{L}' \subset \mathcal{M} = \overline{L}$. According to the definition of the closure $\overline{L}$, we have $L' \cap U \neq \emptyset$. By the definition of $\mathcal{U}$ we have $L' \subset \mathcal{U}$, therefore the shown above inclusion $L \subset \overline{L}'$ implies $\mathcal{M} = \overline{L} \subset \overline{L}'$. Thus $\overline{L}' = \mathcal{M}$ and $\mathcal{M}$ is a minimal set of the foliation (M, F). Therefore $\mathcal{M}$ is a transitive global attractor of $(\mathcal{U}, F_{\mathcal{U}})$. □

5.2. Completeness of Weyl foliations

Recall that a smooth vector field X on $\mathcal{R}$ is *complete*, if X generates a global 1-parametric group of diffeomorphisms of $\mathcal{R}$.

Definition 6. Weyl foliation (M, F) is called *complete*, if every $\mathfrak{g}$-field $X \in \mathfrak{X}_{\widetilde{\mathfrak{M}}}(\mathcal{R})$ is complete.

Thus, by the definition, the completeness of a Weyl foliation (M, F) is equivalent to the completeness of (M, F) considered as a Cartan foliation.

By Corollary 1 a Weyl foliation (M, F) can be considered as a foliation with transverse linear connection. Therefore in accordance with [7, Theorem 4.1] on M there exists a transversally projectable with respect to (M, F) linear connection ∇ and a geodesic invariant transverse q-dimensional distribution $\mathfrak{M}$. It is not difficult to show that a Weyl foliation (M, F) is *complete* if and only if the canonical parameter on every maximal geodesic tangent to $\mathfrak{M}$ of the connection ∇ is defined on $(-\infty, +\infty)$.

5.3. Lemma

The following lemma will be used in the proof of the existence of a global attractor.

Lemma 3. *Let L and L' be any two leaves of a complete Weyl foliation (M, F). Then subsets $\pi^{-1}(L)$ and $\pi^{-1}(L')$ of $\mathcal{R}$ may be connected by some piecewise smooth $\mathfrak{p}$-curve.*

Proof. Let $f : M \to M/F$ be a quotient map onto the leaf space. We denote $[L]$ the leaf L of (M, F) considered as a point of M/F. We say that $[L]$ is equivalent to $[L']$ if there exists a piecewise smooth $\mathfrak{p}$-curve $\sigma : [0, 1] \to \mathcal{R}$ connected $\pi^{-1}(L)$ with $\pi^{-1}(L')$, i.e., $\sigma(0) \in \pi^{-1}(L)$ and $\sigma(1) \in \pi^{-1}(L')$.

Let us show that this relation is indeed an equivalence relation in M/F. The reflexivity and symmetry are obvious. Let us check transitivity of the introduced relation. Let $[L_0] \sim [L_1]$ and $[L_1] \sim [L_2]$, and a $\mathfrak{p}$-curve σ connects $\pi^{-1}(L_0)$ with $\pi^{-1}(L_1)$, a $\mathfrak{p}$-curve σ_1 connects $\pi^{-1}(L_1)$ with $\pi^{-1}(L_2)$. Let $v_0 = \sigma(0) \in \pi^{-1}(L_0)$, $v_1 = \sigma(1) \in \pi^{-1}(L_1)$, $v_2 = \sigma_1(0) \in \pi^{-1}(L_1)$, $v_3 = \sigma_1(1) \in \pi^{-1}(L_2)$. Let $x_i = \pi(v_i), i = 0, ..., 3$. Then $x_1 \cup x_2 \subset L_2$, hence there exists a point $u_0 \in \mathcal{L}(v_2) \cap \pi^{-1}(x_1)$, where $\mathcal{L}(v_2)$ is the leaf of the lifted foliation containing v_2.

Case I: the curve σ_1 is smooth. In this case $\sigma_1(t) = \varphi_t^X(v_2), t \in [0, 1]$, where X is a $\mathfrak{p}$-field. The completeness of the Weyl foliation (M, F) implies that $\mathfrak{p}$-field X is complete. Therefore for any point $v \in \mathcal{R}$ the integral curve $\varphi_t^X(v)$ is defined for all $t \in (-\infty, +\infty)$. Then the $\mathfrak{p}$-curve $\widehat{\sigma}_1(t) := \varphi_t^X(u_0)$ is defined. Since for each fixed t the diffeomorphism φ_t^X is an automorphism of the foliation $(\mathcal{R}, \mathcal{F})$, it is necessary $\widehat{\sigma}_1(1) = \varphi_1^X(u_0) \in \mathcal{L}(v_3) \subset \pi^{-1}(L_2)$.

Case II: the $\mathfrak{p}$-curve σ_1 is piecewise smooth. In this case we consistently apply the previous reasoning to each smooth piece in order to obtain $\mathfrak{p}$-curve $\widehat{\sigma}_1$ with the starting at u_0 and the end at $\widehat{\sigma}_1(1) \in \pi^{-1}(L_2)$.

As points v_1 and u_0 belong to the same fiber $\pi^{-1}(x_1)$, there exists an element $a \in H$ such that $v_1 = u_0 \cdot a$. According to Lemma 1 $\sigma^* := R_a \circ \widehat{\sigma}_1$ is a $\mathfrak{p}$-curve with the origin at v_1. Note that $\sigma^*(0) = v_1 = \sigma(1)$ and $\sigma^*(1) \in \pi^{-1}(L_2)$. Therefore the product of paths $\delta = \sigma \cdot \sigma^*$ is defined. Thus, δ is a $\mathfrak{p}$-curve connecting $\pi^{-1}(L_0)$ with $\pi^{-1}(L_2)$. This means that $[L_0] \sim [L_2]$, i.e. the relation $\sim$ is transitive, hence the introduced relation is an equivalence relation.

We now show that each equivalence class is an open subset of M/F. Consider a point $[L] \in M/F$. Let $A([L])$ be the equivalence class containing $[L]$. In the proof of Lemma 1 for any $x \in L$ and $v \in \pi^{-1}(x)$ we constructed the neighborhood $\mathcal{W}$ at the point v which is adapted with respect to $(\mathcal{R}, \mathbb{F})$. The point v can be connected with any local leaf of $(\mathcal{W}, \mathbb{F}|_{\mathcal{W}})$ by a certain $\mathfrak{p}$-curve. This implies that any two leaves of (M, F) intersected the neighborhood $U = \pi(\mathcal{W})$

are equivalent. Since the projection $f : M \to M/F$ onto the leaf space is open mapping, $f(U)$ is an open subset of M/F contains $[L]$ and $f(U) \subset A([L])$. Thus, the equivalence class $A([L])$ is an open subset of M/F.

Since the complement of $A([L])$ is formed by the union of the remaining equivalence classes each of which is open, then $A([L])$ is a closed subset of M/F. Due to the connectivity of the topological space of M, the leaf space M/F is also connected. Hence a non-empty open-closed subset $A([L])$ coincides with M/F. □

5.4. Proof of Theorem 5

Let (M, F) be a complete Weyl foliation of a signature (k, s) and codimension $q = k + s \geq 2$. Assume that (M, F) admits a leaf $L = L(x)$ with an α-essential holonomy group. Let L' be any other leaf of this foliation. By Lemma 3 there exists a $\mathfrak{p}$-curve σ connecting $\pi^{-1}(L)$ and $\pi^{-1}(L')$. Let $v = \sigma(0) \in \pi^{-1}(x)$, $x \in L$, $v_0 := \sigma(1) \in \pi^{-1}(x_0)$, $x_0 \in L'$.

According to Theorem 4 $\mathcal{M} = \overline{L}$ is a transitive attractor. Let $\mathcal{U} := Attr(\mathcal{M})$ and $\mathcal{V} = \pi^{-1}(\mathcal{U})$. Since the leaf $L = L(x)$ has an α-essential holonomy group, without loss a generality we assume that there exist a point $v \in \pi^{-1}(x)$ and an element $b = \lambda^{-1}A \in H(\mathcal{L}, v)$, where $\lambda \in (0, 1)$. Then there is the ball B of the radius ε with the center in v of the Riemannian manifold $(\mathcal{R}, d)$ such that $B \subset \mathcal{V}$. Therefore there is a natural number k, for which $\lambda^k \cdot l(\sigma) < \varepsilon$. According to the statement 3) of Lemma 1 the length of the curve $\widetilde{\sigma} := R_a \circ \sigma$, where $a = b^k \in H(\mathcal{L}, v) \subset H$, satisfies to the relations $l(\widetilde{\sigma}) = \lambda^k \cdot l(\sigma) < \varepsilon$. Connect the points $u = R_a(v)$ and v by a smooth path h in the leaf $\mathcal{L} = \mathcal{L}(v)$, $h(0) = u$, $h(1) = v$.

As it is known [19, Proposition 3], $\mathfrak{M}$-completeness of the Cartan foliation (M, F) implies that $\mathfrak{M}$ is an Ehresmann connection in sense of Blumenthal and Hebda [3] for this foliation. Here $\mathfrak{M}$ is a q-dimensional distribution on M which is transversal to (M, F). Therefore the induced distribution $\widetilde{\mathfrak{M}} = \pi^*\mathfrak{M}$ is an Ehresmann connection for the foliation $(\mathcal{R}, \mathcal{F})$. Hence there exists the transfer of the $\mathfrak{p}$-curve $\widetilde{\sigma}$ along the leaf path h with respect to the Ehresmann connection $\widetilde{\mathfrak{M}}$. Let $\widehat{\sigma}$ be the result of this transfer. We observe that $\widehat{\sigma}$ is a $\mathfrak{p}$-curve. Then $\widehat{\sigma}(0) = v$ and $l(\widehat{\sigma}) = l(\widetilde{\sigma}) < \varepsilon$ in $(\mathcal{R}, d)$ (see Lemma 1), therefore $\widehat{\sigma}(1) = v_1 \in B$ and $\pi(v_1) \in \mathcal{U}$.

By the property of an Ehresmann connection $\widetilde{\mathfrak{M}}$, the points $v_1 = \widehat{\sigma}(1)$ and $\widetilde{\sigma}(1)$ belong to the same leaf $\mathcal{L}'$ of $(\mathcal{R}, \mathcal{F})$. Since $\widetilde{\sigma} = R_a \circ \sigma$, the points $v_0 = \sigma(1)$ and $\widetilde{\sigma}(1)$ belong to $\pi^{-1}(x_0)$. Therefore the points $\pi(v_1)$ and $\pi(v_0) = x_0$ belong to the leaf $\pi(\mathcal{L}') = L' = L'(x_0)$ and $\pi(v_1) \in L' \cap \mathcal{U}$. Hence the closure $\overline{L'}$ of the leaf L' satisfies to the inclusion $\overline{L'} \supset L$ and $\mathcal{M} = \overline{L}$ is a global transitive attractor of (M, F).

It is necessary to note, that for a proper foliation any minimal set is a closed leaf. Therefore for a proper foliation (M, F) the global attractor $\mathcal{M}$ is a unique closed leaf. □

5.5. Example

Take any pair of natural numbers (k, s). Let $\mathbb{E}^q_k$ be the pseudo-Euclidean space of the signature (k, s), where $q = k + s$, where g_0 is its pseudo-Euclidean metric. Then $g_0(x, x) = -x_1^2 - \ldots - x_k^2 + x_{k+1}^2 + \ldots + x_q^2$ for $x = (x_1, \ldots, x_q) \in \mathbb{E}^q_k$. The linear transformation ψ defined by the following block matrix

$$\frac{1}{e}A = \frac{1}{e}\begin{pmatrix} E_{k-1} & 0 & 0 \\ 0 & A_e & 0 \\ 0 & 0 & E_{s-1} \end{pmatrix}, \quad A_e = \begin{pmatrix} ch1 & sh1 \\ sh1 & ch1 \end{pmatrix},$$

where $\mathbb{E}_{k-1}$ and $\mathbb{E}_{s-1}$ are the unit matrices $(k-1) \times (k-1)$ and $(s-1) \times (s-1)$ accordingly, and ψ is a similarity with the coefficient $\frac{1}{e}$ of the pseudo-Euclidean space $\mathbb{E}^q_k$. The equality

$$n(t, z) := (t - n, \psi^n(z)), \ n \in \mathbb{Z}, (t, z) \in \mathbb{R}^1 \times \mathbb{E}^q_k,$$

defines the free proper discontinues action of the group $\mathbb{Z}$ on the product of manifolds $\mathbb{R}^1 \times \mathbb{E}^q_k$. One-dimensional foliation $F = \{f(\mathbb{R}^1 \times \{z\})|z \in \mathbb{E}^q_k\}$ is induced on the factor-manifold $M := \mathbb{R}^1 \times_{\mathbb{Z}} \mathbb{E}^q_k$. It is a complete Weyl foliation of the signature (k,s), and (M,F) is $(Sim(\mathbb{E}^q_k), \mathbb{E}^q_k)$-foliation.

According to Theorem 3 this foliation is not a pseudo-Riemannian for any metric conformally equivalent to g_0.

The foliation (M,F) has the family of compact leaves continuously depending on $q-1$ parameters, and each compact leaf is diffeomorphic to the circle. Other leaves are diffeomorphic to the real line $\mathbb{R}^1$ and have an essential holonomy groups.

We want to emphasize that there are no leaves with an α-essential holonomy group. It is not difficult to see that (M,F) has not an attractor. $\Box$

6. Riemannian Weyl foliations

6.1. (G,X)-manifolds and (G,X)-foliations

Let X be a connected manifold and G be a group of diffeomorphisms of X. The group G is referred to as act quasi-analytically on X if, for any open subset U in X and an element $g \in G$, the condition $g|_U = id_U$ implies g is the identity transformation of X. We assume that the group G of diffeomorphisms of a manifold X acts on X quasi-analytically.

Definition 7. A foliation (M,F) determined by an X-cocycle $\{U_i, f_i, \{\gamma_{ij}\}\}_{i,j\in J}$ is called a (G,X)-foliation if for any $U_i \cap U_j \neq \emptyset$, $i,j \in J$, there exists an element $g \in G$ such that $\gamma_{ij} = g|_{f_j(U_i \cap U_j)}$.

Definition 8. A manifold B is called a (G,X)-manifold if its zero-dimensional foliation is a (G,X)-foliation.

6.2. Proof of Theorem 6

(i). Let (M,F) be a complete Riemannian Weyl foliation of codimension q, $q \geq 2$, which is not Riemannian. According to [21, Corolarry 5.1] there exists a leaf L with an essential holonomy group. In this case the holonomy group of L is α-essential. Therefore Theorem 5 implies that (M,F) is a transverse similar foliation of the signature $(0,q)$ having a global attractor $\mathcal{M}$.

(ii), (iv). Assume, that (M,F) is modelled on a transverse Riemannian Weyl geometry $(N,[g],\nabla^g)$, where (N,g) is a Riemannian manifold.

Let $q \geq 4$ and W be the Weyl tensor of the type $(1,3)$ of the conformal curvature for Riemannian manifold (N,g). Considering W as a polylinear map $W: \mathfrak{X}N \times \mathfrak{X}N \times \mathfrak{X}N \to \mathfrak{X}N$, we define a norm $\parallel W \parallel (x)$, $x \in N$ by the following a way. Let $\parallel X \parallel (x) := \sqrt{g_x(X,X)}$ for any vector field $X \in \mathfrak{X}(N)$. We put $\parallel W \parallel (x) := \sup\limits_{\|X_i\|(x)\leq 1} \parallel W(X_1,X_2,X_3) \parallel (x)$ $\forall x \in N$, $i = 1,2,3$.

As any Weyl foliation is a conformal foliation, then the holonomy pseudogroup of (M,F) consists from local conformal diffeomorphisms of the Riemannian manifold (N,g). It is well known that the Weyl tensor W is a conformal invariant. Let $\mathfrak{M}$ be a q-dimensional distribution on M transversal to (M,F), i.e. $T_xM = \mathfrak{M}_x \oplus T_xF$ for every $x \in M$. Therefore, on the distribution $\mathfrak{M}$ the transversally projectable Weyl tensor $\widetilde{W}$ is induced. Thus $\widetilde{f}(x) :=\parallel \widetilde{W} \parallel (x)$, $x \in M$, is the base function with respect to (M,F), i.e. a function which is constant on leaves of this foliation. The existence of a global attractor and the continuity of this function imply $\parallel \widetilde{W} \parallel = const$, i.e. the constancy of the function $f(z) :=\parallel W \parallel (z)$, $z \in N$. Let $c =\parallel W \parallel$.

Assume, that $c \neq 0$. As all transformations from the holonomy pseudogroup $\mathcal{H}$ of the foliation (M,F) are local conformal diffeomorphisms, it is not difficult to check up that each transformation from $\mathcal{H}$ preserves the Riemannian metric cg on N, i.e. $\mathcal{H}$ is a pseudogroup of local isometries of the Riemannian manifold (N, cg). Hence (M, F) is a Riemannian foliation. This contradicts the condition of Theorem 6.

Therefore it is necessary that $W \equiv 0$.

If $q = 3$, then $W \equiv 0$ and the replacing W by the Schouten tensor V of the type $(1, 2)$ [16], similarly to the previous case we show the invariance of the Riemannian metric $\| V \|^{\frac{2}{3}} \cdot g$ with respect to the holonomy pseudogroup $\mathcal{H}$ of (M, F) and get the contradiction. Hence $V \equiv 0$.

According to the Weyl-Schouten theorem [16], a q-dimension, $q \geq 3$, the Riemannian manifold (N, g) is conformally flat, i.e. it is locally conformally equivalent to the Euclidean space $\mathbb{E}^q$ and the standard sphere $\mathbb{S}^q$, if and only if, for $q = 3$ the Schouten tensor V is equal to zero and for $q \geq 4$ that the Weyl tensor of conformal curvature W is equal to zero.

It is known that any two-dimensional Riemannian manifold is locally conformally flat.

Thus for $q \geq 2$ the Riemannian manifold (N, g) is conformally flat.

Therefore (N, g) is locally conformal to $\mathbb{S}^q$. Denote by $Conf(\mathbb{S}^q)$ the Lie group of all conformal transformations of $\mathbb{S}^q$. According to the Liouville theorem, for any connected open subsets U, V of $\mathbb{S}^q$ and for each conformal transformation $f : U \to V$ there exists a unique element $\widetilde{f} \in Conf(\mathbb{S}^q)$ such that $f = \widetilde{f}|_U$. Therefore we can consider (M, F) as a conformal foliation modelled on the conformal geometry of the sphere $\mathbb{S}^q$. This means that (M, F) is a transversally homogenious $(Conf(\mathbb{S}^q), \mathbb{S}^q)$-foliation.

The completeness of the Weyl foliation (M, F) implies the completeness of (M, F) considered as $(Conf(\mathbb{S}^q), \mathbb{S}^q)$-foliation. Therefore, by [19, Proposition 3] there exists an Ehresmann connection for (M, F) and we may apply [22, Theorem 2]. According to this theorem there exists a regular covering map $\kappa : \widetilde{M} \to M$ and a simply connected $(Conf(\mathbb{S}^q), \mathbb{S}^q)$-manifold B such that the induced foliation $\widetilde{F} := \kappa^* F$ is formed by fibers of a submersion $r : \widetilde{M} \to M$. Moreover, a group homomorphism

$$\chi : \pi_1(M, x) \to Sim(B)$$

is defined, and the global holonomy group $\Psi := \chi(\pi_1(M, x))$ is isomorphic to the group of covering transformations of the map $\kappa : \widetilde{M} \to M$.

Since (M, F) is a complete conformal foliation, the induced conformal foliation $(\widetilde{M}, \widetilde{F})$ is also complete. This implies completeness of the conformally flat Riemannian manifold B. Therefore B is conformal to the standard sphere $\mathbb{S}^q$ or the Euclidean space $\mathbb{E}^q$.

Observe that the global holonomy group group Ψ has an essential transformation ψ with the fixed point $b = r(\kappa^{-1})(x)$. We emphasize that ψ is the similarity of B with the fixed point b. According to [10, Lemma 2, Chap VI] this is only possible when B is the Euclidean space $\mathbb{E}^q$.

Thus, (M, F) is a a $(Sim(\mathbb{E}^q), \mathbb{E}^q)$-foliation and the statements (ii) and (iv) are proved.

(v). Since the holonomy pseudogroup $\mathcal{H}(M, F)$ is generated by the group Ψ, then $\mathcal{H}(M, F)$ is quasi-analytical. Therefore Theorem 2 implies that the holonomy group of the leaf L is isomorphic to the isotropy subgroup Ψ_z of Ψ at point $z \in pr(\kappa^{-1}(L))$. Hence (v) is proved.

(iii). Since $\mathbb{E}^q$ is contractible, the locally trivial bundle $r : \widetilde{M} \to \mathbb{E}^q$ is trivial, hence $\widetilde{M} = L_0 \times \mathbb{E}^q$ and $r = pr : L_0 \times \mathbb{E}^q \to \mathbb{E}^q$ is the canonical projection.

It is easy to see the restriction $\kappa|_{L_0 \times \{b\}}$, $b \in \mathbb{E}^q$ is the regular covering map onto the corresponding leaf L of (M, F), and the deck transformation group is isomorphic to the isotropy subgroup Ψ_b of Ψ at point b. According to the proved above statement (v) the group Ψ_b is isomorphic to the holonomy group of L. This completes the proof of (iii). □

Acknowledgments

I express my gratitude to Yuri Kordyukov for helpful comments.

This work was supported by the Russian Foundation of Basic Research (grant no. 16-01-00312) and by the fundamental research program of the HSE (project 95) in 2018.

References

[1] Anosov D V 1982 Minimal set *Mathematical encyclopedia* **3** (Moscow: Sov. enc.) pp. 690–91

[2] Blumenthal R 1987 Cartan submersions and Cartan foliations *Ill. Math. J.* **31** pp. 327–43

[3] Blumenthal R and Hebda J J 1984 Ehresmann connections for foliations *Indiana Univ. Math. J.* **33**: 4, pp. 597–12

[4] Boubel C, Mounoud P and Tarquini C 2006 Lorentzian foliations on 3-manifold *Ergod. Th. and Dynam. Sys.* **26** pp. 1339–62

[5] Čap A. and Slovák J 2009 *Parabolic geometries. I. Background and general theory* Mathematical Surveys and Monographs **154** (AMS: Providence, RI) p. x+628

[6] Constantinescu Oana, Crasmareanu M 2012 Sub-Weyl geometry and its linear connections *Int. J. Geom. Methods Mod. Phys.* **9**:5 pp. 15

[7] Dolgonosova A Yu and Zhukova N I 2013 The automorphism group of foliations with transverse linear connection *Cent. Eur. J. Math.* **11**:12 pp. 2076–88

[8] Folland G B 1970 Weyl manifolds *J. Differential Geom.* **4** pp. 145–53

[9] Kobayashi S 1972 *Transformation groups in differential geometry* (Berlin–Heidelberg–New York: Springer Verlag) p. 189

[10] Kobayashi S and Nomizu K 1963 *Foundations of differential geometry: V. 1*, (New York–London: Interscience Publ.) p. 339

[11] Scholz E 2017 The unexpected resurgence of Weyl geometry in late 20-th century physics *ArXiv.org* 1703.03187 pp. 1–92

[12] Sharpe R W 1997 *Differential Geometry: Cartan's Generalization of Klein's Erlangen Prorgam* Grad. texts Math. **188** (New York: Springer) p. 421

[13] Shilnikov L P, Shilnikov A L, Turaev D V and Chua L O 1998 *Methods of qualitative theory in nonlinear dynamics: Part I.*) Singapore–New Jersi–London–Hong Long: World Scientific Publishing Co. Pte, Ltd.) p. 416

[14] Sinukov N S 1979 *Geodesic maps of Riemannian spaces* (Moscow: Science) p. 256 (rus)

[15] Tamura I 1992 *Topology of Foliations: An Introduction* (AMS) p. 198

[16] Udo Hertrich-Jeromin 2003 *Introduction to Moebius Differential Geometry* (New York: Cambridge University Press) p. 426

[17] Weyl H 1952 *Space, Time, Matter* 4 ed. (New York: Dover Publications)

[18] Wojtkowski M P 2002 Weyl Manifolds and Gaussian Thermostats *ICM* **III** pp. 511–23

[19] Zhukova N I 2007 Minimal sets of Cartan foliations *Proc. Steklov Inst. Math.* **256**:1 pp. 105–35

[20] Zhukova N I 2009 Complete foliations with transverse rigid geometries and their basic automorphisms *Bulletin of Peoples' Friendship University of Russia. Ser. Math. Information Sci. Phys.* No 2, pp. 14–35

[21] Zhukova N I 2010 Weyl foliations *Nonlinear dynamics* **6**:1 pp. 219–231 (rus)

[22] Zhukova N I 2012 Global attractors of complete conformal foliations. *Sbornik: Mathematics.* **203**:3 pp. 380–405